高等学校"十三五"规划教材

应用化学专业英语

第二版

黄 忠 谢普会 赵李霞 主编

化学工业出版社

·北京·

《应用化学专业英语》(第二版)共六部分,第一部分为科技英语翻译基础,介绍了科技英语的构词法和翻译特点;第二部分为化学专业英语概论,按无机化学、有机化学、物理化学、分析化学、生物化学、高分子化学、食品化学、农药化学、化工、精细化工几大板块安排内容;第三部分为化学反应和计算;第四部分为化学实验;第五部分为研究生入学考试参考阅读;第六部分为科技论文写作指导。全书共 43 篇课文,编排时对重要的专业词汇进行了注释和音标标注,课后附有练习题以巩固学习效果。书后附录有三套考试样卷及练习参考答案。

《应用化学专业英语》(第二版)可作为高等院校化学类、化工类等各专业本科生的教材,也可供英语爱好者和科技工作者参考。

图书在版编目(CIP)数据

应用化学专业英语/黄忠,谢普会,赵李霞主编. —2 版. —北京:化学工业出版社,2019.3(2025.1重印)
高等学校"十三五"规划教材
ISBN 978-7-122-33586-9

Ⅰ.①应… Ⅱ.①黄… ②谢… ③赵… Ⅲ.①应用化学-英语-高等学校-教材 Ⅳ.①O69

中国版本图书馆 CIP 数据核字(2019)第 000105 号

责任编辑:宋林青　　　　　　　　　　　文字编辑:任睿婷
责任校对:杜杏然　　　　　　　　　　　装帧设计:关　飞

出版发行:化学工业出版社(北京市东城区青年湖南街 13 号 邮政编码 100011)
印　　刷:北京云浩印刷有限责任公司
装　　订:三河市振勇印装有限公司
787mm×1092mm　1/16　印张 19　字数 523 千字　2025 年 1 月北京第 2 版第 7 次印刷

购书咨询:010-64518888　　　　　　　　售后服务:010-64518899
网　　址:http://www.cip.com.cn

凡购买本书,如有缺损质量问题,本社销售中心负责调换。

定　　价:40.00 元　　　　　　　　　　　　　　　　　　　　版权所有　违者必究

《应用化学专业英语》(第二版)
编 写 组

主 编：黄 忠　谢普会　赵李霞
副主编：喻 鹏　胡文斌　尹学琼　万传星　丁志杰
编 者（以姓氏笔画为序）：
　　　　丁志杰　万传星　王志国　尹学琼　刘 辉
　　　　刘光斌　刘佳音　张志云　赵李霞　胡文斌
　　　　黄 忠　龚 磊　喻 鹏　谢普会　潘振良
　　　　戴润英

前　言

　　应用化学专业英语是各高校化学类专业的一门重要必修课程。本书于2010年7月首次出版，几年来应用于多所院校的教学实践中，受到师生的普遍欢迎和好评，同时也发现了书中的不足之处。根据本书使用学校反馈的信息，本书编写组对全书进行了修改、润色，对部分课文内容进行增删、调整。第二版基本保持了原书的结构和框架，主要在以下几个方面进行了修改：

　　① 对课后练习题做了重大调整，题量大幅度增加，并在附录中给出了练习题答案。

　　② 对课后的阅读材料内容进行部分更新且篇幅上尽量选取了短小精炼的文章。

　　③ 对无机化合物命名和有机化合物命名两篇课文内容进行了重大调整。其中，无机化合物命名增加一些例子来说明命名规则，文字表述上也更加详细；有机化合物命名增加了炔、醚、醇、醛、酮、羧酸、酯、胺类等有机化合物的命名方法，形成一个较完整的有机命名系统。

　　④ 常用实验室器具这篇课文中的图片由原来的简笔画图片全部改成直观的实物图片，并进行归类排序方便学生记忆，同时表后增列了常见仪器英语简介，以提高学生对实验操作的英语表述能力。

　　⑤ 研究生入学考试参考阅读部分，由原来的7篇课文精简到现在的3篇，内容上也做了些调整。

　　⑥ 摘要这篇课文，增加了一篇完整的摘要范文。除正文之外，让学生也全面学习摘要中的标题、作者、单位、关键词等英语表述。

　　⑦ 科技英语翻译标准与方法这篇课文，根据需要单独列为第一部分，修改为科技英语翻译基础，共包含9篇课文，涵盖了科技英语翻译的重要方法和技巧。

　　⑧ 化学英文文献检索及网上资源简介、化学专业英语口语简介这两部分，精简成两个附录放在书后，对与大学文献检索和大学英语课程重复的部分进行删减。

　　本次修订分工如下：第一部分、第二部分（第一、第八单元）、第四部分、第六部分（第42课）、附录（1、6）由黄忠（江西农业大学）修订；第二部分（第二、第六单元）、第六部分（第41、43课）由尹学琼（海南大学）修订；第二部分（第三单元）由赵李霞（东北农业大学）修订；第二部分（第四单元）由喻鹏（湖南农业大学）修订；第二部分（第五单元）、附录（3、4）由丁志杰（安徽科技学院）修订；第二部分（第七单元）由万传星（新疆塔里木大学）修订；第二部分（第九单元）由胡文斌（仲恺农业工程学院）修订；第二部分（第十单元）、第三部分、附录（2、5、7、9）由谢普会（河南农业大学）修订；第五部分由潘振良（河南农业大学）修订。江西农业大学的刘光斌、龚磊、戴润英，东北农业大学的刘佳音、张志云，湖南农业大学的刘辉、王志国参加了本书修订的部分工作。全书最后由黄忠修改定稿。

　　本次修订的部分插图没有与课文内容一一对应，纯粹是从专业英语学习角度考虑，敬请读者阅读时参考。本次修订参考了国内外的一些优秀教材，在此谨向这些作者表示深深的谢意。本书第二版的问世离不开广大同仁的大力支持，谨向他们致以诚挚的谢意。由于水平有限，书中难免有疏漏之处，欢迎读者批评指正。

编　者

2018 年 10 月

第一版前言

应用化学专业英语是各高校化学类专业的一门重要必修课程。新形势对专业技术人才的英语水平提出了新的要求,我们在结合多年来积累的教学经验,并参阅国内外相关资料的基础上,将教学设计理念应用于调动学生学习积极性中,编写成这本新颖且实用的教材,以供不同院校的同仁们选择,本书亦可作为化学、化工等领域从事生产、科研等相关工作人员的参考书。

本教材主要包括以下几部分:第一部分应用化学专业基础英语;第二部分化学反应与化学计算;第三部分化学实验;第四部分研究生入学考试专题阅读;第五部分化学论文写作指导;第六部分科技英语简介;第七部分化学英文文献检索及网上资源简介;第八部分化学专业英语口语;第九部分附录(考试参考样卷)。内容由基础到专业,循序渐进,涵盖了"讲、读、写、说、练、查",使学生各方面能力得到训练。

本教材具有以下特点。

① 精读课文篇幅不长,均控制在1000字左右,适合教师安排一次两个课时的内容。

② 版面设计新颖,分左右大小两栏,生词放小栏与课文同步进行,免除了学生阅读时频繁翻书查阅后面的生词表,提高了课堂学习效率。

③ 知识结构安排合理,例如,在有机化学和无机化学两部分,均先讲英文命名法,再依次安排相关内容;无机化学同时又是整本书的开篇,因此用一篇化学科普文章开始第一课,使学生由"大学英语"到"专业英语"有一个良好过渡,符合学习规律。

④ 在课文后安排一段化学趣味小短文,以提高学生兴趣,也可作为师生互动的一个话题,使课堂学习生动活泼起来。

⑤ 在每个学科单元后面增加了一个该学科的主要单词汇总,便于学生课后进一步拓展该学科的英语知识面。

本书编写分工如下:第一部分(第一单元)、第三部分、第六部分、附录(参考样卷ABC)由黄忠(江西农业大学)编写;第一部分(第二、第六单元)、第五部分由薛行华(海南大学)编写;第一部分(第三单元)由付颖(东北农业大学)编写;第一部分(第四、第八单元)由喻鹏(湖南农业大学)和黄忠(江西农业大学)联合编写;第一部分(第五单元)、附录(3、4)由丁志杰(安徽科技学院)编写;第一部分(第七单元)由杨玲(新疆塔里木大学)编写;第一部分(第九单元)由胡文斌(仲恺农业工程学院)和付颖(东北农业大学)联合编写;第一部分(第十单元)、第二部分、第七部分、第八部分、附录(2、5)由谢普会(河南农业大学)编写;第四部分由潘振良(河南农业大学)编写。江西农业大学的刘光斌、龚霞、谭桂霞,东北农业大学的赵李霞,湖南农业大学的张凤,新疆塔里木大学的王丽玲参加了本书编写的部分工作。全书由黄忠修改统稿。

本书的部分插图没与课文内容一一对应，纯粹是从专业英语的学习角度考虑，敬请读者阅读时参考。

本书在编写过程中得到了多方面的关心与帮助，感谢江西农业大学理学院黄长干副院长，化学工业出版社的编辑给予的支持和帮助，感谢参加编写的各兄弟院校的同仁对编委的指导和帮助，并向支持编委编写工作的家人们致以诚挚的谢意。

由于编者学识水平有限，难免有疏漏欠妥之处，恳请同行专家和读者批评指正。

编　者

2010 年 5 月

Contents

Part 1 科技英语翻译基础

Lesson 1　科技英语概论 ·· 2
Lesson 2　科技英语构词法 ··· 7
Lesson 3　单词的译法 ··· 11
Lesson 4　词类转换的译法 ··· 16
Lesson 5　句子成分转换的译法 ··· 21
Lesson 6　词序转变的译法 ··· 24
Lesson 7　被动语态的译法 ··· 27
Lesson 8　后置定语的译法 ··· 31
Lesson 9　长句的译法 ··· 36

Part 2　Introductory Chemistry Speciality English

Unit 1　Inorganic Chemistry ·· 41

Lesson 10　Introduction to Chemistry: The Central, Useful, and Creative Science ········· 41
　　Hands-On Chemistry: Penny Fingers ··· 43
　　【Reading Material】The Extinction of the Dinosaur ························ 45

Lesson 11　The Nomenclature of Inorganic Compound ································· 47
　　Hands-On Chemistry: Burning Money ·· 56
　　【Reading Material】Lucifers and Other Matches ··························· 58

Lesson 12　The Periodic Table ··· 60
　　Hands-On Chemistry: Rubber Egg & Chicken Bones ·················· 63
　　【Reading Material】Out of Oxygen ··· 64

Lesson 13　Chemical Bond ··· 67
　　Hands-On Chemistry: Bending Water ······································· 69
　　【Reading Material】Who Killed Napoleon? ································· 71

Lesson 14　Coordination Chemistry ·· 73
　　Hands-On Chemistry: Relax with Beautiful Bath Salts ················ 76
　　【Reading Material】Nitric Oxide Gas and Biological Signaling ········ 77

Key Terms to Inorganic Chemistry ··· 79

Unit 2　Organic Chemistry ·· 80

Lesson 15　Introduction to Organic Chemistry ··· 80
　　Hands-On Chemistry: Introduction to Countertop Chemistry ········· 84
　　【Reading Material】History of Organic Chemistry ······················· 85

Lesson 16　Nomenclature of Organic Compounds ········ 87
　　　　　Hands-On Chemistry: Dancing Spaghetti ········ 97
　　　　　【Reading Material】IUPAC Nomenclature ········ 98
Lesson 17　Types of Organic Reactions ········ 100
　　　　　Hands-On Chemistry: Remove Tarnish from Silver ········ 101
　　　　　【Reading Material】Harvard University ········ 103
Key Terms to Organic Chemistry ········ 105

Unit 3　Physical Chemistry & Environmental Pollution ········ 106
Lesson 18　Introduction of Physical Chemistry ········ 106
　　　　　Hands-On Chemistry: Atomic Weight Calculation ········ 108
　　　　　【Reading Material】Legal Responsibilities of Undergraduates in Chemistry Department ········ 110
Lesson 19　Chemical Equilibrium and Kinetics ········ 112
　　　　　Hands-On Chemistry: ^{14}C Dating of Organic Material ········ 114
　　　　　【Reading Material】Green Plastics ········ 116
Lesson 20　Environmental Pollution ········ 118
　　　　　Hands-On Chemistry: Etymology of Chemistry ········ 119
　　　　　【Reading Material】The Killer Lake ········ 121
Key Terms to Physical Chemistry ········ 123

Unit 4　Analytical Chemistry ········ 124
Lesson 21　Introduction to Analytical Chemistry ········ 124
　　　　　Hands-On Chemistry: Atmospheric Can-Crusher ········ 126
　　　　　【Reading Material】Experimental Errors ········ 128
Lesson 22　Volumetric Analysis and Qualitative Analysis ········ 129
　　　　　Hands-On Chemistry: What is the Volume of Your Lungs? ········ 132
　　　　　【Reading Material】The Conjugate Acid-Base Pair and Titration ········ 133
Lesson 23　Ultraviolet and Visible Molecular Spectroscopy ········ 135
　　　　　Hands-On Chemistry: Chromatography of Leaves ········ 138
　　　　　【Reading Material】UV / Vis Spectroscopy ········ 139
Lesson 24　Nuclear Magnetic Resonance and Mass Spectroscopy ········ 142
　　　　　Hands-On Chemistry: The Lemon Battery ········ 145
　　　　　【Reading Material】Analytical Chemistry and Society ········ 147
Key Terms to Analytical Chemistry ········ 149

Unit 5　Biochemistry ········ 150
Lesson 25　Introduction to Biochemistry ········ 150
　　　　　Hands-On Chemistry: Sizzle Sources ········ 153
　　　　　【Reading Material】Ada Yonath ········ 155
Key Terms to Biochemistry ········ 157

Unit 6　Polymers ········ 158
Lesson 26　Introduction to Polymers ········ 158
　　　　　Interesting Tidbits: About Polymers ········ 163
　　　　　【Reading Material】The History of Polymers ········ 164

Lesson 27　Discovery of Polyethylene and Nylon ··· 166
　　　　　　Hands-On Chemistry: Clear Slime Polymer ·· 168
　　　　　　【Reading Material】Biopolymers ·· 170
Key Terms to Polymer Science ·· 172

Unit 7　Food Chemistry ·· 173
Lesson 28　Introduction to Food Chemistry: The Science of Studying the Chemical Nature of Foods ············ 173
　　　　　　Hands-On Food Chemistry: Volume Changes During Freezing ··············· 175
　　　　　　【Reading Material】Organic and Natural Foods ···································· 176
Key Terms to Food Chemistry ··· 179

Unit 8　Pesticide Chemistry ·· 180
Lesson 29　Introduction to Pesticide Chemistry ·· 180
　　　　　　Hands-On Chemistry: Cleaning Your Insects ·· 183
　　　　　　【Reading Material】Pesticides and Children ·· 185
Key Terms to Pesticide Chemistry ··· 187

Unit 9　Principles of Chemical Engineering ··· 188
Lesson 30　Introduction to Principles of Chemical Engineering ···························· 188
　　　　　　Hands-On Chemistry: An Explanation to Chemical Engineering ············ 191
　　　　　　【Reading Material】Basic Trend in Chemical Engineering ···················· 192
Lesson 31　Heat Transfer ·· 194
　　　　　　Hands-On Chemistry: Heat Transfer ··· 196
　　　　　　【Reading Material】Applications of Thermodynamics ·························· 197
Key Terms to Chemical Engineering ·· 199

Unit 10　Fine Chemicals ·· 200
Lesson 32　Introduction to Fine Chemicals ··· 200
　　　　　　Hands-On Chemistry: How to Make Disappearing Ink ···························· 203
　　　　　　【Reading Material】Chemists Discover How Antiviral Drugs Bind To and Block Flu Virus ········ 204
Key Terms to Fine Chemistry ··· 206

Part 3　Chemical Reactions and Stoichiometry

Lesson 33　Chemical Reactions ·· 208
　　　　　　Hands-On Chemistry: Any Fun Things I Can Make with Ammonium Nitrate? ········ 210
Lesson 34　Chemical Calculations ·· 211
　　　　　　Hands-On Chemistry: Let Make Chocolate Chip Cookies! ······················ 213

Part 4　Chemical Laboratory

Lesson 35　Common Labware ··· 216
Lesson 36　Work in Chemical Laboratory ··· 222
Lesson 37　An Experiment Case: Synthesis of Aspirin ·· 228

Part 5 Reference Reading for Graduate Admission Examination

Lesson 38　研究生复试英语考试参考样题 ···235
Lesson 39　Letters of Application to Graduate Schools ···237
Lesson 40　诺贝尔化学奖（2014）颁奖晚宴演讲——斯特凡•赫尔 ·······························238

Part 6 Guide for Scientific Paper Writing

Lesson 41　Notes on the Structure of a Scientific Paper ···241
Lesson 42　Style of Writing and Use of English in Essays and Scientific Papers ·······246
Lesson 43　The Abstract ··251

Appendix

Appendix 1　Sample Exams ···257
Appendix 2　Prefixes and Suffixes in Chemistry Speciality English ······························268
Appendix 3　Chemical Abbreviations ··271
Appendix 4　English Name and Pronunciation of Greek Alphabet ·······························272
Appendix 5　English Speaking of Chemical Formula, Equations and Mathematical Expressions ·······273
Appendix 6　Chinese-English Comparison of Common Terms in College Teaching ···275
Appendix 7　Chemical Reference Books and Online Resources in English ···············276
Appendix 8　Key to Exercises and Sample Exams ···277
Appendix 9　The Periodic Table of the Chemical Elements ···290

Part 1
科技英语翻译基础

Lesson 1　科技英语概论

科技英语（English for Science and Technology，简称 EST）是指用于自然科学与工程技术领域的一种英语文体。专业英语（English for Special Science and Technology）是科技英语的一部分，是结合各自专业的科技英语，学好科技英语对专业英语具有很大帮助。大学生进入高年级后开始学习专业英语，对此，相当一部分学生认为，科技英语不过就是"公共英语+科技词汇"，这种观点是很不全面甚至错误的。学好科技英语最好的办法就是先认识科技英语。

一、科技英语的特点

科技英语的特点主要体现在语法、词汇、文体等方面。

1. 语法特点

主要表现为"四多"，即词类转换多、被动语态多、后置定语多、复杂长句多。

（1）词类转换多

词类转换是指在翻译过程中将英文中的某种词类译成汉语中的另一种词类。由于英汉两种语言在表达方式上有很大差别，在翻译时，应进行词类转换，以便符合汉语的表达习惯。例如：

The **operation** of a machine needs some **knowledge** of its performance.

操作机器需要懂得机器的一些性能。

The continuous process can ordinarily be handled in the **less** space.

连续过程通常能节省操作空间。

原文中"operation, knowledge"两个名词均作为"动词"翻译；形容词"less"也作为"动词"翻译。

（2）被动语态多

通常科技人员最关心的是行为、活动、事实本身，至于谁做的无关紧要；运用被动语态不仅文章所描述的内容比主动语态显得更客观，而且形式上也更简洁明了。例如：

Mathematics **is used** in many different fields.

People **use** mathematics in many different fields.

第一句的表达效果更好。

（3）后置定语多

后置定语是指位于其所修饰名词之后的定语。汉语中习惯用前置定语来说明一个名词，而科技英语更多使用后置定语，且定语越长，越易后置。为了强调所修饰的名词通常也会定语后置。例如：

Besides, isomerization processes may also take place **which in turn lead to other fairly complicated reaction.**

此外，还会发生异构化过程，从而相继导致其他复杂反应的发生。

此句中 which 作关系代词，修饰 processes，同时 processes 也作定语从句的主语。

（4）复杂长句多

科技英语在表述一个复杂概念时，为表达清楚，使逻辑严密，往往使用许多长句。例如：

It is important that engineers have an understanding of the physical laws governing these transport processes if they want to understand what is taking place in engineering equipment and to make wise decisions with regard to its economical operation.

如果工程师想搞清楚机械设备中发生了什么并在操作过程做出明智的选择，那么掌握这些决定传递过程的物理定律就显得非常重要。

译成汉语时，需认真分析句子中各成分的关系，按照汉语习惯译成若干个简单句。复杂长句的翻译是科技英语翻译中的一个难点问题。

2. 词汇特点

（1）大量使用专业词汇，经常利用前缀、后缀构成派生词；大量使用缩略语。例如：

tetracycline	四环素［tetra- "四"，cycl(ic) "环"，-ine 某一类化学物质］
all-weather	适应各种气候的，全天候的（all 与 weather 间加连字符）
antiaging	防衰老的（anti 与 aging 间不加连字符）
mascon	质量密集的物质（mass 与 concentration 两词各取前 3 个字母）
gravimeter	比重计（gravity 的前 5 个字母与 meter 合成）
lab	实验室（取 laboratory 的前 3 个字母）
XRD	X 射线衍射（由 X-ray Diffraction 缩略而成）
SEM	扫描电子显微镜（由 Scanning Electron Microscope 缩略而成）

（2）大量使用表示逻辑关系的连接词。常用连接词如下：

① 表示原因：because，as，for，because of，caused by，due to，owing to，as a result of。

② 表示转折：but，yet，however，otherwise，nevertheless。

③ 表示逻辑顺序：so，thus，therefore，furthermore，moreover，in addition to。

④ 表示限制：only，if only，except，besides，unless。

⑤ 表示假设：suppose，supposing，assuming，provided，providing。

3. 文体特点

科技英语总体风格是简洁、客观和精确，行文时陈述句多，感叹句、疑问句少。行文晦涩、感情外露、刻意修辞等方式都会尽量避免。例如：

Since 1940 the chemical industry has grown at a remarkable rate. The lion's share of this growth has been in the organic chemicals sector due to the development and growth of the petrochemicals area since 1950. The explosive growth in petrochemicals in the 1960s and 1970s was largely due to the enormous increase in demand for synthetic polymers such as polyethylene, polypropylene, nylon, polyesters and epoxy resins.

科技论文在说明事理、提出设想、探讨问题和推导公式时，常常涉及各种前提条件和场合。为了避免武断，保持谦虚或谨慎的态度，经常使用虚拟语气，使口吻变得委婉。例如：

If a pound of sand **were broken up and turned** into atomic energy, there **would be** enough power to supply the whole of Europe for a few years.

如果一磅沙子裂变，放出(所有)原子能，那将够整个欧洲用几年。

在说明注意事项时，则常常使用祈使句。例如：

Be careful not to mix the liquids.

注意！不要把这几种液体混起来。

二、科技英语翻译标准

自从中国古代翻译印度佛经以来，翻译界对于翻译的质量标准进行过不少探讨。如隋代的释彦琮认为译者应该：诚心爱法，志愿益人，不惮久时；襟抱平恕，器量虚融，不好专执；耽于道术，澹于名利，不欲高炫。唐玄奘提出翻译"既须求真，又须喻俗"。清末翻译家严复(1853—1921)于1898年在《天演论》的《译例言》中提道："译事三难：信、达、雅"。"信"指的是译文忠实于原文；"达"则指译文通顺，没有语病；"雅"则指译文要自然优美。其中对"雅"的争论最大。鲁迅先生说："凡是翻译，必须兼顾着两面，一则当然求其易解，一则保留着原作的丰姿……"，因而，"雅"当理解为保留原作风格或神韵。而严复本人提出的"雅"是指使用"汉以前字法、句法"要达到"尔雅"的境界，并且坚持"与其伤雅，毋宁失真"的意见，此点在科技英语翻译中是不足取的。

对于科技英语翻译而言，因为原作的语言原本质朴，只要保持其朴素的文风即可，所以翻译时主要应做到"信""达"二字，即译文应准确无误、通顺畅达，使读者易于理解。注意以下两个问题。

（1）科技英语翻译时，应使用专业术语，避免说行外话产生歧义。例如：

The moment the circuit is completed, a current will start flowing toward the coil.

电路被完成这一刻，一个电流将开始流向这个线圈。（误译）

电路一接通，电流就开始流向线圈。

注：要注意对整个句子或语言环境的理解，而不能逐字逐句地翻译。

Heat-treatment is used to **normalize**, to **soften** or to **harden** steels.

热处理被用来使钢正常化，软化或硬化。（误译）

热处理可用来对钢正火、退火或淬火。

（2）科技英语翻译中，刻意追求的不是"雅"，而是科学性和严谨性，这是与非科技英语在翻译标准上的区别。例如：

The machine works properly.

这台机器运转正常。（误译：机器工作正常。）

work：工作、劳动、做事（指人），运转、转动、活动（指机械、身体器官等）。

Force is any push or pull that tends to produce or prevent motion.

力是能产生或阻止运动的任何形式的推或拉。（误译：力是任何倾向于产生或阻止运动的推或拉。）

三、翻译过程

在进行科技英语翻译工作时，首先要根据需要翻译的文献所涉及的专业内容做适当的知识准备，事先读一些有关专著。其次应准备翻译用的工具书，包括普通英汉词典、收录科学技术词汇的英汉词典以及专收该专业词汇的英汉词典。此外，参考一些汉语的科技词典或科技手册也是十分必要的。翻译前，先要通读全文，找出生词及疑难点。通过工具书的帮助解决理解上的困难后，完整、准确地理解全文，然后再一句一句地加以分析、翻译。译出的文字要按照汉语的表达习惯加以修改、组织。最后对文字进行推敲、加工，力求译文准确无误、通顺。概括起来，翻译过程一般经过"理解""表达""校对"三个阶段。

1. 理解阶段

翻译成败的关键是理解。首先把全篇文章阅读一遍，领略大意。有时同一句话，在不同的上下文中意思完全不同，需要结合上下文，逐句推敲，弄清语法结构关系。例如：

The tremendous range of emulsifiers available today permits selection of combinations which make possible emulsification at room temperature.

emulsifiers：乳化剂、乳化器。

究竟是"乳化剂"还是"乳化器"？根据上下文得知应译为"乳化器"。

译文：目前出售的大量不同乳化器使我们能够选择乳化器组，从而在室温下进行乳化。

2. 表达阶段

表达，就是选择恰当的汉语，把已经理解的原作内容叙述出来。在翻译的表达阶段，要注意以下问题。

（1）规范性

科技英语语体讲究论证的逻辑性，语言要规范。翻译时，自然要用符合科技语体规范的汉语来表达。例如：

The rotation of the earth on its own axis causes the change from day to night.

地球绕轴自转，造成昼夜的更替。

Matter is anything having weight and occupying space.

凡物质，都具有质量和占有空间。

（2）逻辑性

科技文章反映的是事物逻辑思维的结果，译者不仅要考虑句中的各种语法关系，更要注意各概念间的逻辑关系。表达的好坏取决于对原文理解的确切程度和对汉语的掌握程度。如果译文仅仅是意思对，但不能用通顺流畅的汉语表达，仍不是一篇好译文。例如：

The homologs of benzene are those containing an alkyl group or alkyl groups in place of one or more hydrogen atoms.

该句话易于理解，但却难于表达。若译作：苯的同系物就是那些被一个或多个烷基取代一个或多个氢原子所形成的产物。该译文尽管意思差不多，但令人感到啰嗦费解。

正确译文：苯的同系物是那些苯环上含有单烷基（取代一个氢）或多烷基（取代多个氢）的物质。

3. 校对阶段

校对阶段是对上两个阶段的内容进行校对，检查译文是否能准确无误地转述原作内容；译文的语言表达是否规范，是否符合汉语习惯。

正确的校对应当是：在誊写工整的译文上进行修改校对，即一校。一校后，再誊写清楚。然后再进行修改校对，如此重复，直到满意为止。

翻译不是一种机械劳动，而是复杂的创造性的脑力劳动，只有熟悉英汉两种语言并掌握了翻译方法之后，才能胜任翻译工作。

Exercises

Translate the following sentences into Chinese.

(1) The power plant is the heart of a ship.
(2) All bodies are known to possess weight and occupy space.
(3) The removal of minerals from water is called softening.
(4) Einstein's relativity theory is the only one which can explain such phenomena.

(5) The most important of the factors affecting plant growth is that it requires the supply of water.
(6) The doctor analyzed the blood sample for anemia.
(7) Many elements in nature are found to be mixtures of different isotopes.
(8) That like charges repel but opposite charges attract is one of the fundamental laws of electricity.
(9) The two units used most frequently in electricity are volt and ampere: this is the unit of voltage and that of current.

Lesson 2　科技英语构词法

一、合成法
将两个或两个以上的词合成为一个词的方法称为合成法（Composition）。这种方法在英语的构词法中十分普遍，在科技英语中占的比重较大。

1. 合成名词

（1）名词+名词（n.+ n.）

由两个或两个以上的名词构成一个合成名词，前面的名词说明后面的名词，合成名词含义由后一个名词表达。例如：

steel pipe　铁管　　　　　　　　　　carbon steel　碳钢
rust resistance　防锈　　　　　　　　water vapor　水蒸气

（2）形容词+名词（adj.+n.）

其含义关系是前者修饰后者。例如：

blueprint　蓝图　　　　　　　　　　periodic table　周期表
atomic weight　原子量

（3）动名词+名词（v.ing+n.）

动名词表示与被修饰词有关的动作，名词表示可用的场所或物品。例如：

launching site　发射场　　　　　　　flying-suit　飞行衣
navigating instrument　导航仪

（4）名词+动名词（n.+v.ing）

paper-making　造纸　　　　　　　　ship-building　造船

（5）其他构成方式

by-product　副产品（介词+名词）　　make-up　化妆品（动词+副词）
out-of-door　户外（副词+介词+名词）　pick-me-up　兴奋剂（动词+代词+副词）

2. 合成形容词

（1）形容词+名词

new-type　新型的　　　　　　　　　long-time　持久的，长期的

（2）名词+分词

① 名词+现在分词：有主动含义，其中名词相当于动作的宾语。

chinese-speaking people　说汉语的人　sound-absorbing material　吸音材料

② 名词+过去分词：有被动含义，其中名词相当于动作的发出者。

man-made satellite　人造卫星

（3）副词+分词（其中副词表示程度、状态）

hard-working　勤劳的　　　　　　　far-ranging　远程的
well-known　著名的　　　　　　　　newly-invented　新发明的

（4）形容词+分词

free-cutting　易切削的　　　　　　　direct-acting　直接作用的

ready-made 现成的   ill-equipped 装备不良的，能力不足的

（5）名词+形容词（此类合成形容词中的名词，有时是比喻的对象）
paper-thin 薄如纸的   colour-blind 色盲的
skin-deep 肤浅的

二、转化法

在现代英语的发展过程中，基本上摒弃了词尾的变化，可以把一个单词直接由一种词类转化为另一种词类，这种构词法称为转化法（Conversion）。其基本特点是保持了原来的词形，但改变了原来的词性，词义基本不变或稍有引申。

1. 名词转化为动词
（1）抽象名词转化为动词

名词	动词
form 形式	form 形成
heat 热	heat 加热
power 动力	power 用动力发动
knowledge 知识	knowledge 了解，知道

（2）物品名称转化为动词

名词	动词
machine 机器	machine 加工
bottle 瓶子	bottle 瓶装
oil 油	oil 加油（如：oil the car）
picture 图画	picture 描绘

2. 形容词转化为动词

形容词	动词
clean 干净的	clean 使干净，清洁
dry 干燥的	dry 干燥，弄干
better 好的	better 改善

3. 副词转化为动词

副词	动词
up 向上	up 提高
back 向后	back 倒车，后退
forward 向前	forward 推进

4. 动词转化为名词

动词	名词
flow 流动	flow 流量（如：the flow of electricity 电量）
stand 站立	stand 支架，看台

三、派生法

派生法（Derivation）就是通过加前缀、后缀构成一个新词。这是化学化工类科技英语中最常用的构词法。据估计，知道一个前缀可帮助人们认识 50 个英语单词。一名科技工作者至少要知道近 50 个前缀和 30 个后缀。这对扩大科技词汇量，增强自由阅读能力，提高翻译质量，加快翻译速度都是大有裨益的。

例如，有机化学中，烷烃就是用前缀(如拉丁或希腊前缀)表示分子中碳原子数再加上

"-ane"作词尾构成的。若将词尾变成"-ene""-yne""-ol""-al""-yl",则分别表示"烯""炔""醇""醛""基"等。以此类推,从而构成千万种化学物质名词。常遇到这样的情况,许多化学化工名词在字典上查不到,但若掌握这种构词法,将其前缀、后缀的含义合在一起即是该词的意义。

更多可参阅附录2化学专业英语常用前后缀。后面的有机物和无机物命名课文中也会详细讨论。

四、压缩法

1. 只取词头字母

EST: English for Science and Technology　科技英语
TOEFL: Test of English as a Foreign Language　非英语国家英语水平考试
IBM: International Business Machines Corporation　国际商务机器公司
CPU: Central Processing Unit　中央处理器
ppm: parts per million　百万分之一
CAD: Computer Aided Design　计算机辅助设计

2. 将单词删去一些字母

h, hr:	hour	小时
y, yr:	year	年
lab:	laboratory	实验室
kg:	kilogram	千克
km:	kilometer	千米
bldg:	building	大楼
corp:	corporation	股份公司
Co. Ltd.:	Corporation Limited	有限公司
CN:	China	中国

五、混成法

混成法(Blending)指将一个词的词头和另一个词的词尾连在一起,构成一个新词。
smog=smoke + fog　烟雾
motel=motor + hotel　汽车旅馆
positron=positive + electron　正电子
medicare = medical + care　医疗保障
sultaine= sulfo + betaine　磺基甜菜碱
modem=modulator + demodulator　调制解调器
aldehyde=alcohol + dehydrogenation　醇+脱氢——醛

六、符号法

&:　and 和
/:　and 或 or,"和"或"或"(例如:M/N——M和N,M或N)
#:　number 号码(例如:#9=No.9= number 9)
$:　dollar 美元,加元
£:　pound 英镑

¥：yuan 元（人民币）

七、字母象形法

字母象形法（Letter Symbolizing）构成词的模式是：大写字母+连字符+名词，用以表示事物的外形、产品的型号、牌号等。英译汉时，注意采用形译法，亦可根据具体情况翻译。

I-bar / I-steel　工字铁　　　　　I-shaped　工字形
T-square　丁字尺　　　　　　　T-beam　T 字梁
T-connection　T 字连接（三通）　U-shaped magnet　马蹄形磁铁
U-pipe　U 形管　　　　　　　　X-ray　X 射线
V-belt　V 带　　　　　　　　　P-N-junction　P-N 交点
n-region　n 区　　　　　　　　 p-region　p 区

Exercises

1. Translate the following terms into English.
   （1）产率　　　（2）碳钢　　　（3）化妆品　　（4）原子量　　（5）周期表
   （6）兴奋剂　　（7）户外　　　（8）烟雾　　　（9）X 射线　　（10）工字钢
2. Translate the following terms into Chinese.
   （1）EST　　　（2）CET　　　（3）rpm　　　（4）SOS　　　（5）CAD
   （6）h　　　　（7）CN　　　　（8）y　　　　（9）U-pipe　　（10）U-shaped magnet
3. Identify the prefixes or suffixes of the following words.
   （1）asexual　　　（2）disproportion　　（3）immaterial　　（4）incomplete
   （5）antibiotic　　（6）counteraction　　（7）semiconductor　（8）kilogram
   （9）hydrocarbon　（10）photochemical　（11）translator　　（12）chemist
   （13）calculation　（14）leakage　　　　（15）resistance　　（16）clockwise

Lesson 3　单词的译法

一、名词单复数的译法

（1）可数名词复数前没有数量词时，一般要把复数含义翻译出来，即在名词前加译"一些、这些、许多"等词，例如：

The teacher may be asked **questions**.
可以向老师提**一些**问题。

Our first electronic **computers** were made in 1958.
我国**首批**电子计算机是 1958 年制成的。

We are **students** of Beijing University.
我们是北京大学的学生。（前面出现了"我们"，后面"学生"就不必加"们"了。）

（2）当名词用复数（可数、不可数）表示泛指或一类时，有时需要译出复数含义，此时往往加译"各种、多种"等总括性词语，有时不需译出复数含义，具体情况应视上下文而定。例如：

Coal, petroleum and natural gas now yield their bond **energies** to man.
煤，石油和天然气现在为人类提供**各种各样的**结合能。

There are **families** of hydrocarbon.
有**好几个**烃族。

cultures：**各种**文化

Properties of **non-metals** vary widely.
非金属的性质差异很大（不需译出复数含义）。

（3）a(an)+名词（可数、不可数）单数表示"一类"时，不用译出不定冠词"a(an)"。例如：

Salts may also be found by the replacement of hydrogen from **an acid** with **a metal**.
盐也能通过用金属置换酸中的氢而获得。

An acid was once defined as a substance that would form hydrogen ions(H^+) in water solution and **a base** as one that would form hydroxide ions(OH^-) in the same.
人们曾把酸定义为在水溶液中能产生氢离子的物质，而碱则是在同样溶液中会产生氢氧根离子的物质。

（4）不可数名词用复数可表示数量之多。

waters　水域
sufferings　灾难
respects　尊敬
regards　问候（指大量的关心）

二、数词复数词组的译法

These books are packed **in tens**.
这些书**每十本装一包**。

They went out **by twos and threes.**
他们三三两两地出去了。

They consulted **tens of** magazines.
他们查阅了几十本杂志。

Automation helps to increase productivity **hundreds of** times over.
自动化使生产率提高了几百倍。

当数词复数用 of 相连时，其最终数字范围应该是几个数字的乘积。

Tens of thousands of foreign friends visit this factory every year.
每年有几万人参观这座工厂。

三、词义引申的译法

词义引申是指对某些英语词汇的含义加以扩展和变通，使其更准确地表达原文的特定意思。词义引申主要包括两方面的内容。

1. 具体化或形象化引申

科技英语中有些字直接译出，要么不符合汉语表达习惯，要么就难以准确地传达原文的含义。在这种情况下，就应该根据特定的语境，用比较具体或形象化的汉语词汇对其词义加以引申。例如：

Other **things** being equal, copper heats up faster than iron.
相同**条件**下，铜比铁热得快。

Steel and cast iron also differ in **carbon**.
钢和铸铁的**含碳量**也不相同。

2. 概括化或抽象化引申

科技英语文章中有些词语的字面意思比较具体或形象，但若直译成汉语，有时会显得牵强，不符合汉语的表达习惯。在这种情况下，就应用较为抽象化的词语加以引申。例如：

Americans every year **swallow** 15000 tons of aspirin, one of the safest and most effective drugs invented by man.
阿司匹林是人类发明的最安全、最有效的药物，美国每年要**消耗**15000吨。

Alloys belongs to **a half-way house** between mixture and compounds.
合金是介于混合物和化合物之间的**一种中间结构**。

Industrialization and environmental degradation seem to go **hand in hand**.
工业化发展似乎**伴随**着环境的退化。

四、词的省译及增译

由于英汉两种语言在表达方式上的不同，在英译汉时，有时需要增加一些词，有时需要减少一些词，以符合汉语的表达习惯。

1. 省译

（1）冠词的省译

一般来说，英语冠词 the 和不定冠词 a 及 an 用作泛指时，常省略不译。另外，定冠词 the 用作特指时，根据汉语表达习惯，有时也可省略不译。

The atom is **the** smallest particle of **an** element.
原子是元素的最小粒子。

Although **the** world is large，man is able to live in only a small part of it.
尽管地球很大，可人类只能在其中很小的一部分地方生活。

（2）代词的省略

英语中表示泛指的人称代词，用作定语的物主代词、反身代词以及用于比较句中的指示代词翻译时，根据汉语的表达习惯常可省略。另外，有些代词可承前省略。例如：

Different metals differ in **their** conductivity.
不同金属具有不同的导电性能。（承前省略）

When the solution in the tank has reached the desired temperature, **it** is discharged.
当罐内溶液达到所要求的温度时，就卸料。

（3）连词的省略

英语中连词使用频率较高，而汉语则不然，因此翻译出来常省略不译。

There are some metals **which** are lighter than water.
有些金属比水轻。

When short waves are sent out and meet an obstacle, they are reflected.
短波发射出后，遇到障碍就反射回来。

（4）动词的省略

有些句子翻译时不用英文中的动词作谓语，而用名词、形容词等作汉译句的谓语。

Friction always opposes the motion whatever its direction **may be**.
不管运动方向怎样，摩擦力的方向总是与运动方向相反。

Then，**came** the development of the microcomputer.
后来，微型计算机发展起来了。

Evidently，semiconductors **have a less conducting capacity** than metal.
显然，半导体的导电能力比金属差。

（5）介词的省略

In the transmission of electric power，a high voltage is necessary.
远距离输电必须用高电压。

Most substances expand **on** heating and contract **on** cooling.
多数物质热胀冷缩。

2. 增译

（1）增译表示时态的词

① 一般现在时，一般无须加减字，但在表示主语特征时，可在动词前面加译"能、可、会"等字。

Extreme temperature **causes** dramatic changes in the properties of rubber.
橡胶处于极限温度时，其性能**会发生**惊人的变化。

Evaporation sometimes **produces** a slurry of crystals in a saturated mother liquor.
有时蒸发能在饱和的母液内**产生**晶浆。

② 一般过去时，有时可在动词前后添加"已、曾、过、了"等字，或于句首加"以前、当时、过去"等时间副词。

Organic compounds **were once thought to** be produced only by living organism.
以前曾认为有机化合物只能从有生命的机体中产生。

The publication of Chemical Engineering **was stopped** during World War Ⅱ, and the blanks **were later filled**.
《化学工程》在第二次世界大战期间**曾停刊过**，不过随后又**填补了**以前的空白。

③ 一般将来时，多数在动词前添加"将、要、会"等字。

They **will** perform the experiment next Monday.
他们**将**于下周一做那个实验。
④ 进行时，可在动词前添加"正、在、正在"，有时在动词后面加译助词"着"。
Even before the Second Five-year plane, China **was already producing** all kinds of lathes, machines, apparatus and instruments.
甚至在第二个五年计划之前，中国就已经**在生产着**各种车床、机器、仪器和设备。
⑤ 完成时，可于动词前面加译时间副词"已（经）、曾（经）"，后面加译助词"了、过、过……了"。
The carbon **had lost** electrons and the oxygen **has gained** electrons in the change.
在变化中，碳**失去了**电子，而氧**获得了**电子。
It was reported that scientist **had worked** at the problem of storing the sun's heat for many years.
据报道，对于贮存太阳能的问题，科学工作者**曾经进行过**多年的**研究**。
（2）增译表示被动语态的词
The mechanical energy **can be changed** back into electrical energy through a generator.
发电机**可以把**机械能再**转变为**电能。
（3）其他词语的增译
① 增加表示动作意义的名词引起的增译："作用、过程、现象、情况、变化"等。
Oxidation will make iron and steel rusty.
氧化**作用**会使钢铁生锈。
The principles of absorption and desorption are basically the same.
吸收和解吸这**两个过程**的原理基本相同。
② 增加表示数量的词（以示强调）引起的增译："两个过程、三个方面、单独、都"等。
Ketones are very closely related to both aldehydes and alcohols.
酮与醛和醇的关系**都**很密切。
The project is the largest public works program undertaken by a state in the history of the United States.
这是美国历史上由一个州**单独**承担的最大的一项公共工程。
③ 增加解说性词语引起的增译。
Transistors are small，efficient and have a long life.
晶体管**体积**小、**效率**高、**寿命**长。
But，neutralization is cumbersome.
但是，中和是个麻烦的**问题**。
④ 增加关联词引起的增译以及增加连贯性的词引起的增译。
Evaporation differs from crystallization in that emphasis is placed on concentrating a solution rather than forming and building crystals.
蒸发和结晶不同，**因为**蒸发着重于将溶液浓缩而不是生成和析出结晶。

Exercises

Translate the following into Chinese.
（1）made of hard-alloy steels
（2）nutritious foods
（3）successes

(4) They will mix freely with other organic compounds and are often soluble in organic solvents.
(5) Properties of non-metals vary widely.
(6) Insects hide themselves in winter.
(7) He has put up a portrait in oils.
(8) When a boy is ill, a doctor usually takes his temperature.
(9) These products are counted by hundreds.
(10) The return of the light into the same medium in which it has been travelling is reflection.
(11) The presence of an acid stronger than nitric acid accelerates the heterolysis into NO^{2+} and OH^-.
(12) The rocket landed on the moon.
(13) From what is stated above, it is learned that the sun's heat can pass through the empty space between the sun and the atmosphere.
(14) The hydrocarbons are all lighter than water, and being almost completely insoluble, float on it.

Lesson 4　词类转换的译法

　　汉、英两种语言的表达方式千差万别，大多数情况下，逐词照译的翻译方式无法表达原文的意思。因此翻译好文章的关键是找到两种语言表达方式的异同，再根据各自的语言规范进行翻译。
　　英语：富于词形变化，词序比较灵活（过去式、过去分词、现在分词，可用于表达不同的时态和语态；定语后置，倒装，疑问句，状语的位置较随意等）。
　　汉语：缺乏词形变化，词序不灵活。
　　因此，英译汉时，译者应大胆摆脱原文表层结构的束缚，根据汉语的习惯，正确表达原意。转换译法是一种可取的、极为常用的翻译方法。转换译法包括词类的转换、句子成分的转换、词序的转换、主被动语态的转换、从句间的转换等。
　　词类转换，就是翻译时改变原文中某些词的词性，以符合汉语的表达习惯，主要涉及名词、动词、形容词、副词及介词等词类转换的翻译方法。

一、名词的转译

　　1. 转译为汉语的动词（n.→v.）
　　（1）具有动作意味的词，常常转译为汉语的动词。
　　Total **determination** of molecular is possible by means of X-ray diffraction.
　　用 X 射线衍射的方法可以全面地**确定**分子结构（谓语部分省译，total 转译为副词，另增译结构）。
　　Such operations require the **transfer** of a substance from the gas stream to the liquid.
　　这样的一些操作要求物质由气流**传递**到液体中去。
　　（2）英语中，某些由动词+er/or 构成的名词，如 teacher，writer，driver 等，当它并不真正表示某个人的职业时，可把该名词转译为汉语的动词。
　　He was a good **calculator**，so we considered the answer correct.
　　他**计算**很好，因此，我们认为这个答案是正确的。
　　I suggested the supports should be located every five meters. He was my good **supporters**.
　　我建议每隔五米设一个支架，他很**支持**我的建议。
　　（3）由动词派生的名词，有时可转译为动词。
　　Laser is one of the most sensational developments in recent years, because of its **applicability** to many fields of science and its **adaptability** to practical uses.
　　激光是近年来轰动一时的科技成就之一，因为它**可用于**各种科学领域，也**适合**各种实际用途。
　　2. 转译为汉语的形容词（n.→adj.）
　　（1）在英语的表达方式中，常常用名词来表达某物的性质，这样的名词一般被转换为汉语的形容词。
　　The laws of thermodynamics are of prime **importance** in the study of heat.
　　研究热时，热力学定律非常**重要**。

16

（2）有时把带有不定冠词的名词转换为汉语的形容词。

The maiden voyage of the newly-built steamship was **a success**.
那艘新轮船的处女航是**成功的**。

He is **a stranger** to the operation of the electronic computer.
他对计算机的操作是**陌生的**。

（3）有些由形容词派生出来的名词，可以转译为形容词。

He found some **difficulties** to design a chemical plant without an electronic computer.
他感到要设计一座化工厂没有计算机是很**困难的**。

The **complication** of mathematical problems made him into difficulties.
复杂的数学问题使他陷入了困境。

3. 转译为汉语的副词（n.→adv.）

Efforts to apply computer techniques have been a **success** in improving pyrolysis techniques.
努力使用计算机技术**成功**促进了热解技术的发展。

二、动词的转译

英语动词，从其词汇意义来讲，可分为两类：状态动词（State Verb）和动作动词（Action Verb），顾名思义，状态动词并不是真正地表示动作，而是表示相对静止的状态。在这类动词中，当表示主语的特征或状态时，为了适应汉语表达的习惯，可翻译为名词。

An acid and a base **react** in a proton transfer reaction. (v.→n.)
酸碱**反应**是质子转移反应。

Many chemical reactions need heat to make them **take place**. (v.→n.)
许多化学反应的**发生**都需要热。

The fibre-optic communication **features** high speed, large capacity and high reliability.
光纤通信的**特点**是速度快、容量大和可靠性高。(v.→n.)

This machine **works** efficiently. (v.→adj.)
这台机器的**工作**效率很高。

The output voltages of the control system **varied** in a wide range. (v.→adj.)
这台控制系统的输出电压**变化**范围很宽。

三、形容词的转译

1. 转译为汉语的名词（adj.→n.）

（1）用作表语或定语的形容词转译为汉语的名词。

Generally speaking, methane series are rather **inert**.
总的来说，甲烷系烷烃的**惰性**很强。

Zirconium is almost as **strong** as steel, but lighter.
锆的**强度**几乎与钢的强度相等，但它比钢轻。

This solar cell is only 7% **efficient**.
这个太阳能电池的**效率**只有 7%。

Contamination leads to **lower** yield.
污染导致成品率的**降低**。

（2）有些形容词加定冠词"the"后表示一类人（或物），这类形容词可称为名词化的形容词。

Electrons move from **the negative** to **the positive**.
电子由负极流向正极。

The social scientist have a keen sense of **the new** and **the old**.
社会科学家对新旧事务有敏锐的感觉。

2. 转译为汉语的动词（adj.→v.）

（1）由动词派生出来的形容词转换为汉语的动词。

As you know that also **present** in solid are numbers of free electrons.
正如我们所知，固体中也**存在**着大量的自由电子。

Figure 5 shows the **rising** power consumption at **increasing** clock frequencies.
图 5 表明，时钟频率增加时，功耗便升高。

（2）有些表示"感觉、欲望、愿望、心理、情感"等心理活动的形容词(短语)，当它们分别和系动词构成谓语时，可把它们转换成汉语的动词。这类词主要有：sure, certain, afraid, aware, glad, sorry, confident, determined, delighted, ashamed, convinced, concerned, careful, thankful, grateful。例如：

They are quite **content** with the data obtained from the experiment.
他们**对**实验中获得的数据非常**满意**。

Scientists are **confident** that all matter is indestructible.
科学家们**深信**，一切物质都是不灭的。

They are **sure** that they will be able to build the factory in a short period of time.
他们**肯定**能在短时期内建成那座工厂。

（3）其他（adj.→v.）

The continuous process can ordinarily be handled in the **less** spaces.
连续过程通常能**节省**操作空间。

There is a large amount of energy **wasted** owing to friction.
由于摩擦而**消耗**了大量的能量。

3. 转译为汉语的副词（adj.→adv.）

（1）当英语名词译为汉语的动词时，修饰该名词的形容词就相应地转换为汉语的副词。

Below 4℃, water is in **continuous** expansion instead of **continuous** contraction.
水在 4℃以下**不断地**膨胀，而不是**不断地**收缩。

（2）习语化的短语动词中心词为名词，当名词被转换为汉语的动词时，修饰中心词的形容词也就相应地转换为副词了。

In assembling color TV sets, they made **full** use of home-made components.
在装配彩色电视机时，他们**充分**利用了国产元件。

But even this illustration does not give an **adequate** picture of the importance of friction.
但是，甚至这个例子还没有**充分地**描绘出摩擦的重要性。

（3）由于汉英两种语言在表达方式上的差异，汉译时，一些用于其他场合的形容词有时也可译为汉语的副词。

At that time, Copernicus was in the **clear** minority.
在那个时候，哥白尼**的确**是属于少数。

四、副词的转译

1. 转译为汉语的形容词（adv.→adj.）

（1）当英语的动词转换为汉语的名词时，原修饰该动词的副词往往转换为汉语的形容词。

The communication system is **chiefly** characterized by its ease with which it can be maintained.

这种通讯装置的**主要**特点是便于维修。

（2）把修饰形容词的副词转换为汉语的形容词。

This film is **uniformly** thin.

该膜薄而**均匀**。

Sulphuric acid is one of **extremely** reactive agents.

硫酸是**强烈的**反应试剂之一。

（3）英语中的少量副词可用作定语，如 here，there，around，home，on，up，off，about 等，汉译时，这类副词可转变为汉语的指示形容词作定语。

The table **above** shows it.

上面表格可说明这一点。

Many plants **here** are of modern construction.

这里的许多工厂是现代建筑。

2. 转译为汉语的名词（adv.→n.）

有些英语副词是由名词派生的，在它们表示"用……方法""在……方面"等意义时，为使译文通顺，可译为汉语的名词。

This crystal is **dimensionally** stable.

这种晶体的**尺寸**很稳定。

The device is shown **schematically** in Fig.8.

图 8 就是这种装置的**简**图。

Oxygen is one of the important element in the physical world, it is very active **chemically**.

氧是物质世界的重要元素之一，它的**化学**性质很活泼。

3. 转译为汉语的动词（adv.→v.）

英语的副词，如 on，off，up，in，out，over，behind，forward 和系动词 be 一起构成合成谓语、作宾补或作状语时，可转译为汉语的动词。

The reaction force to this action force pushes the rocket ship **along**.

这个作用力所产生的反作用力推动宇宙飞船**前进**。

The chemical experiment is **over**.

化学实验**结束**了。

五、介词的转译

有动作意义的介词，如 for，by，in，past，with，over，into，around，across，toward，through 等，汉译时，可转换为汉语的动词。

In this process the solution is pumped **into** a tank.

在这个操作中，溶液用泵打入罐中。

Salts may be formed by the replacement of hydrogen from an acid **with** a metal.

盐可以用金属置换酸中的氢来制备。

Exercises

Translate the following sentence into Chinese.

(1) Most of the TV pictures are influenced by high buildings.

(2) The methyl group on the benzene ring greatly facilitates the nitration of toluene.

(3) They will do their best to build a school for the blind and the deaf.
(4) He is quite familiar with the performance of this transmitter.
(5) There was a large amount of output power wasted due to leakage.
(6) Before the 19th century, most ways of making iron were very crude.
(7) A graph gives a visual representation of the relationship.
(8) They are quite content with the data obtained from the experiment.

Lesson 5　句子成分转换的译法

句子成分转换的译法，即把句子的某一成分（如主语）译成另一成分（如宾语等）。在多数情况下，词类转译必然导致句子成分的转译，如：当英语的动词转译为汉语的名词或副词时，该动词的谓语部分也就相应地转译为汉语的主语、宾语或状语等。例如：

It is because the bounding properties of one pair of electrons in the double bond are not fully satisfied.

这是由于双键上的一对电子不具有充分饱和的键合性质。

分句主语"the bounding properties"转译为宾语；分句定语"one pair of electrons"转译为主语；分句状语"in the double bond"转译为定语；分句表语"are satisfied"转译为定语。

一、主语的转译

1. 转译为宾语

（1）被动句译为汉语的主动句时，可将原文的主语转换为宾语

Water can be shown as containing admixtures.

可以证明水含有杂质。

（2）There be 句型的转译

Organic compounds are not soluble in water because there is no **tendency** for water to separate their molecules into ions.

有机化合物不溶于水，因为水没有将它们的分子离解为离子的**倾向**。

2. 转译为谓语

（1）当主语为动作性名词，如 care，need，attention，inspection，thought，improvement，attempt 等，而谓语动词是系动词，汉译时可将其转换为汉语的谓语。

The following **definitions** apply to the terms used in this specification.

本说明所用的一些术语**定义**如下。

Evaporation **emphasis** is placed on concentrating a solution rather than forming and building crystals.

蒸发**着重**于浓缩溶液，而不是生成和析出晶体。

（2）转换为判断合成谓语

在英语中，当用名词作表语时，主语和表语所表达的内容往往是一致的。汉译时，主语可转换为判断合成谓语，即用判断词"是"和它后面的名词等结合起来，对主语进行判断，共同作谓语。

Matter is anything having weight and occupying space.

凡具有重量和占有空间的东西都是**物质**。

（3）其他

Methane is less than half as heavy as water.

甲烷的重量不到水的一半。（主语→定语）

An atom of oxygen weights 16 times as much as an atom of hydrogen.
氧原子的质量是氢原子质量的 16 倍。（主语→定语）

When a copper plate is put into the sulfuric acid electrolyte, **very few of** its atoms dissolve.
当将铜板置于硫酸电解液中时，几乎没有铜原子溶解。（主语→状语）

二、谓语的转译

当谓语动词为具有名词意义的状态动词时，可转译为主语或宾语。

The past few decades **have been characterized** by a prodigious expansion of the organic chemical industry.
过去数十年的**特征**是有机化学工业得到了惊人的发展。（谓语→主语）

The melting point of alkanes are rather irregular at first, but **tend to** rise somewhat steadily as the molecules become larger.
烷烃的熔点起初很不规则，但随着分子的增大，则有些稳步上升的**趋势**。（谓语→宾语）

三、宾语的转译

TNT has high **explosive power**.
TNT 的**爆炸力**很大。（宾语→主语）

It is assumed that there is little if any leakage through **the condensers**.
可以设想，**电容器**即使漏电，也是很少的。（宾语→主语）

Light makes **vision** possible.
有了光，才能**看见**东西。（宾语→谓语）

四、表语的转译

1. 转译为主语

原文中的主语转换为汉语的判断合成谓语时，已经涉及原文的表语向汉语的主语的转译。

Matter is **anything that occupies space**.
凡占有空间的都是物质。

2. 转译为谓语

"系动词/be+表语"构成的谓语句子，若表语含有一定的动作意味，汉译时常常省译 be，而把表语直接译成谓语。

In this sense, structure analysis **is common to** most organic research.
从这个意义上讲，结构分析普遍应用于大多数的有机研究。

The formula for kinetic energy **is applicable to** any object that is moving.
动能公式适用于任何运动的物体。

3. 转译为定语

在"系动词/be+形容词"的句型中，如果主语原有定语，汉译时可把原文的表语转换为汉语的定语，原主语的定语转换为判断合成谓语。

In their laboratory, few instruments **are valuable**.
他们实验室里，贵重的仪器不多。

五、定语的转译

There are two groups **of metals**: pure metals and their alloys.
金属有两大类：纯金属及其合金。（定语→主语）

Many factors enter into equipment reliability.

涉及设备可靠性的因素很多。（定语→谓语）

Automatic machines **having many advantages** can only do the jobs they have been "told" to do.

自动化机器虽然有很多优点，但它们只能干人们"吩咐"的事。（定语→状语从句）

Benzene can undergo the typical substitution reactions **of halogenation, nitration, sulphonation and Friedel-Crafts reaction.**

苯可以进行典型的取代反应，如卤化、硝化、磺化和傅氏反应。（定语→同位语）

六、状语的转译

1. 转译为主语

在翻译表示地点、方位或某一方面的介词短语所作的状语时，为了使主谓分明，可将其转译为汉语的主语，介词可省译。

When oxides of nitrogen are absorbed **in water** to give nitric acid, a chemical reaction occurs.

当水吸收氮氧化物形成硝酸时，就发生了化学反应。

2. 转译为补语

英语的补语分别说明主语或宾语，故而又可分为主语补足语和宾语补足语。

汉语的补语是动词或形容词的补充说明成分，表示"程度、怎么样、多久、多少次"的意思，特点是谓语和补语之间常用结构助词"得"。

The attractive force between the molecules is **negligibly** small.

分子间的吸引力小得可以忽略不计。

The particles move **faster in the place** where the body is being heated.

物体受热的地方，粒子运动得较快。

Exercises

Translate the following into Chinese.

(1) The melting point is different for different kinds of metals.

(2) When a spring is tightly stretched, it is ready to do work.

(3) TNT is simple and relatively safe to manufacture.

(4) The reagent can be shown as containing admixtures.

(5) The following definitions apply to the terms used in this specification.

(6) Modern industry requires that more and more natural gas be tapped.

(7) Warm-blooded animals have a constant body temperature.

(8) A sketch serves to express one's idea graphically.

(9) This sort of stone has a relative density of 2.7.

(10) Physical changes do not result in formation of new substances, nor do they involve a change in composition.

(11) Many factors enter into equipment reliability.

(12) In size and appearance mercury is very much like moon.

(13) Sodium is very active chemically.

(14) At sea level our atmosphere exerts a pressure of about fifteen pounds per square inch.

Lesson 6　词序转变的译法

词序转变的译法，即译文的词序与原文的词序不同的一种翻译方法。英译汉时，有时词序相同，有时词序不同。是否改变词序，没有一成不变的模式，只取决于一个条件，即译文正确，符合汉语表达习惯。例如：

A six-carbon ring structure in the form of a hexagon with alternate single and double bond was assigned to benzene.

苯被指定为单键和双键交错的由六个碳组成的六角形的环形结构。

该译文的词序基本上是从后往前译。若按原文词序进行翻译，则语句不通顺。疑问句，倒装句，"there be"句型等引起的词序转变不在此讨论。

一、系表结构的词序转变

The commonly employed forms of energy are **kinetic energy and heat energy**.
动能和热能是最常使用的能。（主语、表语词序颠倒）

Non-conductors are **rubber and glass**.
橡胶和玻璃是非导体。（主语、表语词序颠倒）

The alternate double bond arrangement in the six-carbon ring is **aromatics characteristic**.
芳香烃的特点是六个碳原子组成的环上双键交替排列。（主语、表语词序颠倒）

二、宾语的词序转变

（1）宾语有时译在动词（包括谓语、现在分词、动名词）之前，同时加译"将、把、给、使、让、对"等字。例如：

People regarded **the sun** as the chief source of heat and light.
人们把太阳视为主要的热源和光源。（加译"把"，同时把宾语译在谓语前）

We can transform **water** into two gases by passing **an electric current** through it.
让电流通入水中，我们就能把水变为两种气体。

By compressing and cooling **the mixture**, you can separate **one gas** from the other by changing **it** to a liquid.
将该混合物压缩、冷却，使其中一种气体变为液体，就能把它与另一种气体分开。

（2）英语中有直接宾语和间接宾语时，汉语译文中常把一个宾语译在谓语前面，同时加"把、给"等助词。

The figure shows **the readers the relative action** of reciprocating between the work and the tool.

这个图给读者展示了工件和刀具之间往复的相对运动。

或：这个图把工件和刀具之间的运动展

24

示给了读者。
X-ray will show **the doctor** clearly how **the lung suffers**.
X射线会清楚地把肺部损害的程度显示给医生。
双宾语情况下，由于两个宾语是平行的，故两个宾语的位置在汉语译文中可互换。

三、同位语的词序转变

Two factors, **force and distance**, are included in the units of work.
力和距离这两个因素都包括在功的单位内。
The branch of science, **artificial intelligence**, is developing rapidly.
人工智能这门学科正在迅速地发展。

四、定语的词序转变

（1）在英语中，一个或几个形容词作定语，修饰同一个名词时，一般是置于被修饰词的前面，这与汉语是相同的。但是，当以动词不定式短语、分词短语作定语时，通常又置于被修饰的名词之后。在汉语中，不管是以单词作定语还是以词组作定语，一概置于被修饰词之前。按照汉语的表达习惯，英语的后置定语就应译为汉语的前置定语了。

The energy of steam comes from the heat **produced by burning coal**.
蒸汽能来自燃烧煤所产生的热量。
Generally speaking, the fuel **available** is coal.
一般说来，可用的燃料是煤。
Engine revolution should not exceed the maximum **permissible**.
发动机转数不应超过所允许的最大值。

（2）汉语和英语都有两个或两个以上的形容词修饰同一个名词的情况。但在它们的排列顺序上，汉英两种语言差别很大。

英语中多个形容词作定语的排列语序通常是：由次要的到主要的，由程度较弱的到程度较强的，由小范围的到大范围的，由一般的到专有的。

汉语排序与英语的次序恰恰相反，它们是由主要的到次要的，由程度较强的到程度较弱的，由大范围的到小范围的，由专有的到一般的。

因此，在汉译时，应根据汉语的表达习惯，把原文的若干个定语在词序上作必要的、恰当的调整。例如：

the advanced world experience
世界的先进经验
practical social activities
社会实践活动

五、状语的词序转变

Being alloyed with certain metals, aluminum can be strengthened.
铝和某些金属熔合后，强度会增大。（变序，状语译在主语之后）
Many industrial operations can be carried out **in either of two ways which may be called batch and continuous operations.**
许多工业操作可用间歇操作或连续操作来完成。（变序，状语译在谓语之前）
Chemisches Zentralblatt is particularly valuable **for its inclusiveness of the East European and Russian literature.**

《德国化学文摘》也包括东欧和俄罗斯的文献，就这一点来说是特别有价值的。（变序，状语译在表语之前）

Exercises

Translate the following into Chinese.

(1) He studies English very well.
(2) We call such a zinc atom an ion with a double positive charge, Zn^{2+}.
(3) We are all familiar with the fact that nothing in nature will either start or stop moving of itself.
(4) The decimal system of counting has a long history.
(5) In the absence of oxygen, untold numbers of organisms were transformed by heat, pressure, and time into deposits of fossil fuels, coal, petroleum and natural gas.
(6) The test piece shall be of length suitable for the apparatus being used.
(7) Heat is a form of energy into which all other forms are convertible.
(8) The force due to the motion of molecules tends to keep them apart.
(9) An electric current varies directly as the electromotive force and inversely as the resistance.
(10) The electrolytic process for producing hydrogen is not as efficient as the thermochemical process.
(11) Biochemistry is the study of the molecular basis of life.
(12) A continuous increase in the temperature of a gas confined in a container will lead to a continuous increase in the internal pressure within the gas.
(13) Structure analysis is common to most organic research.

Lesson 7　被动语态的译法

据初步统计，被动语态在科技英语中约占 1/3，远高于普通英语中被动句的使用频率。首先，与主动语态相比，被动语态突出所要说明的事物，能减少主观色彩。其次，被动句用行为、活动、作用、事实等作主语，能立即引起读者的注意（例如：结果、数据、实验现象等）。最后，通常被动句比主动句更为简洁明了。被动句有两种译法，即仍译成被动句和译成主动句。

一、仍译成被动句

（1）译文可加译"被、供、由、让、给、受、遭、得到、为……所、加以……、予以……"等助词，再译谓语。这些助词在汉语中都表示被动的意思。但根据具体情况，有时也可不加任何词而直接译出。

The atomic theory **was** not **accepted** until the last century.
原子学说直到上个世纪才**为**人们**所接受**。（或：**才被接受**）
Other evaporation materials and processes will **be discussed** briefly.
其他一些蒸发材料和蒸发过程将简单地**加以**讨论。
Purified hydrogen is **passed** over a liquid metal halide.
让纯化氢气通过液态金属卤化物。
Considerable use **is made of** these data.
这些资料**得到**充分的利用。
The electromagnetic disturbances **are caused by** lighting discharges.
电磁干扰**是由**雷电放电**引起的**。

（2）用 as 引出主语补足语的被动语态词组。
科技英语中，有不少用 as 引出主语补足语的被动语态词组，它们在汉语中已有固定译法。例如：

be accepted as　　被承认为　　　　be accounted as　　被认为是
be adopted as　　　被用作为　　　　be classified as　　被划分为
be conceived as　　被想象为　　　　be considered as　　被认为是
……

Heat **is regarded as** a form of energy.
热**被看作是**能的一种形式。

二、译成汉语主动句

（1）译成汉语的无主语句
这种译法一般是将英语句中的主语译成汉语中的宾语，即先译谓语，后译主语（主谓变序）。也就是说，将被动句的谓语译成主动形式，被动句中的主语译成此谓语的宾语，不译出主语。

Measures **have been taken** to diminish friction.
已经采取了一些措施来减少摩擦。

If the product is a new compound, the structure must **be proved** independently.
如果产物是新化合物，则必须单独**鉴定**其结构。

Now headings, sub-headings and tables of contents in English **are provided**.
现在**提供**英文的（主）标题，副标题和目录表。

The resistance can **be determined** provided that the voltage and current are known.
只要知道电压和电流，**就能确定**电阻。

（2）加译"人们、我们、大家、有人"等主语（主语不明确是谁）

The metal, iron in particular, **is known** to be an important material in engineering.
大家知道，金属，特别是铁，是工程方面的重要材料。

If one or more electrons are removed, the atom **is said** to be positively charged.
如果原子失去了一个或多个电子，**我们就说**这个原子带正电荷。

Potassium and sodium **are seldom met** in their natural state.
我们很少见到自然状态的钾和钠。

（3）用英语句中的动作者作汉语中的主语

The complicated problem will be solved **by them**.
他们会解决这个复杂的问题。

The molecules are held together **by attractive forces**.
引力把分子聚集在一起。

Even when the pressure stays the same, great changes in air density are caused **by changes in temperature**.
即使压力不变，**气温的变化**也能引起空气密度的巨大变化。

（4）将英语句中的一个恰当成分译成汉语句中的主语

The students have been answered all the **questions**.
学生们的**问题**已全部得到回答。

The oxides of nitrogen are absorbed **in water** to give nitric acid.
水吸收氮氧化物形成硝酸。

（5）在原主语前加译"把、将、使、给"等词

Borax was added to beeswax to enhance its emulsifying power.
将硼砂加到蜂蜡中以增加蜂蜡的乳化力。

The mechanical energy can be changed back into electrical energy through a generator.
发电机可以**把机械能**再转变为电能。

（6）将被动语态译成"是……的、对……进行"等

The slurry **is filtered** to recover the electrolyte solution.
对泥浆**进行过滤**可回收电解液。

The higher degree of unsaturation **is associated** with somewhat greater chemical reactivity.
高不饱和度**是**与高化学反应活性相关**的**。

（7）被动意义译成主动意思

Batch operations **are** frequently **found** in experimental and pilot-plant operations.
间歇操作**常见于**实验室操作及中试操作。

The kinds of activities which engage the organic chemist **may be grouped** in the following way.
有机化学工作者所从事的活动可以按照下列方法**归类**。

（8）主谓合译

把主语和谓语合在一起翻译，是运用转换法翻译被动语态句子的一种形式。它应用于翻译某些及物动词短语，如 make use of，pay attention to 等，它们的中心词 use，attention 等在被用作被动语态句的主语时，仍按原动词短语的含义进行翻译，因此形成了主谓合译。

Attention should be paid to the study of proteins.
应该注意蛋白质的研究。
Stress must be laid on the development of the electronics industry.
必须强调电子工业的发展。
类似的短语有：

make references to	提到，参考	make mention to	提及
take care of	照顾，照料	look down on(upon)	轻视，看不起
make fun of sb	取笑某人	put an end to	结束，终止
lose sight of	忽略，看不见	set fire to	点燃，使……燃烧
take advantage of sth	趁机利用		

三、科技英语中常用的被动句型

常见的句型是 It is done that……，其中 it 作形式主语，that 引导主语从句；另一种常见句型是 be done as，该种被动结构引出主语补足语。这些句型多半在汉语中已有习惯的译法。例如：

It can be seen that	可见，能够看见	It can be foreseen that	可以预料
It cannot be denied that	无可否认	It has been objected that	有人反驳
It has been viewed that	讨论了	It has been proved that	已经证实
It is preferred that	最好		

四、主动形式表示被动含义的句子

（1）英语中有一些动词，例如：wash，wear，find，build，divide，blow，catch，work，consist，look 等，在某些场合下，能以主动态表示被动含义。汉译这类句子时，往往采用顺译法，以相应的汉语主动句表示。

The integrated circuit **finds** wide application in electrical engineering.
集成电路在电气工程方面**得到了**广泛的应用。
The new laboratory **is building**.
新实验室**在建设中**。

（2）有时，在英语的某些形容词后以动词不定式或动名词的主动式表示被动含义。

A single piece of silicon is hardly big enough **to pick up**.
单个的硅芯片小得**捡**不起来。
The problem is worth **studying**.
这个问题值得**研究**。

（3）有的形容词在作表语时表示被动含义。

Atoms are not visible to the naked eyes.
原子是肉眼看不见的。

（4）有的介词短语在作表语或定语时，表示被动含义。

Our electronic workshop is under construction.
我们的电子车间正在建设中。

This law under discussion is true of all gases.
所讨论的定律对所有气体都是适用的。

Exercises

Translate the following sentences into Chinese.
(1) The laws of thermodynamics will be discussed in the next articles.
(2) Since methane is its first member, it is also known as the methane series.
(3) Radio waves are also known as radiant energy.
(4) Much greater magnification can be obtained with the electron microscope.
(5) Silver is known to be the best conductor.
(6) Large quantities of lasers are used by optic communication in the generation of light emitting.
(7) When the solution in the tank has reached the desired temperature, it is discharged.
(8) The workers were seen repairing the generator.
(9) Absorption process is therefore conveniently divided into two groups: physical process and chemical process.
(10) An aldehyde is prepared from the dehydrogenation of an alcohol and hence the name.
(11) In the reaction both the acid and the base are neutralized forming water and salt.
(12) Electrons are known to be minute negative charges of electricity.
(13) Temperature is changed quickly from room temperature to 125℃ and is held there for at least 15 minutes.

Lesson 8　后置定语的译法

按定语所处的位置可将定语分为前置定语和后置定语。前置定语，即定语位于它所修饰的成分之前，一般较短，因而较简单。后置定语，是指位于名词或代词之后的定语，通常比较复杂。然而，科技英语的准确性与严密性，需要频繁使用后置定语来表达，这也是科技英语中复杂长句多的一个原因。因此，尽管定语是句子的次要成分，却是影响译文质量好坏的重要因素。后置定语包括介词短语、形容词、代词、数词、名词、非限定动词、副词、同位语和从句等。

一、介词短语作后置定语

In general, ethers are good solvents **for fats, waxes and resins.**
醚通常是脂肪、蜡和树脂的良好溶剂。

The gas **from coke ovens** is washed with water to remove ammonia.
焦炉煤气用水洗涤可除去氨。

The presence **of a substituent group** in benzene exerts a profound control over both orientation and the ease **of introduction of the entering substituent**.
苯中已有取代基对进入基团的位置及进入的难易程度有决定作用。

二、形容词（或其短语）作后置定语

Hydrocarbons that do not contain the maximum number of hydrogen atoms **possible** are called unsaturated hydrocarbons.
不含有最大可能氢原子数的烃称为不饱和烃。
（误：不含有最大氢原子数的烃可能称为不饱和烃。）

Acetylene is hydrocarbon especially **high** in heat value.
乙炔是热值特别高的烃。

An engineer has to know the properties of the materials **available.**
工程师必须了解现有的各种材料的性能。

三、非限定动词（动词不定式、现在分词、过去分词）作后置定语

翻译这类句子时，应特别注意现在分词和过去分词所表示的意义不同。现在分词，往往表示动作正在发生，表示的动作具有主动意义。过去分词，往往表示动作已经完成。及物动词（vt.）的过去分词表示被动意义；不及物动词（vi.）的过去分词不具有被动含义，仅表示动作已经完成。例如：

the evaporated water，若 evaporate 为 vt.，则译为"被蒸发的水"；若 evaporate 为 vi.，

则译为"蒸发了的水"。

substituted group，若 substitute 为 vt.，则译为"被取代的基团"；若 substitute 为 vi.，则译为"取代了的基团"，也就是"取代基"。

由此可以看出，动词是 vt. 还是 vi.，其代表的意思有时正相反。实际翻译时，应根据上下文决定是哪种译意。例如：

Man was not the first living thing **to communicate through the use of sound**.
第一个用声音进行联络的生物并不是人类。（不定式短语作后置定语，修饰 living thing）

California has a statewide plan **to balance the distribution of water**.
加利福尼亚有一个在全州性的用水分配平衡计划。（不定式短语作后置定语，修饰 plan）

It is this nitronium ion that reacts with the hydrocarbon **being nitrated**.
就是这个硝镦离子与**要被硝化的**烃反应。（现在分词短语作后置定语，修饰 hydrocarbon）

The homologs of benzene are those containing **an alkyl group or alkyl groups in place of one or more hydrogen atoms**.
苯的同系物是一个或多个氢原子被烷基取代的物质。（现在分词短语作后置定语）

The functional group of a ketone consists of a carbon atom **connected by a double bond to an oxygen atom**.
酮的官能团是碳氧双键。（过去分词短语作后置定语）

四、定语从句作后置定语

同被动语态一样，定语从句在科技英语中出现的频率也很高。科技文献的特点是需要概念清楚，逻辑性强，因而常用能表达完整意思的定语从句来修饰名词、代词或相当于名词的词，以便明确、完整地表达该词的意义。定语从句是最复杂的一种从句，不仅结构复杂、含义繁多，有时还具有补充、转折、因果、目的、条件、让步等意义，是汉译时最难处理的一种从句。

定语从句包括限制性定语从句和非限制性定语从句两种。由关系代词(that，which，who，whom，whose，but，as)或关系副词(where，when，why，as，wherein，whereon)引导。定语从句除译作"……的"外，还有许多翻译技巧和方法。概括起来分为"合译法"和"分译法"。

1. 合译法

所谓合译法，就是把从句融合于主句中，把定语从句置于所修饰的名词前，译成汉语的单句。

（1）翻译成定语

凡是结构比较简单，在逻辑上对先行词有明确的限制和确定作用的定语从句，如果翻译时置于所修饰的名词前不影响意思表达而且符合汉语表达习惯的，一般译成"……的"，置于被修饰的词之前。例如：

When fuels burn, they return once more to the simple materials out **of which they were made**.
燃料燃烧时，又转变成**原来构成它们的**那些简单的物质。

One may already have some idea of the material **with which organic chemistry deals**, but he may not know the extent **to which organic chemistry touches on our life**.
也许有人已经对**有机化学中的**物质有所了解，但是他可能并不知道有机化学深入我们日常生活已经到了什么程度。

（2）翻译成谓语

有些定语从句是全句的中心，如要突出从句的内容，汉译时，可将从句顺序译成简单句中的谓语，而把主句压缩成主语。

A semiconductor is a material **that is neither a good conductor nor a good insulator.**

半导体材料**既不是良导体，也不是好的绝缘体**。

Strength, hardness and plasticity of metals are the properties **that make them so useful for industry**.

金属的强度，硬度和可塑性等性能**使得它们在工业上非常有应用价值**。

2. 分译法

如果定语从句较长，结构较复杂，或意思上独立性较强，在逻辑意义上有分层叙述及转折等作用时，一般可把定语从句译成并列句。

（1）译成表示同等关系的并列分句

有的定语从句，起着对先行词进一步说明的作用，往往译成并列分句，有时加译"其、它、这"等词。

An acid is a compound **whose solutions can produce hydrogen ions.**

酸是这样一种化合物，**其溶液能产生氢离子**。

The first higher homolog of benzene is toluene **which is the raw material for the manufacture of the explosive.**

苯的第一个高级同系物是甲苯，**它是制造炸药的原料**。

（2）译成表示转折的并列句子

当定语从句在意思上与主句相对照或语气转折时，可译成转折句，加译"而、但、却、可是"等。

Mild oxidation of a primary alcohol gives an aldehyde which **may be further oxidized to an organic acid.**

伯醇的氧化生成醛，**而醛能进一步氧化生成有机酸**。

Copernicus, **whose theory was not made public until after his death**, believed that the sun was the center of the solar system.

哥白尼认为太阳是太阳系的中心，**但他的理论直到他死后才公布于众**。

（3）如果关系代词 which 与其在主句所修饰的词能够按顺序说明，表达清楚，可省略不译

Live-wire operation is an advanced technique, **by which wireman can repair transmission without cutting off the current.**

带电作业是一项先进技术，**线路工人不必切断电流就能修理输电线**。

When the free electrons in a conductor move in one direction only, the current thus established is called a direct current, **which is often abbreviated to DC.**

当导体的自由电子只以一个方向运动时，产生的电流称为直流电，**常缩写为 DC**。

（4）译成状语从句

当定语从句在意义上相当于表示原因、结果、目的、条件、让步等的状语从句，可译成状语从句，加译"因为、虽然、只要、因而"等词。

① 译成原因状语从句

Iron is not as strong as steel **which is an alloy of iron with some other element.**

铁的强度不如钢高，**因为钢是铁和一些其他元素形成的合金**。

The water should be free from dissolved salts **which will cause deposits on the tubes and lead to overheating.**

水中应不含溶解盐，**因为它会沉积在管壁上，导致管壁过热**。

② 译成结果状语从句

Copper, **which is used widely for carrying electricity**, offers very little resistance.
铜的电阻很小，所以广泛用来输电。

Molecules have perfect elasticity, **in consequence of which they undergo no loss of energy after a collision**.
分子的弹性很好，所以碰撞后并没有能量损失。

③ 译成目的状语从句

The workers oil the moving parts of these machines regularly, **the friction of which may be greatly reduced**.
工人们定期给这些机器的传动部分加油，以便大大减小摩擦。

He wishes to write an article **that will attract public attention to the matter**.
他想写一篇论文以引起公众对这个问题的关注。

④ 译成条件状语从句

For any substance **whose formula is known**, a mass corresponding to the formula can be computed.
不管什么物质，只要知道其分子式，就能求出与分子式对应的质量。

Alloys, **which contains a magnetic substance**, generally also have a magnetic property.
金属如果含有磁性物质，一般也会具有磁性。

⑤ 译成让步状语从句

The problem, **which is very complicated**, has been solved.
这个问题虽然很复杂，但已经解决了。

Photographs are taken of stars, **the light of which is too faint to be seen by eyes at all.**
虽然很多星体的光线非常微弱，眼睛根本看不到，但它们的照片还是可被拍下来的。

⑥ 译成时间状语从句

Water, **the temperature of which reaches 100 ℃**, begins to boil.
当水温达到100℃时，开始沸腾。

Electricity **which is passed through the thin tungsten wire inside the bulb** makes the wire very hot.
电通过灯泡里的钨丝时，会使钨丝变热。

Exercises

Translate the following sentences into Chinese.
(1) All of the carbon-carbon bonds in benzene are alike and have properties intermediate between those of a single and a double bond.
(2) What do you have to say?
(3) Sewage sludge is an organic material containing a large variety of carbon-based molecules.
(4) An engine derives power from the heat produced by burning fuel.
(5) This discovery explains the diminished unsaturation properties shown by the ring structure.
(6) Of all the forms of energy that we use, electrical energy is the most convenient.
(7) The elements themselves were changing, which was something that had always been thought impossible by all but the old alchemists.
(8) There are some metals which possess the power to conduct electricity and ability to be magnetized.

(9) Matter is composed of molecules that are composed of atoms.
(10) The considerable success of these studies has brought a theoretical unity to the whole field of organic chemistry which has the effect of making its principles easier to teach and to learn.
(11) As has been pointed out above, solids do not expand as much as gases and liquids.
(12) Do you have any idea how fast sound travels?

Lesson 9 长句的译法

在表达一些较复杂的概念时，英、汉两种语言特点差别较大。英语的特点是：利用各种修饰语构成较长的简单句，利用连词将简单句连在一起构成更长的复合句。汉语的特点是：尽量分成几个简短的句子来说明一个概念，不常使用长句。根据两种语言特点的差异，在翻译时，**应将英语长句分解成几个汉语短句**，然后再根据逻辑次序和意思轻重，重新安排句子结构。翻译的基本步骤为：

① 通读全句，以确定句子种类（简单句、并列句、复合句）。

② 如为简单句，则应先分析出主语、谓语、宾语、表语等主要成分，再分析定语、状语等次要成分，并弄清主次成分之间的关系，同时注意时态、语气和语态的变化。

③ 如为复合句，则应先找出主句，再确定从句及其之间的关系。其包含的各从句，再分别按简单句分析。

长句的使用，在科技英语中非常普遍。英语长句的翻译是一个综合运用语言和翻译技巧的过程。翻译时，首先抓住全句的中心内容，弄清各部分之间的语法关系及逻辑关系，分清上下层次及前后联系，然后根据汉语特点、习惯和表达方式，正确译出原文意思。

长句的译法包括顺译、倒译、分译三种方法。

一、顺译法

英语长句的叙述层次与汉语相同时，可以按照英语原文的顺序，依次译出。例如：

Objectionable hydrogen sulfide is removed from such a gas or from naturally occurring hydrocarbon gases by washing with various alkaline solutions in which it is absorbed.

要从这样的煤气或天然存在的烃类气体中除去有害物质硫化氢，就要用能吸收硫化氢的各种碱性溶液来洗涤。

本句为简单句，与汉语结构相似；被动语态 is removed 采用"主谓变序"的译法，译成"除去有害物质硫化氢"。

Gas absorption is an operation in which a gas mixture is contacted with a liquid for the purposes of preferentially dissolving one or more components of the gas and to provide a solution of these in the liquid.

气体吸收是这样一种操作：让气体混合物与液体接触，以使气体中的一种或多种组分优先溶解在液体中，并提供由这些组分和液体所形成的溶液。

本句为复合句，定语从句可译成并列句，被动语态 is contacted 加译"让"，译成"让气体混合物与液体接触"。

There were running in the garden a group of children aged from seven to twelve.

在花园里有一群七岁到十二岁大小的孩子在跑着。

（a group of children aged from seven to twelve 为主语部分，因较长，放在句子后面，前面用 there 引导。）

There have been opened up to the vast and excellent science, in which my work is the beginning, ways and means by which other minds more acute than mine will explore its remote corners.

通往伟大而美好科学的途径和手段已经展现在眼前。在科学领域中，我的研究只是个开端，比我敏锐的其他人所采用的途径和手段将会探索科学领域的遥远境地。

本句为复合句。翻译好本句的关键是找对主句的主语，若主句主语搞不清，翻译必错。There 句型的翻译方法：当句子主语有较长的定语修饰时，可以将主语及修饰部分放在后面，前面用 there 引导，以使句子平衡。本句中主句主语为 ways and means, 因其后有较长的定语从句（by which…）修饰，故将主语放在句子后部，前面用 there 引导。

The continuous process although requiring more carefully designed equipment than the batch process, can ordinarily be handled in less space, fits in with other continuous steps more smoothly, and can be conducted at any prevailing pressure without release to atmospheric pressure.

虽然连续过程比间歇过程要求更为周密设计的设备，但连续过程通常能节约操作空间，能较顺利地适应其他连续操作步骤，并能在任何常用的压力下进行，而不必向大气释放压力。

（本句中 less 为形容词，通过词类转换译法译成"节省"；最后一个分句通过词意引申译法译成"……并能在任何常用的压力下进行，而不必暴露在大气中"。）

二、倒译法

有时英语长句的叙述层次与汉语相反，翻译时，应根据汉语习惯，改变原文语序，进行翻译。这种译法为倒译法，又称变序译法。倒译法常在下列情况下采用：

① 主句后面，带有很长的状语（特别是原因和方式状语）或状语从句（特别是原因、条件、让步状语从句）。

② 主句后面，带有很长的定语、定语从句，或宾语从句。

This is why the hot water system in a furnace will operate without the use of a water pump, if the pipes are arranged so that the hottest water rises while the coldest water runs down again to the furnace.

如果把管子装成这个样子，使最热的水上升，而最冷的水再往下回流到锅炉里去，那么，锅炉中的热水系统不用水泵就能循环，道理就在于此。（采用倒译法，先译 if 引导的条件状语从句）

Theoretically, women were not supposed to be recruited into the continental army, but if a woman was a good soldier, no one made an issue of sex at a time when the army was so short of soldiers that boys not yet in their teens were also being recruited in violation of rules.

从理论上说，妇女不准被征召参加大陆军，但当时兵源严重不足，连那些不足 13 岁的男童都被违反规定征召入伍，假如一个妇女是一个好兵，又有谁会把性别看成是一个问题呢？（采用倒译法，先译 when 引导的时间状语从句）

We learn that sodium or any of its compounds produces a spectrum having a bright yellow double line by noticing that there is no such line in the spectrum of light when sodium is not present, but that if the smallest quantity of sodium be thrown into the flame or other sources of light, the bright yellow line instantly appears.

我们注意到，如果把非常少量的钠投入到火焰或其他光源中时，立即出现一条亮黄色的双线，当没有钠存在时，光谱中就没有这样的双线。由此我们知道钠或钠的任何化合物所产生的光谱都带有一条亮黄色的双线。

（从 by noticing that 开始一直到句尾，为一个既长又复杂的句子，作主句的方式状语，状语中又含从句；采用倒译法，先译较长的方式状语，再译主句。）

37

还有一种倒译法，即在几句话中，为翻译表达清楚，将后面的句子先行译出。此种倒译属于句子之间的倒译，不同于上面所讲的一个句子的倒译。例如：

Normally, in evaporation the thick liquor is the valuable product and the vapor is condensed and discarded. Mineral-bearing water is often evaporated to give a solid-free product for boiler feed, for special process requirements, or for human consumption. This technique is often called water distillation, but technically it is evaporation. In one specific situation, however, the reverse is true.

通常情况下，蒸发时，黏稠液是有价值的产物，而蒸汽被冷凝下来并且扔掉。**但是在某一特定的情况下，恰恰相反。**含有矿物质的水常被蒸发以获得不含固体的产物，以供锅炉加水、特殊过程或供人类使用。此项技术通常称为水的蒸馏，而在技术上却是蒸发。

①原文中最后一句提前译出，见译文中黑体字部分，便于理解原文意思；②几个被动句有的加译"被"，有的采用"被动意义译成主动意思"的翻译方法。

三、分译法

当长句中的主句与从句或分词短语及介词短语等各种修饰词之间的关系不是很密切，各具有相对的独立性时，为符合汉语习惯，有时需将原文的某一短语或从句先行单独译出，并借助适当的总括性词语或其他语法手段将前后句联系到一起；或将几个并列成分先概括地合译在前面，而后分别加以叙述；或将原文中不好处理的成分拆开，译成相应的句子或另一独立句子，这种翻译方法称为分译法。

例如：

The diode consists of a tungsten filament, which gives off electrons when it is heated, and a plate toward which the electrons migrate when the field is in the right direction.

二极管由一根钨丝和一块极板组成：钨丝受热时便放出电子，当电场方向为正时，这些电子便向极板移动。（先将两个并列的名词 filament 和 plate 分出去合译在前面，然后再分别叙述。）

Our object is to draw attention to those areas responsible for the quantum jump in sophistication, improved cosmetic attributes and safety of today's products.

我们的目的是提请人们注意那些使下列方面发生跃变的领域，即当代化妆品所采用的先进技术方面、质量改进方面以及安全性方面。

①添加总括性词语"下列方面"先行译出，而后具体叙述。②to draw attention to 不能译作"我们注意"。

Another development, by itself not strictly scientific in nature, is the growing awareness that the energy resources at present available to human technology are limited, and that photosynthesis is the only large-scale process on earth by which a virtually inexhaustible source of energy, i.e., the radiation energy of the sun, is collected and converted into a form of energy that is not only used by plants, but by all forms of life, including man.

另一发展就其本质而言，不是严格科学的，而是人们正日益认识到的事实，即人类技术领域目前可以利用的能源是有限的，在地球上只有光合作用能将实际上是取之不尽用之不竭的能源——太阳的辐射能聚集起来，转换为不仅植物，而且包括人在内的所有生命都可以利用的能量形式的大规模过程。（先加译总括性词语"下面这个事实"，再分别叙述。）

实际翻译时，往往是顺译、倒译和分译互相融合，常常很难说出是哪种单独的译法。分节讲述的目的是掌握基本要点，使之能灵活运用于复杂的实际翻译中。

Exercises

Translate the following sentences into Chinese.

(1) How these two things—energy and matter—behave, how they interact one with the other, and how people control them to serve themselves make up the substance of two basic physical sciences, physics and chemistry.

(2) This simple fact shows that the more of the force of friction is got rid of, the further will the ball travel, and we are led to infer that, if all the impeding force of gravitation could be removed, there is no reason why the ball, once in motion, should ever stop.

(3) Aluminum remained unknown until the nineteenth century, because nowhere in nature is it found free, owing to its always being combined with other elements, most commonly with oxygen, for which it has a strong affinity.

(4) Fire is a chemical reaction in which atoms of the gas oxygen combine with atoms of certain other elements, such as hydrogen or carbon.

(5) Manufacturing process may be classified as unit production with small quantities being made and mass production with large numbers of identical parts being produced.

(6) The versatility of chemical engineering originates in training to the practice of breaking up a complex process into individual physical steps, called unit operations, and into the chemical reactions.

(7) The two pairs of electrons of oxygen may be shared with one carbon alone forming a double bond($>$C=O), or with two separate carbons (—C—O—C—), or with one carbon and one hydrogen forming only single bonds(—C—O—H).

(8) Designs which avoid the need for hazardous materials, or use less of them, or use them at lower temperatures and pressures, or dilute them with inert materials will be inherently safe and not require elaborate safety systems.

Part 2
Introductory Chemistry Speciality English

Unit 1

Inorganic Chemistry

Lesson 10 Introduction to Chemistry: The Central, Useful, and Creative Science

1. productive
   [prə'dʌktɪv, pro-]
   有生产力的
2. segment
   ['sɛgmənt]
   部分
3. account for 占
4. statistic
   [stə'tɪstɪk]
   统计量
5. take for granted
   认为……理所当然
6. toxic ['taksɪk]
   有毒的

7. property
   ['prapəti]
   性质
8. undergo
   [ˌʌndɚ'go]
   经历
9. transformation
   [ˌtrænsfɚ'meʃən]
   （原子结构等）蜕变

Chemistry is one of the oldest sciences, and it has certainly been one of the most **productive** in improving human life. The industries that use chemistry to manufacture products—the chemical process industries—are the largest **segment** of manufacturing in advanced societies; in the United States, for example, they **account for** more than 30% of all manufacturing. This **statistic** does not include the output of related industries, such as electronics, automobiles, or agriculture, that use the products of the chemical process industries.

Much of what is done in designing and producing chemical products such as modern medicines is unknown to the average person or **taken for granted**. Most people are not aware that it is chemistry that creates such useful substances. We often hear about "**toxic** chemicals" or "chemical pollution" without hearing about the absolutely central role that chemistry plays in human well-being. Here I attempt to present a more balanced picture.

What Is Chemistry?

Chemistry is the science that tries to understand the **properties** of substance and the changes that substances **undergo**. It is concerned with substances that occur naturally —the minerals of the earth, the gases of the air, the water and the salts of the seas, the chemicals found in living creatures —and also with new substances created by humans❶. It is concerned with natural changes—the burning of a tree that has been struck by lightning, the chemical changes that are central to life—and also with new **transformations** invented and created by chemists.

---

❶ It is concerned with…and also with… 它与……有关，也与……有关；同样下一句也是这个意思。

Vocabulary	
10. prerecorded times	史前时代
11. quotation [kwo'teʃən]	引用（语）
12. investigate [ɪn'vɛstɪˌget]	调查，研究
13. enormously [ɪ'nɔrməslɪ]	巨大地
14. in the last century	在上个世纪
15. electronics [ɪˌlɛk'trɑnɪks]	电子学
16. metallurgy	冶金
17. mineralogy [ˌmɪnə'rɑlədʒɪ]	矿物学
18. among others	除了别的以外
19. ultimate ['ʌltəmɪt]	最终的，根本的
20. utilize ['jutlˌaɪz]	利用
21. shelter ['ʃɛltɚ]	避难所
22. brutish ['brutɪʃ]	如野兽般的
23. harmony ['hɑrmənɪ]	和谐，协调
24. aspiration [ˌæspə'reʃən]	渴望
25. isolate ['aɪsəˌlet]	分离
26. determine [dɪ'tɚmɪn]	测定

Chemistry has a very long history. In fact, human activity in chemistry goes back to **prerecorded times**.

What Do Chemists Do?

As the **quotation** at the head of this chapter indicates, chemists are involved in two different types of activities. Some chemists **investigate** the natural world and try to understand it, while other chemists create new substances and new ways to perform chemical changes that do not occur in nature. Both activities have gone on since the first appearance of humans on earth, but the pace has increased **enormously in the last century** or so.

Why Do Chemists Call Their Discipline the "Central Science"

Chemistry touches many other scientific fields. It makes major contributions to agriculture, **electronics**, biology, medicine, environmental science, computer science, engineering, geology, physics, **metallurgy**, and **mineralogy**, **among** many **others**. It does not ask the physicists' question: What is the **ultimate** nature of all matter? Instead it asks the chemists' questions: Why do the substances of the world differ in their properties? How can we control and most effectively **utilize** these properties?

Interesting and exciting as the physics question is, answers to the chemists' questions allow us to create new medicines, make new materials for **shelter** and clothing and transportation, invent new ways to improve and protect our food supply, and improve our lives in many other ways as well. Thus we see chemistry as "central" to the human effort to move above the **brutish** existence of our caveman ancestors into a world where we can exist not only **in harmony with**❶ nature, but also in harmony with our own **aspirations**.

What Makes Chemistry the "Useful Science" and the "Creative Science"

The two questions are linked. Some chemists explore the natural world and find useful chemical substances not known before. This exploration has been carried out extensively by examining the chemicals found in plants and animals on land, and it still goes on. Now there is also a major search for new chemicals from plants and animals in the seas. Once these chemicals are **isolated** and their chemical structures are **determined**, the creativity of chemists takes over.

---

❶ in harmony with 与……和谐相处。

27. destructive [dɪˈstrʌktɪv] 破坏性的
28. synthesize [ˈsɪnθɪˌsaɪz] 合成
29. readily [ˈrɛdəli, ˈrɛdli] 轻而易举地
30. relative [ˈrɛlətɪv] 相关物
31. pay off 取得成功，使得益
32. fierce [fɪrs] 猛烈的
33. repel [rɪˈpɛl] 抵制
34. synthetically 综合地，合成地
35. microorganism [ˌmaɪkroˈɔrgəˌnɪzəm] 微生物
36. yeast [jist] 酵母
37. mold [mold] 真菌
38. antibiotic [ˌæntɪbaɪˈɑtɪk] 抗生素
39. mating 交尾
40. irreverently 无礼地
41. alchemist [ˈælkəmɪst] 炼金术士

42. fundamental [ˌfʌndəˈmɛntl] 基本的
43. in a sequence 顺次
44. penny [ˈpɛni] 便士
45. finger [ˈfɪŋgɚ] 用手指拨弄

❶ 新物质过去常常是通过化学家们自嘲地称之为"摇"和"烘焙"的两种方式制备的。

Normally we would not continue to harvest the living sources of useful new drugs, for instance—this could be too **destructive** and too costly. Instead chemists devise ways to **synthesize** the newly discovered compounds, to create them from other simpler materials, so they can be **readily** available. Sometimes the original chemical structures are altered by creative synthesis, to see whether the properties of a novel **relative** of the natural compound are even better.

There is a reason that the search for useful natural chemicals often **pays off**. The natural world is not the peaceful place we dream of—there are **fierce** battles for survival. Insects eat plants, and some plants have developed chemicals that **repel** those insects. When we learn what those chemicals are, we can make them **synthetically** and use them to help protect our food plants. Bacteria can infect plants, animals, and other **microorganisms** such as **yeasts** and **molds**, not just humans. Some organisms have developed powerful **antibiotics** to protect themselves. Most of the effective antibiotics in human use have come from the exploration of nature's chemistry, although sometimes the medicines we use are versions improved by chemists.

Insects also use chemicals to signal each other for **mating**. As we learn what those chemicals are, we can make them and attempt to control the populations of harmful insects.

The most creative act in chemistry is the design and creation of new molecules. How is this done? This question will be more fully addressed in other articles, but a brief answer is given here. New chemicals used to be made by what chemists **irreverently** refer to as "shake and bake"❶: Heat up some mixture and see what happens, as in the earlier examples of making metals and glass. The **alchemists** of the past devoted themselves to heating up various mixtures in the vain hope to turn lead into gold. They did not succeed, but they did create some interesting new chemical processes and new substances.

Syntheses are now normally designed using the **fundamental** principles that chemists have discovered. As many as 30 or more predicted chemical steps are sometimes needed, **in a sequence**, to permit the synthesis of a complicated molecule from available simple chemicals. This could not be done without a clear understanding of chemical principles.

Hands-On Chemistry: Penny Fingers

Pennies dated 1982 or earlier are nearly pure copper, each having a mass of about 3.5grams. Pennies dated after 1982 are made of copper-coated zinc, each having a mass of about 2.9grams. Hold a pre-1982

46. index finger
食指
47. forearm
[fɔr'ɑrm, for-] 前臂
48. inertia
[ɪ'nɚʃə] 惯性
49. subtle
['sʌtl] 精细的
50. well-tuned sense
察觉
51. threshold
['θreʃ,old, -,hold]极限
52. stack
[stæk] 堆叠

penny on the tip of your right **index finger** and a post-1982 penny on the tip of your left index finger. Move your **forearms** up and down to feel the difference in **inertia** — the difference of 0.6grams (600 milligrams) is **subtle** but not beyond a set of **well-tuned senses**. If one penny on each finger is below your **threshold**, try two pre-1982s **stacked** on one finger and two post-1982s stacked on the other. Share this activity with a friend.

Hands-On Chemistry Insights: Penny Fingers

You have to be moving the pennies up and down in order to optimize your threshold of detection. What you are sensing here is the difference in inertia. Recall that inertia is an object's resistance to any change in its motion. If you minimize the motion, you minimize your ability to detect any difference in inertia the two coins may have. Switch pennies between your left and right index fingers (with your eyes closed) to confirm your ability to detect a difference. With or without the motion, the pennies exert a downward pressure that your nerve endings sense. To feel this pressure, repeat the experiment with the pre-1982s on one index finger and the post-1982s on the other index finger, but this time keep your hands still. How many do you need to stack before you can sense a difference in pressure? If you did this pressure experiment on the moon, would you need to stack more or fewer? Why?

Exercises

1. Translate the following from Chinese into English.
    (1) 有毒化学品   (2) 化学污染   (3) 物性   (4) 自然变化   (5)科学领域
    (6) 分离   (7) 测定   (8) 合成   (9) 基本原理

2. Match each of the significant terms in the left hand column with the most appropriate explanation in the right hand column.

    Significant Terms   **Explanation**
    (1) matter   A. study of living matter
    (2) theory   B. the branch of science that deals with matter, with the changes that matter can undergo, and with the laws that describe these changes
    (3) chemistry   C. a change in the composition of matter
    (4) principles of chemistry   D. description of the elements and their compounds, their physical states, and how they behave
    (5) chemical change   E. the chemistry of all of the elements and their compounds, with the exception of compounds of carbon with hydrogen and of their derivatives
    (6) descriptive chemistry   F. everything that has mass and occupies space

(7) inorganic chemistry	G. the chemistry of compounds of carbon with hydrogen and of their derivatives
(8) organic chemistry	H. study of natural laws and processes other than those peculiar to living matter
(9) qualitative analysis	I. a change in which the composition of the matter involved is unaltered
(10) quantitative analysis	J. explanations of chemical facts, for example, by theories and mathematics
(11) physical sciences	K. the identification of substances, often in a mixture
(12) physical change	L. the determination of the amounts of substances in a mixture
(13) biological sciences	M. unifying principle or group of principles that explains a body of facts or phenomena

3. Define each of the following terms.
   (1) experiment   (2) law   (3) hypothesis   (4) theory
4. Which statement below is true?
   (a) Chemistry is not concerned with describing and explaining the different forms of matter.
   (b) Energy is not a concern of chemistry.
   (c) Chemistry is concerned with describing and explaining processes such as burning a log or dissolving a piece of copper in an acid.
   (d) Chemistry does not play a role in other sciences.
   (e) None of the above
5. Write a letter to a friend or your parents, telling your chemistry-studying life.
6. A topic for speaking.
   Chemistry is everywhere. —Do you agree with the motto? Why? Why not?

【Reading Material】

The Extinction of the Dinosaur

Dinosaurs dominated life on earth for millions of years and then disappeared very suddenly. To solve the mystery, **paleontologists**（古生物学家）studied fossils and skeletons found in rocks in various layers of earth's crust. Their findings enabled them to map out which species existed on earth during specific geologic periods. They also revealed no dinosaur skeletons in rocks formed immediately after the **Cretaceous**（白垩纪）period, which dates back some 65 million years. It is therefore assumed that the dinosaurs became extinct about 65 million years ago.

Among the many hypothesis put forward to account for their disappearance were **disruptions**（中断）of the food chain and a dramatic change in climate caused by violent volcanic eruptions. However, there was no convincing evidence for any one hypothesis until 1977. It was then that a group of paleontologists working in Italy obtained some very puzzling data at a site near **Gubbio**（古比奥）. The chemical analysis of a layer of clay deposited above sediments formed during the

Cretaceous period (and therefore a layer that records events occurring after the Cretaceous period) showed a surprisingly high content of the element iridium (Ir). Iridium is very rare in earth's crust but is comparatively abundant in **asteroids** (小行星).

This investigation led to the hypothesis that the extinction of dinosaurs occurred as follows. To account for the quantity of iridium found, scientists suggested that a large asteroids several miles in diameter hit earth about the time the dinosaurs disappeared. The impact of the asteroid on earth's surface must have been so tremendous that it literally vaporized a large quantity of surrounding rocks, soils, and other objects. The resulting dust and **debris** (岩屑) floated through the air and blocked the sunlight for months or perhaps years. Without ample sunlight most plants could not grow, and the fossil record confirms that many types of plants did indeed die out at this time. Consequently, of course, many plant-eating animals **perished** (消亡), and then, in turn, meat-eating animals began to starve. **Dwindling** (逐渐减少的) food sources would obviously affect large animals needing great amounts of food more quickly and more severely than small animals. Therefore, the huge dinosaurs, the largest of which might have weighed as much as 30 tons, vanished due to lack of food.

Questions

1. How does the study of dinosaur extinction illustrate the scientific method?
2. Suggest two ways that would enable you to test the asteroid collision hypothesis.
3. In your opinion, is it justifiable to refer to the asteroid explanation as the theory of dinosaur extinction?
4. Available evidence suggests that about 20 percent of the asteroid's mass turned to dust and spread uniformly over earth after settling out of the upper atmosphere. This dust amounted to about $0.02 g/cm^2$ of earth's surface. The asteroid very likely had a density of about $2 g/cm^3$. Calculate the mass (in kilograms and tons) of the asteroid and its radius in meters, assuming that it was a sphere. (The area of earth is $5.1 \times 10^{14} m^2$) (Source: Consider a Spherical Cow—A Course in Environmental Problem Solving by J Harte. University Science Books, Mill Valley. CA 1988.)

Lesson 11　The Nomenclature of Inorganic Compound

1. informal
   [ɪnˈfɔrml]
   非正式的
2. common names
   俗名
3. quartz
   [kwɔrts]
   石英
4. systematic
   [ˈsɪstəˈmætɪk]
   系统的
5. nomenclature
   [ˈnomənˌkletʃɚ]
   系统命名法
6. cation
   [ˈkætaɪən]
   阳离子
7. parentheses
   [pəˈrenθəsiz]
   圆括号
8. transition metal
   过渡金属
9. oxidation number
   氧化数
10. ferrous [ˈfɛrəs]
    亚铁的
11. ferric [ˈfɛrɪk]
    三价铁的
12. appendix
    [əˈpɛndɪks]
    附录
13. quicklime
    [ˈkwɪklaɪm]
    生石灰
14. halogens
    [ˈhælədʒənz]
    卤族
15. oxoanion
    含氧阴离子

Many compounds were given **informal**, **common names** before their compositions were known. Common names include water, salt, sugar, ammonia, and **quartz**. A **systematic** name, on the other hand, reveals which elements are present and, in some cases, the arrangement of atoms. The systematic naming of compounds, which is called chemical **nomenclature**, follows the simple rules described in this section.

Names of Cations

The name of a monatomic cation is the same as the name of the element forming it, with the addition of the word ion, as in sodium ion for Na^+. When an element can form more than one kind of cation, such as Cu^+ and Cu^{2+} from copper, we use the oxidation number, the charge of the cation, written as a Roman numeral in **parentheses** following the name of the element. Thus, Cu^+ is a copper(Ⅰ) ion and Cu^{2+} is a copper(Ⅱ) ion. Similarly, Fe^{2+} is an iron(Ⅱ) ion and Fe^{3+} is an iron(Ⅲ) ion. Most **transition metals** form more than one; so unless we are given other information, we need to include the oxidation number in the names of their compounds.

Some older systems of nomenclature are still in use. For example, some cations were once **denoted** by the endings -ous and -ic for the ions with lower and higher charges, respectively. To make matters worse, these endings were in some cases added to the Latin form of the element's name. Thus, iron(Ⅱ) ions were called **ferrous** ions and iron(Ⅲ) ions were called **ferric** ions (see **Appendix** 11-3). We do not use this system in this text, but you will sometimes come across it and should be aware of it.

Names of Anions

Monatomic anions, such as the Cl^- ions in sodium chloride and the O^{2-} ions in **quicklime** (CaO), are named by adding the suffix -ide and the word ion to the first part of the name of the element (the "stem" of its name), as shown in the list of anions in Table 11-1; thus, S^{2-} is a sulfide ion and O^{2-} is an oxide ion. There is usually no need to give the charge, because most elements that form monatomic anions form only one kind of ion. The ions formed by the **halogens** are collectively called halide ions and include fluoride (F^-), chloride (Cl^-), bromide (Br^-), and iodide(I^-) ions.

Polyatomic ions include the **oxoanions**, which are ions that contain oxygen (see Table 11-1). If only one oxoanion of an element exists, its name is formed by adding the suffix -ate to the stem of the name of the element, as in the carbonate ion, CO_3^{2-}. Some elements can form two types of oxoanions, with different numbers of oxygen atoms, so we need

Table 11-1 Common anions and their parent acid

Anion	Parent acid	Anion	Parent acid
fluoride ion, F^-	hydrofluoric acid, HF* (hydrogen fluoride)	cyanide ion, CN^-	hydrocyanic acid, HCN* (hydrogen cyanide)
chloride ion, Cl^-	hydrochloric acid, HCl* (hydrogen chloride)	nitrite ion, NO_2^-	nitrous acid, HNO_2
		nitrate ion, NO_3^-	nitric acid, HNO_3
bromide ion, Br^-	hydrobromic acid, HBr* (hydrogen bromide)	phosphate ion, PO_4^{3-}	phosphoric acid, H_3PO_4
		hydrogen phosphate ion, HPO_4^{2-}	
iodide ion, I^-	hydroiodic acid, HI* (hydrogen iodide)	dihydrogen phosphate ion, $H_2PO_4^-$	
		sulfite ion, SO_3^{2-}	sulfuous acid, H_2SO_3
oxide ion, O^{2-}	water, H_2O	hydrogen sulfite ion, HSO_3^-	
hydroxide ion, OH^-		sulfate ion, SO_4^{2-}	sulfuric acid, H_2SO_4
sulfide ion, S^{2-}	hydrosulfuric acid, H_2S* (hydrogen sulfide)	hydrogen sulfate ion, HSO_4^-	
hydrogen sulfide ion, HS^-		hypochlorite ion, ClO^-	hypochlorous acid, HClO
acetate ion, CH_3COO^-	acetic acid, CH_3COOH	chlorite ion, ClO_2^-	chlorous acid, $HClO_2$
carbonate ion, CO_3^{2-}	carbonic acid, H_2CO_3	chlorate ion, ClO_3^-	chloric acid, $HClO_3$
hydrogen carbonate ion (bicarbonate ion), HCO_3^-		perchlorate ion, ClO_4^-	perchloric acid, $HClO_4$

* The name of the aqueous solution of the compound. The name of the compound itself is in parentheses.

names that distinguish them. Nitrogen, for example, forms both NO_2^- and NO_3^-. In such cases, the ion with the larger number of oxygen atoms is given the suffix -ate, and that with the smaller number of oxygen atoms is given the suffix -ite. Thus, NO_3^- is nitrate, and NO_2^- is nitrite. The nitrate ion is an important source of nitrogen for plants and is included in some fertilizers (such as ammonium nitrate, NH_4NO_3).

16. hypochlorite
    [ˌhaɪpəˈklɔraɪt]
    次氯酸根

17. perchlorate
    [pɜˈklɔret]
    高氯酸根

Some elements—particularly the halogens—form more than two kinds of oxoanions. The name of the oxoanion with the smallest number of oxygen atoms is formed by adding the prefix **hypo-**❶ to the -ite form of the name, as in the **hypochlorite** ion ClO^-. The oxoanion with more oxygen atoms than the -ate for oxoanion is named with the prefix **per-**❷ added to the -ate form of the name. An example is the **perchlorate** ion, ClO_4^-. The rules for naming polyatomic ions are summarized in Appendix 11-1.

Table 11-2 Prefixes used for naming compounds

Prefix	Meaning	Prefix	Meaning
mono-	1	hepta-	7
di-	2	octa-	8
tri-	3	nona-	9
tetra-	4	deca-	10
penta-	5	undeca-	11
hexa-	6	dodeca-	12

❶ **Hypo** comes from the Greek word for "under".

❷ **Per** is the Latin word for "all over", suggesting that the element's ability to combine with oxygen is finally satisfied.

Hydrogen is present in some anions. Two examples are HS^- and HCO_3^-. The names of these anions begin with "hydrogen". Thus, HS^- is the hydrogen sulfide ion, and HCO_3^-, is the hydrogen carbonate ion. You might also see the name of the ion written as a single word, as in hydrogencarbonate ion. In an older system of nomenclature, which is still quite widely used, an anion containing hydrogen is named with the prefix bi-, as in bicarbonate ion for HCO_3^-. If two hydrogen atoms are present in an anion, as in $H_2PO_4^-$, then the ion is named as a dihydrogen anion, in this case as dihydrogen phosphate ion.

Names of Ionic Compounds

An ionic compound is named with the cation name first, followed by the name of the anion; the word ion is omitted in each case. The oxidation number of the cation is given if more than one charge is possible. However, if the cation comes from an element that exists in only one charge state, then the oxidation number is omitted. Typical names include potassium chloride (KCl), a compound containing K^+ and Cl^- ions; and ammonium nitrate (NH_4NO_3), which contains NH_4^+ and NO_3^- ions. The cobalt chloride that contains Co^{2+} ions ($CoCl_2$) is called cobalt(Ⅱ) chloride.

Some ionic compounds form crystals that **incorporate** a **definite** proportion of molecules of water as well as the ions of the compound itself. These compounds are called hydrates. For example, copper(Ⅱ) sulfate normally exists as blue crystals of composition $CuSO_4 \cdot 5H_2O$. The **raised dot** in this formula is used to separate the water of **hydration** from the rest of the formula. This formula indicates that there are five H_2O molecules for each $CuSO_4$ formula unit. Hydrates are named by first giving the name of the compound, then adding the word hydrate with a Greek prefix indicating how many molecules of water are found in each formula unit (Table 11-2). For example, the name of $CuSO_4 \cdot 5H_2O$, the common blue form of this compound, is copper(Ⅱ) sulfate pentahydrate. When this compound is heated, the water of hydration is lost and the blue crystals **crumble** to a white powder (Figure 11-1). The white powder is $CuSO_4$ itself. When we wish to emphasize that the compound has lost its water of hydration, we call it anhydrous. Thus, $CuSO_4$ is anhydrous copper(Ⅱ) sulfate.

18. incorporate
 [ɪnˈkɔːpəreɪt]
 包含
19. definite
 [ˈdɛfənɪt]
 一定的
20. raised dot
 凸点
21. hydrate
 [ˈhaɪdret]
 水合物
22. hydration
 [haɪˈdreʃən]
 水合
23. crumble
 [ˈkrʌmbl]
 破碎

Figure 11-1

[Blue crystals of copper(Ⅱ) sulfate pentahydrate $CuSO_4 \cdot 5H_2O$ lose water above 150℃ and form a white anhydrous powder ($CuSO_4$). The color is restored when water is added; and, in fact, anhydrous copper sulfate has such a strong attraction for water that it is usually colored a very pale blue from reaction with the water in air.]

Toolbox 11-1

How to Name Ionic Compounds

Conceptual Basis

The aim of chemical nomenclature is to be simple but unambiguous. Ionic and molecular compounds use different procedures; so it is important first to identify which type of compound is present. To name an ionic compound, we name the ions present and then combine the names of the ions.

Procedure

Step 1 Identify the cation and the anion (see Table 11-1 or Appendix 11-1, if necessary). To determine the oxidation number of the cation, decide what cation charge is required to cancel the total negative charge of the anions.

Step 2 Name the cation. If the metal can have more than one oxidation number (most transition metals and metals in some groups), give its charge as a Roman numeral.

Step 3 If the anion is monatomic, change the ending of the element's name to -ide. For an oxoanion:

① Give the ion with the larger number of oxygen atoms the suffix -ate and that with the smaller number of oxygen atoms the suffix -ite.

② For elements that form more than two oxoanions, add the prefix hypo- to the name of the oxoanion with the smallest number of oxygen atoms. Add the prefix per- to the oxoanion with the highest number of oxygen atoms.

For other polyatomic anions, find the name of the ion in Table 11-1 or Appendix 11-1. If hydrogen is present, name it as a second cation.

Step 4 If water molecules appear in the formula, the compound is a hydrate. Add the word hydrate with a Greek prefix corresponding to the number of water molecules in front of H_2O. For examples of how these rules are applied, see Example 11-1.

Ionic compounds are named by starting with the name of the cation (with its oxidation number if more than one charge is possible), followed by the name of the anion; hydrates are named by adding the word hydrate, **preceded** by a Greek prefix indicating the number of water molecules in the formula unit.

24. precede
[prɪ'sid]
先于

Example 11-1 Sample exercise: Naming ionic inorganic compounds
Use the rules in Toolbox 11-1 to name (1) $CrCl_3 \cdot 6H_2O$ and (2) $Ba(ClO_4)_2$.
Solution

	(1) $CrCl_3 \cdot 6H_2O$	(2) $Ba(ClO_4)_2$
Step 1 Identify the cation and anion.	Cr^{3+}, Cl^-	Ba^{2+}, ClO_4^-
Step 2 Name the cation, giving the charge of the transition metal cation as a Roman numeral. Note that in part (1) there are three Cl^- ions, so the charge on Cr must be +3.	chromium(III)	barium
Step 3 Name the anion and combine the name of the ions with the cation first.	chromium(III) chloride	barium perchlorate
Step 4 If H_2O is present, add hydrate with a Greek prefix.	chromium(III) chloride hexahydrate	

Table 11-3 Common names for some simple molecular compounds

Formula*	Common name
NH_3	ammonia
N_2H_4	hydrazine
NH_2OH	hydroxylamine
PH_3	phosphine
NO	nitric oxide
N_2O	nitrous oxide
C_2H_4	ethylene
C_2H_2	acetylene

* For historical reasons, the molecular formulas of binary hydrogen compounds of group VA elements are written with the group VA element first.

Names of Inorganic Molecular Compounds

Many simple inorganic molecular compounds are named by using the Greek prefixes in Table 11-2 to indicate the number of each type of atom present. Usually, we do not use a prefix if only one atom of an element is present; for example, NO_2, is nitrogen dioxide. An important exception to this rule is carbon monoxide, CO. Most of the common binary molecular compounds——molecular compounds built from two elements have at least one element from group VIA or VIIA. These elements are named second, with their endings changed to -ide:

phosphorus trichloride, PCl_3, dinitrogen oxide, N_2O
sulfur hexafluoride, SF_6 dinitrogen pentoxide, N_2O_5

Some exceptions to these rules are the phosphorus oxides and compounds that are generally known by their common names. The phosphorus oxides are distinguished by the oxidation number of phosphorus, which is calculated as though phosphorus were a nonmetal and the oxygen present as O^{2-}. Thus, P_4O_6 is named phosphorus(Ⅲ) oxide as though it was $(P^{3+})_4(O^{2-})_6$ and P_4O_{10} is named phosphorus(Ⅴ) oxide as though it was $(P^{5+})_4(O^{2-})_{10}$. These compounds, though, are molecular. Certain binary molecular compounds, such as NH_3, and H_2O, have widely used common names (Table 11-3).

Both the names and the molecular formulas of compounds formed between hydrogen and nonmetals in group ⅥA or ⅦA are written with the H atom first: for example, the formula for hydrogen chloride is HCl and that for hydrogen sulfide is H_2S. Note that, when these compounds are dissolved in water, many act as acids and are named as acids. Binary acids are named by adding the prefix hydro- and changing the ending of the element name to -ic acid, as in hydrochloric acid for HCl in water and hydrosulfuric acid for H_2S in water. An aqueous solution, a solution in water, is indicated by (aq) after the formula. Thus, HCl, the compound itself, is hydrogen chloride, and HCl(aq), its aqueous solution, is hydrochloric acid.

25. oxoacid
[ɒkˈsoʊsɪd]
含氧酸

An **oxoacid** is an acidic molecular compound that contains oxygen. Oxoacids are the parents of the oxoanions in the sense that an oxoanion is formed by removing one or more hydrogen ions from an oxoacid molecule (see Table 11-1). In general, -ic oxoacids are the parents of -ate oxoanions and -ous oxoacids are the parents of -ite oxoanions. For example, the molecular compound H_2SO_4, sulfuric acid, is the parent acid of the sulfate ion, SO_4^{2-}. Similarly, the parent acid of the sulfite ion, SO_3^{2-}, is the molecular compound H_2SO_3, sulfurous acid.

Toolbox 11-2
How To Name Simple Inorganic Molecular Compounds
Conceptual Basis
The aim of chemical nomenclature is to be simple but unambiguous. A systematic name specifies the elements present in the molecule and the numbers of atoms of each element.
Procedure
Determine the type of compound and then apply the corresponding rules.
(1) Binary Molecular Compounds Other Than Acids
The compound is generally not an acid if its formula does not begin with H.
Step 1 Write the name of the first element, followed by the name of the second, with its ending changed to -ide.
Step 2 Add Greek prefixes to indicate the number of atoms of each element. Mono- is usually omitted.
(2) Acids
An inorganic acid has a formula that typically begins with H;

oxoacids have formulas that begin with H and end in O. We distinguish between binary hydrides, such as HX, which are not named as acids, and their aqueous solutions, HX(aq), which are.

Step 1 If the compound is a binary acid in solution, add "hydro···ic acid" to the root of the elements name.

Step 2 If the compound is an oxoacid, derive the name of the acid from the name of the polyatomic ion that it produces. In general, -ate ions come from -ic acids; -ite ions come from -ous acids.

Retain any prefix, such as hypo- or per-.

26. retain
[rɪ'ten]
保留

Example 11-2 Sample exercise: Naming inorganic molecular compounds
Use the rules in Toolbox 11-2 to write the systematic names of the compounds (1) N_2O_4; (2) ClO_2; (3) HI(aq); (4) HNO_2.
Solution (1) The molecule N_2O_4 has two nitrogen atoms and four oxygen atoms; so its name is dinitrogen tetroxide. (2) A ClO_2 molecule has one chlorine atom and two oxygen atoms; so its name is chlorine dioxide. (3) HI(aq) is a binary acid formed when hydrogen iodide dissolves in water. The second element present in HI(aq) is iodine; so the name of HI(aq) is hydroiodic acid. Note that the name of the molecular compound HI is hydrogen iodide. (4) HNO_2 is an oxoacid, the parent acid of the nitrite ion, NO_2^-. Therefore, the name of HNO_2 is nitrous acid.

Example 11-3 Writing the formula of a binary compound from its name
Write the formulas of (1) cobalt(Ⅱ) chloride hexahydrate; (2) diboron trisulfide; (3) silicon tetrachloride.
Strategy First check to see whether the compounds are ionic or molecular. Many compounds that contain a metal are ionic. Write the symbol of the metal first, followed by the symbol of the nonmetal. If the charge on the cation is not indicated by a Roman numeral in parentheses, it is governed by its position in the periodic table. The charge on the anion is governed by its group in the periodic table. Subscripts are chosen to balance charges. Compounds of two nonmetals are normally molecular. Their formulas are written by listing the symbols of the elements in the same order as in the name and giving them subscripts corresponding to the Greek prefixes used.
Solution (1) The combination of metal and nonmetal tells us that the compound is ionic. The Ⅱ in cobalt(Ⅱ) signifies that the metal cobalt has a charge of +2 in the compound; chlorine is in group ⅦA, so the chloride ion has a charge of −1. We can balance the charge of the +2 cation with two −1 anions; so the formula is $CoCl_2$. This compound is a hexahydrate, so we add six water molecules, the complete formula is therefore $CoCl_2 \cdot 6H_2O$. The subscripts for the individual atoms in a molecular compound are taken from the Greek prefixes, so we write (2) B_2S_3, and (3) $SiCl_4$.

Binary molecular compounds are named by using Greek prefixes to indicate the number of atoms of each element present; the element named second has its ending changed to -ide.

Vocabulary	
27. coordination compound 配位化合物	
28. elaborate [ɪˈlæbəret] 复杂的	
29. polydentate [ˌpɒlɪˈdenteɪt] 多齿配位体	
30. alphabetical [ˌælfəˈbetɪkl] 按字母顺序的	
31. denote [dɪˈnot] 代表	
32. originate [əˈrɪdʒəˌnet] 来自	
33. distinct [dɪˈstɪŋkt] 清楚的	

Name of d-Metal Complexes and Coordination Compounds

The following rules are adequate for most common complexes; more **elaborate** rules are needed if the complex contains more than one metal atom.

Procedure

(1) When naming a d-metal complex, name the ligands first and then the metal atom or ion.

(2) Neutral ligands, such as $H_2NCH_2CH_2NH_2$ (ethylenediamine), have the same name as the molecule, except for H_2O (aqua), NH_3 (ammine), CO (carbonyl), and NO (nitrosyl).

(3) Anionic ligands end in -o; for anions that end in -ide (such as chloride), -ate (such as sulfate), and -ite (such as nitrite), change the endings as follows:

$$\text{-ide} \longrightarrow \text{-o} \quad \text{ate} \longrightarrow \text{-ato} \quad \text{-ite} \longrightarrow \text{-ito}$$

Examples: chloro, cyano, sulfato, carbonato, sulfito, and nitrito.

(4) Greek prefixes indicate the number of each type of ligand in the complex ion:

$$\text{di-2, tri-3, tetra-4, penta-5, hexa-6}\cdots$$

If the ligand already contains a Greek prefix (as in ethylenediamine) or if it is **polydentate** (able to attach at more than one binding site), then the following prefixes are used instead:

$$\text{bis-2, tris-3, tetrakis-4}\cdots$$

(5) Ligands are named in **alphabetical** order, ignoring the Greek prefix that indicates the number of each one present. (Notice that Cl_2, in the coordination sphere of the second complex in rule 7 represents two chloride ion ligands, named as a dichloro, and not a Cl_2 molecular ligand.)

(6) The chemical symbols of anionic ligands precede those of neutral ligands in the chemical formula of the complex (but not necessarily in its name). Thus, Cl^- precedes H_2O and NH_3.

(7) A Roman numeral **denotes** the oxidation number of the central metal ion:

$[FeCl(H_2O)_5]^+$ pentaaquachloroiron(Ⅱ) ion
$[CrCl_2(NH_3)_4]^+$ tetraamminedichlorochromium(Ⅲ) ion
$[Co(en)_3]^{3+}$ tris(ethylenediamine)cobalt(Ⅲ) ion

(8) If the complex has an overall negative charge(an anionic complex), the suffix -ate is added to the stem of the metal's name. If the symbol of the metal **originates** from a Latin name, then the Latin stem is used. For example, the symbol for iron is Fe, from the Latin *ferrum*. Therefore, any anionic complex of iron ends with *-ferrate* followed by the oxidation number of the metal in Roman numerals:

$[Fe(CN)_6]^{4-}$ hexacyanoferrate(Ⅱ) ion
$[Ni(CN)_4]^{2-}$ tetracyanonickelate(Ⅱ) ion

(9) The name of a coordination compound (as **distinct** from a

complex cation or anion) is built in the same way as that of a simple compound, with the (possibly complex) cation named before the (possibly complex) anion:

NH$_4$[PtCl$_3$(NH$_3$)]   ammonium amminetrichloroplatinate
[Cr(OH)$_2$(NH$_3$)$_4$]Br   tetraamminedihydroxochromium(Ⅲ) bromide

This procedure is illustrated in Example 11-4.

Example 11-4 (1) Name the coordination compound [Co(NH$_3$)$_3$(H$_2$O)$_3$]$_2$(SO$_4$)$_3$.
(2) Write the formula of sodium dichlorobis(oxalato)platinate(Ⅳ).

Solution (1) First, we note that there are three SO$_4^{2-}$ ions for every two complex ions, and so the charge on the complex cation must be +3 to ensure charge neutrality of the compound. Therefore, the complex cation is [Co(NH$_3$)$_3$(H$_2$O)$_3$]$^{3+}$. Because all the ligands are neutral, the cobalt must be present as cobalt(Ⅲ). The complex ion contains three ammonia molecules (ammine) and three water molecules (aqua). It follows that the name of the cation is tri-amminetriaquacobalt(Ⅲ), and the compound is triamminetriaquacobalt(Ⅲ) sulfate. (2) Two chloride ligands, each with a charge of −1, and two oxalate ions, each with charge −2, are attached to a platinum(Ⅳ) ion, and so the charge on the complex is −2. The symbol for oxalate is ox, so the formula for the complex anion is [PtCl$_2$(ox)$_2$]$^{2-}$ and that for the compound that it forms is Na$_2$[PtCl$_2$(ox)$_2$].

APPENDIX 11-1   The Nomenclature of Polyatomicions

Formula	Name	Formula	Name	Formula	Name
NO$_2^-$	nitrite	ClO$^-$	hypochlorite	Cr$_2$O$_7^{2-}$	dichromate
NO$_3^-$	nitrate	ClO$_2^-$	chlorite	S$_2$O$_3^{2-}$	thiosulfate
CN$^-$	cyanide	ClO$_3^-$	chlorate	S$_2^{2-}$	disulfide
SCN$^-$	thiocyanate	ClO$_4^-$	perchlorate	SiO$_3^{2-}$	silicate
HCO$_3^-$	hydrogen carbonate, bicarbonate	CO$_3^{2-}$	carbonate	PO$_4^{3-}$	phosphate
HSO$_3^-$	hydrogen sulfite, bisulfite	SO$_4^{2-}$	sulfate	PO$_3^{3-}$	phosphite
HSO$_4^-$	hydrogen sulfate, bisulfate	SO$_3^{2-}$	sulfite	SiO$_4^{4-}$	silicate
H$_2$PO$_4^-$	dihydrogen phosphate	CrO$_4^{2-}$	chromate	NH$_4^+$	ammonium
HPO$_4^{2-}$	hydrogen phosphate, biphosphate	O$_2^{2-}$	peroxide	H$_3$O$^+$	hydronium
MnO$_4^-$	permanganate	OH$^-$	hydroxide	CH$_3$COO$^-$	acetate

34. detergent
[dɪ'tɜrdʒənt]
洗涤剂

35. beverage
['bɛvərɪdʒ]
饮料

36. antacid
[ænt'æsɪd]
抗酸药

APPENDIX 11-2   Common Names of Chemicals

Many chemicals have acquired common names, sometimes as a result of their use over hundreds of years and sometimes because they appear on the labels of consumer products, such as **detergents**, **beverages**, and **antacids**. The following substances are just a few that have found their way into the language of everyday life.

Formula	Chemical name	Common name
SiO$_2$	silicon dioxide	quartz
CaO	calcium oxide	lime

(paper currency is on fire yet is not consumed by the flames)

		to be continued
Formula	Chemical name	Common name
$Ca(OH)_2$	calcium hydroxide	slaked lime
NaOH	sodium hydroxide	lye
K_2CO_3	potassium carbonate	potash
$NaHCO_3$	sodium bicarbonate	baking soda, bicarbonate soda
NaCl	sodium chloride	table salt
HCl	hydrochloric acid	muriatic acid
N_2O	dinitrogen monoxide	laughing gas
$Na_2S_2O_3$	sodium thiosulfate	hypo

APPENDIX 11-3   Name of Some Common Cations with Variable Charge Numbers

Modern nomenclature includes the oxidation number of elements with variable oxidation states in the names of their compounds, as in cobalt(II) chloride. However, the traditional nomenclature, in which the suffixes -ous and -ic are used, is still encountered. The table translates from one system into the other for some common elements.

Element	Cation	Old-style name	Modern name
cobalt	Co^{2+}	cobaltous	cobalt(II)
	Co^{3+}	cobaltic	cobalt(III)
copper	Cu^+	cuprous	copper(I)
	Cu^{2+}	cupric	copper(II)
iron	Fe^{2+}	ferrous	iron(II)
	Fe^{3+}	ferric	iron(III)
lead	Pb^{2+}	plumbous	lead(II)
	Pb^{4+}	plumbic	lead(IV)
manganese	Mn^{2+}	manganous	manganese(II)
	Mn^{3+}	manganic	manganese(III)
mercury	Hg_2^{2+}	mercurous	mercury(I)
	Hg^{2+}	mercuric	mercury(II)
tin	Sn^{2+}	stannous	tin(II)
	Sn^{4+}	stannic	tin(IV)

Hands-On Chemistry: Burning Money

This is a neat "magic trick" that illustrates the process of combustion, the **flammability** of alcohol, and the special qualities of the material used to make currency.

Materials: dollar bill (higher **denomination** if you're brave), tongs, matches or a lighter, salt, solution of 50% alcohol and 50% water (you can mix 95% alcohol with water in a 1:1 ratio, if desired).

Procedure: (1) Prepare the alcohol and water solution. You can mix

37. flammability
    [,flæmə'bɪləti]
    易燃

38. denomination
    [dɪ,nɑmɪ'neʃn]
    币值

50mL of water with 50mL of 95%~100% alcohol. (2) Add a pinch salt or other colorant to the alcohol/water solution, to help produce a visible flame. (3) Soak a dollar bill in the alcohol/water solution so that it is thoroughly wet. (4) Use **tongs** to pick up the bill. Allow any excess liquid to drain. Move the damp bill away from the alcohol/water solution. (5) Light the bill on fire and allow it to burn until the flame goes out.

39. tong [tɔŋ]
钳子

Hands-On Chemistry Insights: Burning Money

A combustion reaction occurs between alcohol and oxygen, produ- cing heat and light (energy) and carbon dioxide and water.

$$C_2H_5OH + 3O_2 \longrightarrow 2CO_2 + 3H_2O + energy$$

When the bill is soaked in an alcohol/water solution, the alcohol has a high vapor pressure and is mainly on the outside of the material (a bill is more like fabric than paper, which is nice, if you've ever accidentally washed one). When the bill is lit, the alcohol is what actually burns. The temperature at which the alcohol burns is not high enough to evaporate the water, which has a high specific heat, so the bill remains wet and isn't able to catch fire on its own. After the alcohol has burned, the flame goes out, leaving a slightly damp dollar bill.

Exercises

1. Write the formula of
   (1) aluminum oxide (2) strontium phosphate (3) aluminum carbonate (4) lithium nitride
2. Name the following ionic compounds. Write both the old and the modern names wherever appropriate.
   (1) $Ca_4(PO_4)_2$, the major inorganic component of bones (2) SnF_2 (3) V_2O_5 (4) Cu_2O
3. Write the formula of
   (1) titanium dioxide (2) silicon tetrachloride (3) carbon disulfide
   (4) sulfur tetrafluoride (5) lithium sulfide (6) antimony pentafluoride
   (7) dinitrogen pentoxide (8) iodine heptafluoride
4. Name each of the following binary molecular compounds.
   (1) SF_6 (2) N_2O_5 (3) NI_3 (4) XeF_4 (5) $AsBr_3$ (6) ClO_2
5. The following aqueous solutions are common laboratory acids. What are their names?
   (1) $HCl\,(aq)$ (2) $H_2SO_4(aq)$ (3) $HNO_3(aq)$ (4) $CH_3COOH(aq)$
   (5) $H_2SO_3(aq)$ (6) $H_3PO_4(aq)$
6. Write the formula for the ionic compound formed from
   (1) sodium and oxide ions (2) potassium and sulfate ions (3) silver and fluoride ions
   (4) zinc and nitrate ions (5) aluminum and sulfide ions
7. Name each of the following compounds.
   (1) Na_2SO_3 (2) Fe_2O_3 (3) FeO (4) $Mg(OH)_2$ (5) $NiSO_4 \cdot 6H_2O$
   (6) PCl_5 (7) $Cr(H_2PO_4)_3$ (8) As_2O_3 (9) $RuCl_2$
8. Name each of the following complex ions and determine the oxidation number of the metal.
   (1) $[Fe(CN)_6]^{4-}$ (2) $[Co(NH_3)_6]^{3+}$ (3) $[Co(CN)_5(H_2O)]^{2-}$ (4) $[Co(SO_4)(NH_3)_5]^+$

9. Write the formula for each of the following coordination compounds.
   (1) potassium hexacyanochromate(Ⅲ)   (2) pentaamminesulfatocobalt(Ⅲ) chloride
   (3) tetraamminediaquacobalt(Ⅲ) bromide   (4) sodium bis(oxalato)diaquaferrate(Ⅲ)
10. Which of the following ligands can be polydentate? If the ligand can be polydentate, give the maximum number of places on the ligand that can bind simultaneously to a single metal center.
   (a) HN(CH$_2$CH$_2$NH$_2$)$_2$ (b) CO$_3^{2-}$ (c) H$_2$O   (d) oxalate

【Reading Material】

Lucifers and Other Matches

Samuel Jones, an Englishman, patented one of the first kinds of matches in 1828. It consisted of a glass **bead**（有孔珠子）containing sulfuric acid surrounded by a coating of sugar with an oxidizing agent. You ignited the match by breaking the bead using a pair of **pliers**（钳子）or, if you were more daring, your teeth. This action released the acid, which ignited an **exothermic**（放热的）reaction in the surrounding **combustible**（易燃的）materials.

Later Jones began to market a **friction**（摩擦）match discovered, but not patented, by John Walker. Walker, who had been experimenting with explosives, discovered this match one day when he tried to remove a small glob of a dried mixture from a stick. He rubbed the stick on the floor and was surprised when it **burst**(突然发生)into flame. Jones called his matches "Lucifers". They were well named: when lighted, they gave off a shower of **sparks**（火星）and smoky fumes with the **acrid**（刺鼻的） odor of sulfur dioxide. Jones had every box **inscribed**（写）with the warning "Persons whose lungs are **delicate**（娇弱的）should by no means use Lucifers."

A few years later, a Frenchman, Charles Sauria, invented the white phosphorus match, which became an immediate success. When rubbed on a rough surface, the match lighted easily, without hazardous sparks, and smelled better than Lucifers. The match head contained white phosphorus, an oxidizing agent, and glue. White phosphorus is a yellowish-white, **waxy**（蜡状的）substance, often sold in the form of sticks looking something like fat **crayons**(蜡笔). Unlike crayons, though, white phosphorus ignites **spontaneously**（自发地）in air (so it is generally stored under water). The glue in the match mixture had two purposes: it protected the white phosphorus from air, and it held the match mixture firmly together. The white phosphorus match had one serious drawback. White phosphorus is quite poisonous. Workers in match factories often began to show the **agonizing symptoms**(痛苦症状) of "**phossy jaw**(磷毒性颌骨坏死)", from white phosphorus poisoning, in which the **jawbone**（下颌骨）**disintegrates**（碎裂）. The manufacture of white phosphorus matches was **outlawed**（宣布不合法）in the early 1900s.

The head of the strike-anywhere match, which you can buy today at any grocery store, contains the relatively nontoxic tetraphosphorus trisulfide, P$_4$S$_3$, and potassium chlorate, KClO$_3$. By rubbing the match head against a surface, you create enough heat by friction to ignite the match material. Tetraphosphorus trisulfide then burns in air in a very exothermic reaction.

$$P_4S_3(s) + 8O_2(g) \longrightarrow P_4O_{10}(s) + 3SO_2(g) \quad \Delta H = -3677 kJ$$

Safety matches have a head containing mostly an oxidizing agent and require a striking surface containing nonpoisonous red phosphorus.

Questions

1. Is phosphorus essential for life?
2. Elemental phosphorus exists in two major forms, what are they?

Lesson 12 The Periodic Table

(Dmitri Mendeleev is credited with developing the first periodic table of the elements.)

1. recur
[rɪˈkɚ] 重现
2. reorganize
[riˈɔrgəˌnaɪz]
重新组织
3. substantiate
[səbˈstænʃiˌet]
证实
4. credited with
被认为有功劳
5. sequentially
按顺序地
6. electron configuration
电子构型
7. specify
[ˈspɛsəˌfaɪ]
具体说明
8. representative
[ˌrɛprɪˈzɛntətɪv]
典型的，有代表性的
9. designation
[ˌdɛzɪgˈneʃən]
名称
10. octet [akˈtɛt]
八电子
11. inert [ɪˈnət]
惰性的
12. at a time
依次，每次

Dmitri Mendeleev published the first periodic table in 1869. He showed that when the elements were ordered according to atomic weight, a pattern resulted where similar properties for elements **recurred** periodically. Based on the work of physicist Henry Moseley, the periodic table was **reorganized** on the basis of increasing atomic number rather than on atomic weight. The revised table could be used to predict the properties of elements that had yet to be discovered. Many of these predictions were later **substantiated** through experimentation. This led to the formulation of the periodic law, which states that the chemical properties of the elements are dependent on their atomic numbers.

Dmitri Mendeleev is **credited with** developing the first periodic table of the elements.

Elements in the periodic table are arranged in periods (rows) and groups (columns). Each of the seven periods is filled **sequentially** by atomic number. Groups include elements having the same **electron configuration** in their outer shell, which results in group elements sharing similar chemical properties. The electrons in the outer shell are termed valence electrons. Valence electrons determine the properties and chemical reactivity of the element and participate in chemical bonding. The Roman numerals found above each group **specify** the usual number of valence electrons. There are two sets of groups. The group A elements are the **representative** elements, which have s or p sublevels as their outer orbitals. The group B elements are the nonrepresentative elements, which have partly filled d sublevels (the transition elements) or partly filled f sublevels (the lanthanide series and the actinide series). The Roman numeral and letter **designations** give the electron configuration for the valence electrons (e.g., the valence electron configuration of a group VA element will be s^2p^3 with 5 valence electrons).

Periodic Properties of the Elements

The properties of the elements exhibit trends. These trends can be predicted using the periodic table and can be explained and understood by analyzing the electron configurations of the elements. Elements tend to gain or lose valence electrons to achieve stable **octet** formation. Stable octets are seen in the **inert** gases, or noble gases, of group VIII of the periodic table. In addition to this activity, there are two other important trends. First, electrons are added one **at a time** moving from left to right across a period. As this happens, the electrons of the outermost shell

13. shield…from…
   屏蔽……免受……
14. affinity
   [əˈfɪnɪti]
   亲和力

experience increasingly strong nuclear attraction, so the electrons become closer to the nucleus and more tightly bound to it. Second, moving down a column in the periodic table, the outermost electrons become less tightly bound to the nucleus. This happens because the number of filled principal energy levels (which **shield** the outermost electrons **from** attraction to the nucleus) increases downward within each group. These trends explain the periodicity observed in the elemental properties of atomic radius, ionization energy, electron **affinity** and electronegativity.

Atomic Radius

The atomic radius of an element is half of the distance between the centers of two atoms of that element that are just touching each other. Generally, the atomic radius decreases across a period from left to right and increases down a given group. The atoms with the largest atomic radii are located in group Ⅰ and at the bottom of groups.

Moving from left to right across a period, electrons are added one at a time to the outer energy shell. Electrons within a shell cannot shield each other from the attraction to protons. Since the number of protons is also increasing, the effective nuclear charge increases across a period. This causes the atomic radius to decrease.

Moving down a group in the periodic table, the number of electrons and filled electron shells increases, but the number of valence electrons remains the same. The outermost electrons in a group are exposed to the same effective nuclear charge, but electrons are found farther from the nucleus as the number of filled energy shells increases. Therefore, the atomic radii increase.

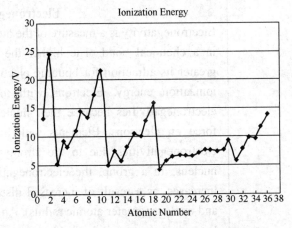

(This graph displays the periodic trend of ionization energy versus atomic number.)

Ionization Energy

The ionization energy, or ionization potential, is the energy required to completely remove an electron from a gaseous atom or ion. The closer and more tightly bound an electron is to the nucleus, the more difficult it will be to remove, and the higher its ionization energy will be. The first ionization energy is the energy required to remove one electron from the parent atom. The second ionization energy is the energy required to remove a second valence electron from the **univalent** ion to form the **divalent** ion, and so on. Successive ionization energies increase. The second ionization energy is always greater than the first ionization energy. Ionization energies increase moving from left to right across a period (decreasing atomic radius). Ionization energy decreases moving down a group (increasing atomic radius). Group I elements have low ionization energies because the loss of an electron forms a stable octet.

Electron Affinity

Electron affinity reflects the ability of an atom to accept an electron. It is the energy change that occurs when an electron is added to a gaseous atom. Atoms with stronger effective nuclear charge have greater electron affinity. Some **generalizations** can be made about the electron affinities of certain groups in the periodic table. The group IIA elements, the **alkaline** earths, have low electron affinity values. These elements are relatively stable because they have filled s subshells. Group VIIA elements, the **halogens**, have high electron affinities because the addition of an electron to an atom results in a completely filled shell. Group VIII elements, noble gases, have electron affinities near zero, since each atom possesses a stable octet and will not accept an electron readily. Elements of other groups have low electron affinities.

Electronegativity

Electronegativity is a measure of the attraction of an atom for the electrons in a chemical bond. The higher the electronegativity of an atom, the greater its attraction for bonding electrons. Electronegativity is related to ionization energy. Electrons with low ionization energies have low electronegativities because their nuclei do not **exert** a strong attractive force on electrons. Elements with high ionization energies have high electronegativities due to the strong pull exerted on electrons by the nucleus. In a group, the electronegativity decreases as atomic number increases, as a result of increased distance between the valence electron and nucleus (greater atomic radius). An example of an electropositive (i.e., low electronegativity) element is **cesium**; an example of a highly electronegative element is fluorine.

Summary of Periodic Table Trends

Moving from left to right: atomic radius decreases, ionization energy increases, electronegativity increases.

Moving from top to bottom: atomic radius increases, ionization energy decreases, electronegativity decreases.

Hands-On Chemistry: **Rubber Egg & Chicken Bones**

A mad scientist can make a toy out of just about anything, including a boiled egg. Soak an egg in a common kitchen ingredient, vinegar, to dissolve its shell and make the egg rubbery enough that you can bounce it on the floor like a ball. Soaking chicken bones in vinegar will soften them so that they will become rubbery and flexible.

(A mad scientist can do much more with an egg than just eat it.)

Rubber Egg Materials   hard-boiled egg, glass or jar (big enough to hold the egg), vinegar.

How to Turn the Egg into a Bouncy Ball

(1) Place the egg in the glass or jar.

(2) Add enough vinegar to completely cover the egg.

(3) Watch the egg. What do you see? Little bubbles may come off the egg as the acetic acid in the vinegar attacks the calcium carbonate of the eggshell. Over time the color of the egg may change as well.

(4) After 3 days, remove the egg and gently rinse the shell off of the egg with tap water.

(5) How does the boiled egg feel? Try bouncing the egg on a hard surface. How high can you bounce your egg?

(6) You can soak raw eggs in vinegar for 3~4 days, with a slightly different result. The eggs' shell will become soft and flexible. You can gently squeeze these eggs, but it's not a great plan to try to bounce them on the floor.

How to Make Rubbery Chicken Bones

If you soak chicken bones in vinegar (the thinner bones work best), the vinegar will react with the calcium in the bones and weaken them so that they will become soft and rubbery, as if they had come from a rubber chicken. It is the calcium in your bones that makes them hard and strong. As you age, you may deplete the calcium faster than you replace it. If too much calcium is lost from your bones, they may become brittle and susceptible to breaking. Exercising and eating a diet that includes calcium-rich foods can help prevent this from happening.

Exercises

1. Which element of periodic group I A is not an alkali metal?
2. Which of the following neighbors of fluorine in the periodic table has properties most like those of fluorine: oxygen, neon, or chlorine?
3. (1) Which two elements are in group VI of period 5?

   (2) Which element(s) is (are) in group III of period 2?

4. Using the periodic table, determine: (1) How many elements whose names start with the letter S are transition elements? (2) How many elements whose symbols start with the letter T are transition elements?

5. Would you expect manganese or selenium to act more like a typical metal?

6. The formula of an oxide of phosphorus is P_2O_5. Make an educated guess as to which one of the following compounds exists: As_2O_5; Ga_2O_5; S_2O_5.

7. Using a periodic table, obtain the symbol corresponding to each of the following elements, and name the element.

   (1) element in group ⅡA, period 5   (2) element in group ⅣA, period 3
   (3) element in group ⅦB, period 4   (4) element in group ⅣB, period 4
   (5) element in group ⅤA, period 4.

8. Which metals are in group ⅤA?

9. For the element phosphorus, obtain the atomic number, atomic weight, group, period, and determine whether the element is a metal, nonmetal, or metalloid.

10. Use the periodic table to identify each of the following.
    (1) the sixth element of the second transition series   (2) the element of the fifth period that is also in group ⅣB   (3) the third lanthanide   (4) the second transition element   (5) The seventh actinide metal   (6) the first element of group Ⅷ   (7) the fourth halogen   (8) the first alkali metal (9) the last coinage metal

11. Elements in the same family in the periodic table_____.
    (a) have the same number of protons   (b) have the same number of electrons
    (c) have similar chemical properties   (d) are in the same horizontal row
    (e) None of the above

12. Which elements in group ⅥA are metals?
    (a) all of them   (b) oxygen (O), sulfur (S), and selenium (Se)
    (c) tellurium (Te)   (d) polonium (Po)   (e) None of the above

【Reading Material】

Out of Oxygen

In September 1991 four men and four women enter the world's largest glass **bubble** (泡泡), known as Biosphere Ⅱ, to test the idea that human could design and build a totally **self-contained** (自给自足的) **ecosystem** (生态系统), a model for some future **colony** (殖民地) on another planet. Biosphere Ⅱ (earth is considered Biosphere Ⅰ) was a 3-acre mini-world, completed with a tropical rain forest, **savanna** (无树平原), **marsh** (湿地), desert and working farm that was intended to be fully sufficient. This unique experiment was to continue for 2 to 3 years, but almost immediately there were signs that the project could be in **jeopardy** (危险).

Soon after the bubble had been sealed, sensors inside the facility showed that the concentration of oxygen in Biosphere Ⅱ's atmosphere had fallen from its initial level of 21 percent (by volume), while the amount of carbon dioxide had risen from a level of 0.035 percent (by volume), or 350 ppm (parts per million). **Alarmingly** (让人担忧地), the oxygen level continued to fall at a rate of about 0.5 percent a month and the level of carbon dioxide kept rising, forcing the crew to turn on **electrically powered** (电动的) chemical **scrubbers** (洗涤器), similar to those on submarines, to remove some of the excess CO_2. Gradually the CO_2 level stabilized around 4000 ppm, which is high but not dangerous. The loss of oxygen did not stop, though. By January 1993 (16 months into the experiment), the oxygen concentration had dropped to 14 percent, which is equivalent to the O_2 concentration in air at an elevation of 4360m (14300 ft). The crew began having trouble performing normal tasks. For their safety it was necessary to **pump** (泵) pure oxygen into Biosphere II.

With all the plants present in Biosphere II, the production of oxygen should have been greater as a consequence of **photosynthesis** (光合作用). Why had the oxygen concentration declined to such a low level? A small part of the loss was blamed on unusually cloudy weather, which had slowed down plant growth. The possibility that iron in the soil was reacting with oxygen to form iron(III) oxide or rust was **ruled out** (被排除) along with several other explanations for lack of evidence. The most **plausible** (似乎可信的) hypothesis was that microbes (microorganisms) were using oxygen to metabolize the excess organic matter that had been added to the soils to promote plant growth. This turned out to be the case.

Identifying the cause of oxygen **depletion** (消耗) raised another question. **Metabolism** (新陈代谢) produces carbon dioxide. Based on the amount of oxygen consumed by the microbes, the CO_2 level should have been at 40000 ppm, 10 times what was measured. What happened to the excess gas? After ruling out leakage to the outside world and reactions between CO_2 with compounds in the soils and in water, scientists found that the **concrete** (混凝土) inside Biosphere II was consuming large amounts of CO_2!

Concrete is a mixture of sand and **gravel** (碎石) held together by a binding agent which is a mixture of calcium silicate hydrates and calcium hydroxide. The calcium hydroxide is the key ingredient in the CO_2 mystery. Carbon dioxide **diffuses** (扩散) into the porous structure of concrete then reacts with calcium hydroxide to form calcium carbonate and water:

$$Ca(OH)_2(s) + CO_2(g) \longrightarrow CaCO_3(s) + H_2O(l)$$

Under normal conditions, this reaction goes on slowly. But CO_2 concentrations in Biosphere II were much higher than normal, so the reaction proceeded much faster. In fact, in just over 2 years, $CaCO_3$, had accumulated to a depth of more than 2 cm in Biosphere II's concrete. Some 10000m² of exposed concrete was hiding 500000 to 1500000 moles of CO_2.

The water produced in the reaction between $Ca(OH)_2$ and CO_2 created another problem: CO_2 also reacts with water to form carbonic acid (H_2CO_3), and hydrogen ions produced by the acid promote the corrosion of the **reinforcing** (加固) iron bars in the concrete, thereby weakening its structure. This situation was dealt with effectively by painting all concrete surfaces with an **impermeable** (不可渗透的) coating.

In the meantime the decline in oxygen (and hence also the rise in carbon dioxide) slowed, perhaps because there was now less organic matter in the soils and also because new lights in the agricultural areas may have **boosted** (促进) photosynthesis. The project was **terminated** (终止) prematurely and, as of 1996, the facility was transformed into a science education and research center.

The Biosphere II experiment is an interesting project from which we can learn a lot about earth and its **inhabitants** (居民). If nothing else, it has shown us how complex earth's ecosystems are and how difficult it is to **mimic** (仿生) nature, even on a small scale.

Questions

1. What solution would you use in a chemical scrubber to remove carbon dioxide?

2. Photosynthesis converts carbon dioxide and water to carbohydrates and oxygen gas, while metabolism is the process by which carbohydrates react with oxygen to form carbon dioxide and water. Using glucose ($C_6H_{12}O_6$) to represent carbohydrates, write equations for these two processes.

3. Why was diffusion of O_2 from Biosphere II to the outside world not considered a possible cause for the depletion in oxygen?

4. Carbonic acid is a diprotic acid. Write equations for the stepwise ionization of the acid in water.

5. What are the factors to consider in choosing a planet on which to build a structure like Biosphere II?

Lesson 13　Chemical Bond

1. linkage
['lɪŋkɪdʒ]
连接
2. electrostatic
[ɪˌlɛktrɔ'stætɪk]
静电的
3. electrodynamic
电力学的
4. coulombic
库仑力的
5. repulsion
[rɪ'pʌlʃən]
斥力
6. quantum
['kwɑːntəm]
量子
7. categorize
['kætɪgəˌraɪz]
分类
8. discretely
离散地
9. mimic ['mɪmɪk]
模拟
10. relinquish
[rɪ'lɪŋkwɪʃ]
松开，放开
11. autonomous
[ɔ'tɑnəməs]
独立的
12. entity ['ɛntɪti]
实体
13. harmoniously
[hɑr'monɪəsli]
和谐地
14. buoy ['bui, bɔɪ]
激励
15. sufficient
[sə'fɪʃənt]
充分的
16. collapse
[kə'læps]
崩溃
17. neutral
['nutrəl, 'nju-]
中性的
18. three-dimensional
立体的
19. array [ə'reɪ]
排列
20. lattice ['lætɪs]
晶格

Chemical bond is the term used to describe the **linkages** between atoms joined together to form molecules or crystals. Chemical bonds are the result of electromagnetic interactions that may be either **electrostatic** or **electrodynamic** in nature or a combination of the two. Electrostatic bonding forces result from **coulombic** attraction or **repulsion** between charged particles whereas electrodynamic bonding forces result from the sharing of electrons and are described by **quantum** mechanical theories of the valence bond and of molecular orbitals.

Chemical bonding is **categorized** within five different classes of chemical bonds: ionic, covalent, coordinate covalent, metallic, and hydrogen. Actual bonds may have properties that aren't so **discretely** categorized, so a given bond could be defined by more than one of these terms.

Chemical bonding theory explains one aspect of the relational nature of physical existence and **mimics** in some ways the essential nature of bonds holding people together in families, groups, societies, or nations. In the formation of a chemical bond between two atoms (or ions), for example, each of the two **relinquishes** a part of its energy of **autonomous** existence as a contribution toward the energy of the chemical bond formed between them. As a consequence, the two bound **entities** each exists with a lower individual energy than when they were autonomous and the bond is stable. Similarly, two people interacting **harmoniously** are naturally bonded and may feel themselves **buoyed** by the relationship, as though their existence requires less energy. Conversely, separating two chemically-bonded atoms (or ions) requires the input of enough energy to return to each entity energy **sufficient** for autonomous existence. Likewise with harmoniously bonded people, an input stronger than the bond between them is needed to break their relationship.

The Ionic Bond

The ionic bond refers to the electrostatic forces holding the ions together, but not **collapsing** together, in an ionic solid. Ionic solids are electrically **neutral** and contain both cations (positively charged ions), and anions (negatively charged ions). These ions are held together in an ordered **three-dimensional array**, a crystal **lattice**, by coulombic attractive forces acting between the net positive or negative charges on the ions and by repulsive forces associated with the electron shells of each ion. The lattice arrangement means that each positive ion is surrounded most closely by negative ions and each negative ion is surrounded most closely by positive

21. exemplify
[ɪgˈzɛmpləˌfaɪ]
例证

22. lithium
[ˈlɪθiəm]
锂

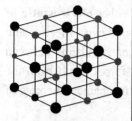

[The Sodium Chloride Crystal Structure is an example of a crystal lattice held together by electrostatic forces. Each atom has six nearest neighbors, with **octahedral geometry**. This arrangement is known as cubic close packed (ccp). Light blue = Na^+ ion, Dark green = Cl^- ion]

23. octahedral
[ˌɑktəˈhidrəl]
八面体的

24. geometry
[dʒiˈɑmɪtri]
几何排列

25. collective
[kəˈlɛktɪv]
集体的

26. soluble
[ˈsɑljəbəl]
可溶解的

:N≡N:
:Xe:

(Lewis dot diagrams are used to show how elements bond together in a molecule.)

ions. In this way, the lattice minimizes the repulsive forces between the similarly charged ions and maximizes the attractive forces between the oppositely charged ions. While ions in principle can be monatomic or polyatomic, the ions in most crystals are monatomic.

The formation of an ionic bond is **exemplified** in the reaction between the metal, **lithium**, and the non-metal, fluorine. Metals loose electrons to form cations, and non-metals gain electrons to form anions. The resulting ionic solid is held together by ionic bonds formed due to the electrostatic forces acting between charged ions.

$$Li + F \longrightarrow Li^+F^-$$

In this example, lithium has one valence electron, which is held rather loosely and is easily lost. Fluorine has seven valence electrons and it tends to gain one electron in order to give a stable octet of electrons. If the electron moves from lithium to fluorine each ion acquires the stable electronic configuration of a noble gas. The bonding energy (from the electrostatic forces acting within the crystal structure) is large enough that the net energy of the bonded ions is lower than their **collective** energy in the unbonded state. Reactions that form ionic crystals occur only if the overall energy change for the reaction is favorable [the bonded atoms (or ions) have a lower energy than the free ones]. The larger the energy change, the stronger the bond, because the energy lost by the atoms (or ions) is imparted into the energy of the bond itself.

The electrostatic forces acting within an ionic solid are quite large. Such bonds are stronger than hydrogen bonds, but similar in strength to covalent bonds. Consequently ionic solids have relatively high melting points and boiling points. At the same time, however, many ionic solids are water **soluble** because the ions are readily attracted by water's polar molecules. In the solid state, ionic solids do not conduct electricity, as the ions are fixed in place and their electrons are securely bound to each of the ions. When ionic solids are melted or dissolved in water, the ions are free to move and in these states ionic compounds conduct electricity via the movement of ions themselves rather than of electrons.

The Covalent Bond

Covalent bonding is a form of chemical bonding characterized by the sharing of one or more pairs of electrons between two atoms, producing a mutual attraction that holds the resultant molecule or polyatomic ion together. Atoms tend to share electrons in such a way that their outer electron shells satisfy the octet rule. Such bonds are always stronger than the intermolecular hydrogen bond and similar in strength to or stronger than the ionic bond.

Covalent bonding most frequently occurs between atoms of non-metals with similar electronegativities. In contrast, metals with their

27. roam [rom]
漫游, 徘徊

28. indispensable
[ˌɪndɪˈspɛnsəbəl]
不可缺少的

easily-removed electrons being somewhat free to **roam** in the material are more likely to form metallic bonds when confronted with another species of similar electronegativity.

However, covalent bonding in metals and, particularly between metals and organic compounds is particularly important, especially in industrial catalysis and process chemistry, where many **indispensable** reactions depend on covalent bonding with metals.

Hands-On Chemistry: Bending Water

Static electricity can be a problem whenever the humidity is low. It causes shocks and makes dust stick to surfaces, and it can literally make your hair stand on end. In this experiment, you will see that it also can move things around.

What You Will Need: a nylon comb, a water faucet.

Adjust the faucet to produce a small stream of water. The stream should be about 1.5 millimeters (1/16 inch) in diameter.

Run the comb through your hair several times. Slowly bring the teeth of the comb near the stream of water, about 8 to 10 centimeters (3 or 4 inches) below the faucet. When the teeth of the comb are about an inch or less away from the stream, the stream will bend toward the comb.

Move the comb closer to the stream. How does the distance between the stream and the comb affect how much the stream bends?

Run the comb through your hair several more times. Does the comb bend the stream more now? Change the size of the stream by adjusting the faucet. Does the size of the stream affect how much the stream bends? If you have other combs, you can try these to see if some bend the stream more than others.

Hands-On Chemistry Insights: Bending Water

Static electricity is the accumulation of an electrical charge in an object. The electrical charge develops when two objects are rubbed against one another. When the objects are rubbed together, some electrons (charged components of atoms) jump from one object to the other. The object that loses the electrons becomes positively charged, while the object that they jump to becomes negatively charged. The nature of the objects has a large effect on how many electrons move. This determines how large an electrical charge accumulates in the objects. Hair and nylon are particularly good at acquiring charge when they are rubbed together.

A charged object attract small particles, such as dust. The charge in the object causes a complementary charge to develop in something close to it. The complementary charge is attracted to the charged object. If the complementary charge forms on something tiny, such as dust particles, these tiny particles move to the charged object. This is why your television screen becomes dusty faster than the television cabinet. When a television operates, electrons fly from the back to the screen. These electrons cause the screen to become charged. The charge on the screen attracts dust.

The comb attracts the stream of water in the same way. The charge on the comb attracts the molecules of water in the stream. Because the molecules in the stream can be moved easily, the stream bends toward the comb. When you comb your hair with a nylon comb, both the comb and your hair become charged. The comb and your hair acquire opposite charges. Because the individual hairs acquire the same charge, they repel each other. Perhaps you noticed that after running the nylon comb through your hair, the hairs on your head stood on end. This is a result of your hairs repelling each other because they are charged. Static electricity is more of a problem when humidity is low. When humidity is high, most surfaces are coated with a thin film of water. When objects coated by a film of water are rubbed together, the water prevents electrons from jumping between the objects.

Exercises

1. What are the two principle types of chemical bonds?
2. Define the following terms.
    (1) ionic bond   (2) covalent bond   (3) bonding pair   (4) lone pair   (5) electronegativity
3. Which bond in each of the following pairs is more polar?
    (1) Na—O, O—O   (2) O—O, H—O   (3) F—O, F—H
4. For each of the following pairs of elements, state whether the compound formed is likely to be ionic or covalent.
    (1) Ba, O   (2) C, F   (3) In, Cl   (4) P, Br
5. Select the molecule from each pair that is polar.
    (1) HF, F_2 (2) CF_4, NF_3 (3) SiH_4, H_2O
6. How many electrons are there in the valence shell of the sodium atom?
    (a) 1   (b) 2   (c) 4   (d) 7   (e) None of the above
7. How many electrons are there in the valence shell of the chlorine atom?
    (a) 1   (b) 2   (c) 4   (d) 7   (e) None of the above
8. How many electrons are there in the valence shell of the sodium ion?
    (a) 1   (b) 2   (c) 4   (d) 7   (e) None of the above
9. How many electrons are there in the valence shell of the chloride ion?
    (a) 1   (b) 2   (c) 4   (d) 7   (e) None of the above
10. An electron pair that is shared between two atoms is called_____.
    (a) a rebonding pair   (b) a bonding pair   (c) a lone pair
    (d) a double bond   (e) None of the above
11. An electron pair that is entirely on one atom in a molecule is called_____.
    (a) a rebonding pair   (b) a bonding pair   (c) a lone pair
    (d) a double bond   (e) None of the above
12. Which one of the following elements is more electronegative than sulfur (S)?
    (a) Te   (b) Al   (c) Si   (d) P   (e) None of the above
13. Which one of the following elements is more electronegative than oxygen (O)?
    (a) C   (b) N   (c) Cl   (d) F   (e) None of the above
14. The molecular shape of the phosphine molecule, PH_3, is_____.
    (a) linear   (b) angular   (c) trigonal planar
    (d) trigonal pyramidal   (e) None of the above

15. The molecular shape of the carbon disulfide molecule, CS_2, is_____.
    (a) linear (b) angular (c) trigonal planar (d) trigonal pyramidal (e) None of the above
16. The molecular shape of the silane molecule, SiH_4, is_____.
    (a) linear (b) angular (c) trigonal planar (d) trigonal pyramidal (e) None of the above
17. The molecular shape of the hydrogen sulfide molecule, H_2S, is_____.
    (a) linear (b) angular (c) trigonal planar (d) trigonal pyramidal (e) None of the above
18. Which of the following is a polar molecule?
    (a) CO_2 (b) NH_3 (c) CH_4 (d) SiF_4 (e) None of the above

【Reading Material】

Who Killed Napoleon?

After his defeat at Waterloo in 1815, Napoleon was **exiled**(流放) to St Helena, a small island in the Atlantic Ocean, where he spent the last six years of his life. In the 1960s, samples of his hair were analyzed and found to contain a high level of **arsenic** (砷), suggesting that he might have been poisoned. The prime suspects are the governor of St Helena with whom Napoleon did not get along, and the French royal family who wanted to prevent his return to France.

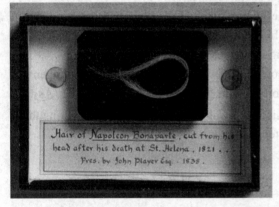

Elemental arsenic is not that harmful. The commonly used poison is actually arsenic (Ⅲ) oxide. As_2O_3, a white compound that dissolves in water, is tasteless, and if **administered** (给药) over a period of time, is hard to detect. It was once known as the "**inheritance** (遗产) powder" because it could be added to grandfather's wine to **hasten** (使加紧) his **demise** (让位) so that his grandson could inherit the estate!

In 1832 the English chemist James Marsh devised a procedure for detecting arsenic. This test, which now **bears** (带有…标记) Marsh's name, combines hydrogen formed by the reaction between zinc and sulfuric acid with a sample of the suspected poison. If As_2O_3 is present, it reacts with hydrogen to form a toxic gas, arsine (AsH_3). When arsine gas is heated, it decomposes to form arsenic, which is recognized by its metallic **luster** (光泽). The Marsh test is an effective **deterrent** (威慑) to murder by As_2O_3, but it was invented too late to do Napoleon any good, in fact, he was a victim of **deliberate** (蓄意的) arsenic poisoning.

Doubts about the **conspiracy** (阴谋) theory of Napoleon's death developed in the early 1990s, when a sample of the wallpaper from his drawing room was found to contain copper arsenate ($CuHAsO_4$), a green **pigment** (颜料) that was commonly used at the time Napoleon lived. It has been suggested that the damp climate on St Helena promoted the growth of **molds** (霉) on the wallpaper. To rid themselves of arsenic, the molds could have converted it to trimethyl arsine [$(CH_3)_3As$], which is a **volatile** (易挥发的) and highly poisonous compound. Prolonged exposure to these vapors would have ruined Napoleon's health and would also account for the presence of arsenic in his body, though it may not have been the primary cause of his death. This **provocative** (挑衅的) theory is supported by the fact that Napoleon's regular guests suffered from **gastroin-**

testinal disturbances (胃肠道功能紊乱) and other symptoms(症状) of arsenic poisoning and that their health all seemed to improve whenever they spent hours working outdoors in the garden, their main hobby on the island.

We will probably never know whether Napoleon died from arsenic poisoning, intentional or accidental, but this exercise in historical **sleuthing**(调查) provides a fascinating example of the use of chemical analysis. Not only is chemical analysis used in **forensic science** (法医学), but it also plays an essential part in **endeavors** (努力) ranging from pure research to practical applications, such as quality control of commercial products and medical **diagnosis** (诊断).

Questions

1. The arsenic in Napoleon's hair was detected using a technique called neutron activation. When As-75 is bombarded with high-energy neutrons, it is converted to the radioactive As-76 isotope. The energy of the γ rays emitted by the radioactive isotope is characteristic of arsenic, and the intensity of the rays establishes how much arsenic is present in a sample. With this technique, as little as 5ng (5×10^{-9}g) of arsenic can be detected in 1g of material. (1) Write symbols for the two isotopes of As, showing mass number and atomic number. (2) Name two advantages of analyzing the arsenic content by neutron activation instead of a chemical analysis.

2. Arsenic is not an essential element for the human body. (1) Based on its position in the periodic table, suggest a reason for its toxicity. (2) In addition to hair, where else might one look for the accumulation of the element if arsenic poisoning is suspected?

3. The Marsh test for arsenic involves the following steps: (1) The generation of hydrogen gas when sulfuric acid is added to zinc. (2) The reaction of hydrogen with arsenic (Ⅲ) oxide (As_2O_3), to produce arsine. (3) Conversion of arsine to arsenic by heating. Write equations representing these steps and identify the type of the reaction in each step.

Lesson 14 Coordination Chemistry

1. coordination [kouˌɔːdɪˈneɪʃn] 配位
2. ligand [ˈlɪɡənd] 配位体
3. polydentate 多配位基的
4. chelate [ˈkiːleɪt] 螯合的
5. sphere [sfɪr] 球
6. originally [əˈrɪdʒənəli] 最初, 原先
7. reversible [rɪˈvɜːsəbəl] 可逆的
8. evolve [ɪˈvɑlv] 进化
9. irreversible [ˌɪrɪˈvɜːsəbəl] 不可逆的
10. Prussian blue 普鲁士蓝
11. copper vitriol 胆矾
12. octahedral [ˌɑktəˈhidrəl] 八面体的
13. chirality 手性
14. ammine [ˈæmɪn] 氨络合物
15. inexplicable [ɪnˈɛksplɪkəbəl] 无法说明的
16. lone electron pair 孤电子对

Coordination chemistry is the science concerned with the interactions of organic and inorganic ligands with metal centres. It studies the physical and chemical properties, syntheses and structures of **coordination** compounds.

In chemistry, a coordination complex or metal complex, is a structure consisting of a central atom or ion (usually metallic), bonded to a surrounding array of molecules or anions (ligands, complexing agents). The atom within a **ligand** that is directly bonded to the central atom or ion is called the donor atom. **Polydentate** (multiple bonded) ligands can form a **chelate** complex. A ligand donates at least one pair of electrons to the central atom/ion.

Compounds that contain a coordination complex are called coordination compounds. The central atom or ion, together with all ligands form the coordination **sphere**.

Coordination refers to the "coordinate covalent bonds" (dipolar bonds) between the ligands and the central atom. **Originally,** a complex implied a **reversible** association of molecules, atoms, or ions through such weak chemical bonds. As applied to coordination chemistry, this meaning has **evolved**. Some metal complexes are formed virtually **irreversibly** and many are bound together by bonds that are quite strong.

History

Coordination complexes were known - although not understood in any sense - since the beginning of chemistry, e.g. **Prussian blue** and **copper vitriol**. The key breakthrough occurred when Alfred Werner proposed in 1893 that Co(Ⅲ) bears six ligands in an **octahedral** geometry. In 1914, he resolved the first coordination complex, called hexol, into optical isomers, overthrowing the theory that **chirality** was necessarily associated with carbon compounds.

The theory allows one to understand the difference between coordinated and ionic chloride in the cobalt **ammine** chlorides and to explain many of the previously **inexplicable** isomers.

Structure of Coordination Complexes

The ions or molecules surrounding the central atom are called ligands. Ligands are generally bound to said central atom by a coordinate covalent bond (donating electrons from a **lone electron pair** into an empty metal orbital), and are thus said to be coordinated to the atom.

17. ambiguous
   [æm'bɪgjuəs]
   不确定的
18. lanthanide
   ['lænθə,naɪd]
   镧系元素
19. actinide
   ['æktə,naɪd]
   锕系元素
20. accommodate
   [ə'kɑmə,det]
   容纳

[cisplatin, PtCl₂(NH₃)₂
A platinum atom with four ligands]

21. polyhedron
   [,pɑlɪ'hidrən]
   多面体
22. deviate
   ['divi,et]
   偏离
23. distortion
   [dɪ'stɔrʃən]
   扭曲，变形
24. linear 线性的
25. trigonal planar
   三角平面的
26. tetrahedral
   [,tɛtrə'hidrəl]
   四面体的
27. square planar
   正方平面型的
28. trigonal bipyramidal
   三角双锥
29. square pyramidal
   四棱锥的

Geometry

In coordination chemistry, a structure is first described by its coordination number, the number of ligands attached to the metal (more specifically, the number of σ-type bonds between ligand(s) and the central atom). Usually one can count the ligands attached, but sometimes even the counting can become **ambiguous**. Coordination numbers are normally between two and nine, but large numbers of ligands are not uncommon for the **lanthanides** and **actinides**. The number of bonds depends on the size, charge, and electron configuration of the metal ion and the ligands. Metal ions may have more than one coordination number.

Typically the chemistry of complexes is dominated by interactions between s and p molecular orbitals of the ligands and the d orbitals of the metal ions. The s, p, and d orbitals of the metal can **accommodate** 18 electrons (see 18-Electron rule; for f-block elements, this extends to 32 electrons). The maximum coordination number for a certain metal is thus related to the electronic configuration of the metal ion (more specifically, the number of empty orbitals) and to the ratio of the size of the ligands and the metal ion. Large metals and small ligands lead to high coordination numbers, e.g. $[Mo(CN)_8]^{4-}$. Small metals with large ligands lead to low coordination numbers, e.g. $Pt[P(CMe_3)]_2$. Due to their large size, lanthanides, actinides, and early transition metals tend to have high coordination numbers.

Different ligand structural arrangements result from the coordination number. Most structures follow the points-on-a-sphere pattern (or, as if the central atom were in the middle of a **polyhedron** where the corners of that shape are the locations of the ligands), where orbital overlap (between ligand and metal orbitals) and ligand-ligand repulsions tend to lead to certain regular geometries. The most observed geometries are listed below, but there are many cases which **deviate** from a regular geometry, e.g. due to the use of ligands of different types (which results in irregular bond lengths; the coordination atoms do not follow a points-on-a-sphere pattern), due to the size of ligands, or due to electronic effects (e.g. Jahn-Teller **distortion**):

Linear for two-coordination
Trigonal planar for three-coordination
Tetrahedral or **square planar** for four-coordination
Trigonal bipyramidal or **square pyramidal** for five-coordination
Octahedral (orthogonal) or trigonal prismatic for six-coordination
Pentagonal bipyramidal for seven-coordination
Square antiprismatic for eight-coordination
Tri-capped trigonal prismatic (Triaugmented triangular prism) for nine coordination.

Some exceptions and provisions should be noted:

Vocabulary	
30. indistinct [ˌɪndɪˈstɪŋkt] 不清楚的	
31. alternative [ɔlˈtɚnətɪv] 可选择的	
32. stabilization 稳定性	
33. mutable [ˈmjutəbəl] 易变的	
34. stereoisomerism 立体异构	
35. orientation [ˌɔriɛnˈteʃən] 方向, 方位	
36. adjacent [əˈdʒesənt] 邻近的	
37. superimposable 镜像重合	
38. optically active 光学活性	
39. polarized light 偏振光	
40. bidentate [baɪˈdɛnˌtet] 两齿的	

The idealized descriptions of 5-, 7-, 8-, and 9- coordination are often **indistinct** geometrically from **alternative** structures with slightly different L-M-L (ligand-metal-ligand) angles. The classic example of this is the difference between square pyramidal and trigonal bipyramidal structures.

Due to special electronic effects such as (second-order) Jahn-Teller **stabilization**, certain geometries are stabilized relative to the other possibilities, e.g. for some compounds the trigonal prismatic geometry is stabilized relative to octahedral structures for six-coordination.

Isomerism

The arrangement of the ligands is fixed for a given complex, but in some cases it is **mutable** by a reaction that forms another stable isomer. There exist many kinds of isomerism in coordination complexes, just as in many other compounds.

Stereoisomerism

Stereoisomerism occurs with the same bonds in different **orientations** relative to one another. Stereoisomerism can be further classified into:

— **cis-trans Isomerism and facial-Meridional Isomerism**

cis-trans isomerism occurs in octahedral and square planar complexes (but not tetrahedral). When two ligands are mutually **adjacent** they are said to be cis, when opposite each other, trans. When three identical ligands occupy one face of an octahedron, the isomer is said to be facial, or fac. In a fac isomer, any two identical ligands are adjacent or cis to each other. If these three ligands and the metal ion are in one plane, the isomer is said to be meridional, or mer. A mer isomer can be considered as a combination of a trans and a cis, since it contains both trans and cis pairs of identical ligands.

— **Optical Isomerism**

Optical isomerism occurs when the mirror image of a compound is not **superimposable** with the original compound. It is so called because such isomers are **optically active**, that is, they rotate the plane of **polarized light**. The symbol Λ (lambda) is used as a prefix to describe the left-handed propeller twist formed by three **bidentate** ligands, as shown. Similarly, the symbol Δ (delta) is used as a prefix for the right-handed propeller twist.

— **Structural Isomerism**

Structural isomerism occurs when the bonds are themselves different. Linkage isomerism is only one of several types of structural isomerism in coordination complexes (as well as other classes of chemical compounds). Linkage isomerism occurs with ambidentate ligands which can bind in more than one place. For example, NO_2 is an ambidentate ligand, it can bind to a metal at either the N atom or at an O atom.

41. lump [lʌmp]
    块糖
42. stir [stɚ]
    拌入，搅动
43. rosemary
    [ˈrozˌmɛri]
    艾菊
44. lavender
    [ˈlævəndɚ]
    薰衣草
45. mint [mɪnt]
    薄荷
46. evenly [ˈivənlɪ]
    均匀地
47. lather [ˈlæðɚ]
    肥皂泡

Hands-On Chemistry: Relax with Beautiful Bath Salts

What You'll Need   1 cup of washing soda, a plastic bag, a rolling pin (or something similar that can crush **lumps**), a bowl, a spoon for **stirring**, essential oil, food coloring.

Instructions   Take the cup of washing soda and put it into a plastic bag. Crush the lumps with a rolling pin or similar object. Empty the bag into a bowl and stir in 5 or 6 drops of your favorite essential oil such as **rosemary**, **lavender** or **mint**. Stir in a few drops of food coloring until the mixture is **evenly** colored. Put the mixture into clean dry containers and enjoy as you please.

What's Happening?   Bath salts are typically made from Epsom salts (magnesium sulfate), table salt (sodium chloride) or washing soda (sodium carbonate). The chemical make up of the mixture makes it easy to form a **lather**. Bath salts are said to improve cleaning and deliver an appealing fragrance when bathing.

Exercises

1. A complex has the composition $Co(NH_3)_4(H_2O)BrCl_2$. Conductance measurements show that there are three ions per formula unit, and precipitation of AgCl with silver nitrate shows that there are two Cl^- ions not coordinated to cobalt. What is the structural formula of the compound?
2. Another complex studied by Werner had a composition corresponding to the formula $PtCl_4$: $2KCl$. From electrical-conductance measurements, he determined that each formula unit contained three ions. He also found that silver nitrate did not give a precipitate of AgCl with this complex. Write a formula for this complex that agrees with this information.
3. Identify the type of isomers represented by each of the following pairs.
    (1) $[Cu(NH_3)_4][PtCl_4]$ and $[Pt(NH_3)_4][CuCl_4]$
    (2) $[Cr(OH)_2(NH_3)_4]Br$ and $[CrBr(OH)(NH_3)_4]OH$
4. Give the coordination number of the transition-metal atom in each of the following complexes.
    (1) $Au(CN)_4^-$ (2) $[Co(NH_3)_4(H_2O)_2]Cl_3$ (3) $[Au(en)_2]Cl_3$ (4) $Cr(en)_2(C_2O_4)^+$
5. Determine the oxidation number of the transition element in each of the following complexes.
    (1) $K_2[Ni(CN)_4]$ (2) $Mo(en)_3^{3+}$ (3) $Cr(C_2O_4)_3^{3-}$ (4) $[Co(NH_3)_5(NO_2)]Cl_2$
6. Consider the complex ion $Cr(NH_3)_2Cl_2(C_2O_4)^-$.
    (1) What is the oxidation state of the metal atom?
    (2) Give the formula and name of each ligand in the ion.
    (3) What is the coordination number of the metal atom?
    (4) What would be the charge on the complex if all ligands were chloride ions?
7. Write the IUPAC name for each of the following coordination compounds.
    (1) $K_3[FeF_6]$ (2) $Cu(NH_3)_2(H_2O)_2^{2+}$ (3) $(NH_4)_2[Fe(H_2O)F_5]$ (4) $Ag(CN)_2^-$

8. Give the IUPAC name for each of the following.
   (1) Fe(CO)$_5$ (2) Rh(CN)$_2$(en)$^{2+}$ (3) [Cr(NH$_3$)$_4$SO$_4$]Cl   (4) MnO$_4^-$
9. Write the structural formula for each of the following compounds.
   (1) potassium hexacyanomanganate(Ⅲ)
   (2) sodium tetracyanozincate(Ⅱ)
   (2) tetraamminedichlorocobalt(Ⅲ) nitrate
   (4) hexaamminechromium(Ⅲ) tetrachlorocuprate(Ⅱ)

【Reading Material】

Nitric Oxide Gas and Biological Signaling

In 1998, the Nobel committee awarded its prize in physiology or medicine to three scientists for the **astounding** (使人震惊的) discovery that nitric oxide gas, NO, **functions as** (具有……功能) the signaling agent between biological cells in a wide variety of chemical processes. Until this discovery, biochemists had thought that the major chemical reactions in a cell always involved very large molecules. Now they discovered that a simple gas, NO, could have a central role in cell chemistry.

Prize winners Robert Furchgott and Louis Ignarro, independently, **unraveled** (解开) the role of nitric oxide in blood-pressure **regulation** (调节). Cells in **the lining of arteries** (动脉内壁) detect increased blood pressure and respond by producing nitric oxide. NO rapidly diffuses through the artery wall to cells in the surrounding muscle tissue. In response, the muscle tissue relaxes, the **blood vessel** (血管) expands, and the blood pressure drops.

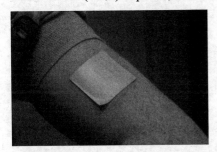

In a related discovery, another prize winner, Ferid Murad, explained how **nitroglycerin** (硝化甘油) works to **alleviate** (缓和) the intense chest pain of an **angina** (心绞痛) attack which results from reduced blood flow to the heart muscle as a result of partial **blockage** (堵塞) of arteries by **plaque** (斑块). **Physicians** (内科医生) have **prescribed** (开药) nitroglycerin for angina for more than a century, knowing only that it works. Murad found that nitroglycerin breaks down in the body to form nitric oxide, which relaxes the arteries, allowing greater blood flow to the heart. Alfred Nobel, were he alive today, would no doubt be **stunned** (惊呆) by this news. Nobel, who established the prizes bearing his name, made his fortune in the nineteenth century from his invention of **dynamite** (炸药), a mixture of nitroglycerin with clay that **tamed** (抑制) the **otherwise** (别的) **hazardous** (危险的) explosive. When Nobel had heart trouble, his physician recommended that he eat a small quantity of nitroglycerin, he refused. In a letter, he wrote "It is ironical that I am now ordered by my physician to eat nitroglycerin." Today, a patient may take either nitroglycerin pills (containing tenths of a milligram of compound in a stabilized mixture) for occasional use or can use a chest **patch** (贴片) to **dispense** (渗透) nitroglycerin continuously to the skin, where it is absorbed.

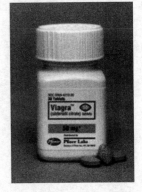

Research papers on the biological role of nitric oxide now number in the tens of thousands. For example, scientists have found that white blood cells use nitric oxide in a kind of chemical warfare. These cells emit concentrated clouds of NO that surround bacterial or tumor cells, killing them by interfering with certain cell processes. Researchers have also discovered that nitric oxide plays a role in penile **erection** (勃起) . Pharmacologists found that the drug **Viagra** (伟哥) assists in the action of NO in **dilating** (膨胀) arteries, leading to erection.

Questions

1. Why did the Nobel committee award its prize in physiology or medicine to three scientists in 1998?
2. What is the physicochemical property of nitric gas?
3. How will you prepare the nitric gas in the laboratory?

Key Terms to Inorganic Chemistry

1. the ideal-gas equation 理想气体状态方程
2. real gases: deviation from ideal behavior 真实气体：对理想气体行为的偏离
3. the van der Waals equation 范德华方程
4. system and surroundings 系统与环境
5. state and state functions 状态与状态函数
6. process and phase 过程和相
7. the first law of thermodynamics 热力学第一定律
8. heat and work 热与功
9. endothermic and exothermic processes 吸热与放热过程
10. enthalpies of reactions 反应热
11. hess's Law 盖斯定律
12. enthalpies of formation 生成焓
13. reaction rates 反应速率
14. reaction order 反应级数
15. rate constants 速率常数
16. activation energy 活化能
17. reaction mechanisms 反应机理
18. homogeneous catalysis 均相催化剂
19. heterogeneous catalysis 非均相催化剂
20. the equilibrium constant 平衡常数
21. the direction of reaction 反应方向
22. spontaneous processes 自发过程
23. entropy (standard entropy) 熵（标准熵）
24. the second law of thermodynamics 热力学第二定律
25. entropy changes 熵变
26. standard free-energy changes 标准自由能变
27. acid-bases 酸碱
28. the dissociation of water 水离解
29. the pH scales pH值范围
30. proton-transfer reactions 质子转移反应
31. conjugate acid-base pairs 共轭酸碱对
32. relative strength of acids and bases 酸碱的相对强度
33. Lewis acids and bases 路易斯酸碱
34. hydrolysis of metal ions 金属离子的水解
35. buffer solutions 缓冲溶液
36. the common-ion effects 同离子效应
37. buffer capacity 缓冲容量
38. formation of complex ions 配离子的形成
39. solubility 溶解度
40. the solubility-product constant K_{sp} 溶度积常数
41. precipitation and separation of ions 离子的沉淀与分离
42. selective precipitation of ions 离子的选择沉淀
43. oxidation-reduction reactions 氧化还原反应
44. oxidation number 氧化数
45. balancing oxidation-reduction equations 氧化还原反应方程的配平
46. half-reaction 半反应
47. galvani cell 原电池
48. cell EMF 电池电动势
49. standard electrode potentials 标准电极电势
50. oxidizing and reducing agents 氧化剂和还原剂
51. quantum numbers 量子数
52. electron spin 电子自旋
53. atomic orbital 原子轨道
54. many-electron atoms 多电子原子
55. energies of orbital 轨道能量
56. the pauli exclusion principle 泡林不相容原理
57. electron configurations 电子构型
58. the periodic table 周期表
59. row and group 行和族
60. isotopes, atomic numbers, and mass numbers 同位素，原子数，质量数
61. periodic properties of the elements 元素周期律
62. radius of atoms 原子半径
63. ionization energy 电离能
64. electron affinities 亲电性
65. effective nuclear charge 有效核电荷
66. valence bond theory 价键理论
67. covalence bond 共价键
68. orbital overlap 轨道重叠
69. multiple bonds 重键
70. hybrid orbital 杂化轨道
71. molecular geometries 分子空间构型
72. polarity molecules 极性分子
73. charges, coordination numbers, and geometries 电荷数、配位数及几何构型
74. structural isomerism 结构异构
75. peroxides and superoxides 过氧化物和超氧化物

Unit 2

Organic Chemistry

Lesson 15 Introduction to Organic Chemistry

The Chemistry of Carbon

1. derogatory
   [dɪˈrɑgəˌtɔri]
   减损的，毁损的
2. term [tɚm]
   术语
3. valency
   [ˈvelənsi]
   价,化合价

What is Organic Chemistry?

The word "organic" is one of the most overused in the English language. People use it as a **derogatory** term in phrases like "don't eat that, it's not organic". Of course, there is a precise scientific definition of the word. In science, organic can be a biological or chemical **term**. In biology it means any thing that is living or has lived. The opposite is non-organic. In chemistry, an organic compound is one containing carbon atoms. The opposite term is inorganic. Organic chemistry is a chemistry subdiscipline involving the scientific study of the structure, properties, and reactions of organic compounds.

Molecules

All substances are made up of molecules which are collections of atoms. All the molecules in existence are made up of about a hundred different kinds of atoms. For example, a water molecule is composed of two atoms of hydrogen and one atom of oxygen. We write its formula as H_2O. A molecule of sulphuric acid contains two atoms of hydrogen, one atom of sulphur and four atoms of oxygen. Its formula is H_2SO_4. These are simple molecules containing only a few atoms. Most inorganic molecules are small.

Molecules with Carbon

Most atoms are only capable of forming small molecules. However one or two can form larger molecules. By far the best atom for making large molecules with is carbon. Carbon can make molecules that have tens, hundreds, thousands even millions of atoms! A single carbon atom is capable of combining with up to four other atoms. We say it has a **valency** of 4. Sometimes a carbon atom will combine with fewer atoms. The carbon atom is one of the few that will combine with itself. In other words carbon combines with other carbon atoms. This means that carbon atoms can form chains and rings onto which other atoms can be attached. This leads to a huge number of different compounds. Organic chemistry is essentially the chemistry of carbon. Carbon compounds are classified according to how the carbon atoms are arranged and what other groups of atoms are attached.

Hydrocarbons

The simplest organic compounds are made up of carbon and hydrogen atoms only. Even these run into thousands! Compounds of carbon and hydrogen only are called hydrocarbons.

Alkanes

The simplest hydrocarbon is **methane**, CH_4. This is the simplest member of a series of hydrocarbons. Each **successive** member of the series has one more carbon atom than the preceding member. This is shown in Table 15-1.

Table 15-1 A few common alkanes

Formula	Structure	Name / Uses
CH_4	CH_4	methane - gas used for cooking
C_3H_8	$CH_3CH_2CH_3$	propane - heating fuel
C_4H_{10}	$CH_3CH_2\,CH_2CH_3$	butane - lighter / camping fuel

CH$_4$

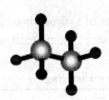

CH$_3$CH$_3$

There is a series of these compounds with this general formula: C_nH_{2n+2}. This series of compounds are called alkanes. The lighter ones are gases and used as fuels. The middle ones (7 carbons to 12 carbons) are liquids used in petrol (gasoline). The higher ones are waxy solids. Candle wax is a mixture of alkanes. After butane, the names of these compounds are from the Greek for the number of carbon atoms followed by the **suffix** -ane. So, **decane** would have the formula: $C_{10}H_{22}$. **Polythene** is a very large alkane with millions of atoms in a single molecule. Apart from being **flammable**, alkanes are stable compounds found underground.

In the alkanes, all four of the carbon valency bonds are taken up with links to different atoms. These types of bonds are called single bonds and are generally stable and resistant to attack by other chemicals. Alkanes contain the maximum number of hydrogen atoms possible. They are said to be saturated. The alkanes are not the only hydrocarbons.

Alkenes

Another series of hydrocarbon compounds is called the alkenes. These have a general formula: C_nH_{2n}. Alkenes have fewer hydrogen atoms than the alkanes. The extra valencies left over occur as double bonds between a pair of carbon atoms. The double bonds are more reactive than single bonds making the alkenes chemically more reactive. The simplest alkenes are listed in Table 15-2. These compounds are named in a similar manner

13. alkyne
   ['æl,kaɪn]
   炔烃

14. ethyne
    ['εθ,aɪn]
    乙炔

15. cyclohexane
    环己烷

16. hexagonal
    [hɛk'sægənəl]
    六角形的,六边形的

17. symmetrical
    [sɪ'mɛtrɪkəl]
    对称的

18. naphthalene
    ['næfθə,lin, 'næp-]
    萘(球),卫生球

19. oxalic acid
    草酸

20. rhubarb
    ['ru,barb]
    大黄

21. hydroxymethanoic
    acid 羟基甲酸

to the alkanes except that the suffix is -ene.

Table 15-2   A few simple alkenes

Formula	Structure	Name / Uses
C_2H_4	$CH_2\!=\!CH_2$	ethene - used as an industrial starter chemical
C_3H_6	$CH_3\!-\!CH\!=\!CH_2$	propene
C_4H_8	$CH_3\!-\!CH_2\!-\!CH\!=\!CH_2$	1-butene

Alkynes

The third series are the alkynes. These have the following formula: C_nH_{2n-2}. Alkynes have two carbon atoms joined by a triple bond.

This is highly reactive making these compounds unstable. Examples of alkynes are listed in Table 15-3. These highly reactive substances have many industrial uses. Again the naming of these compounds is similar to the alkanes except that the suffix is -yne.

Table 15-3   Examples of alkynes

Formula	Structure	Name / Uses
C_2H_2	$HC\!\equiv\!CH$	**ethyne** - better known as acetylene which is used for welding underwater
C_3H_4	$CH_3C\!\equiv\!CH$	propyne
C_4H_6	$CH_3CH_2C\!\equiv\!CH$	butyne

Carbon Rings

Alkanes, alkenes and alkynes all contain carbon atoms in linear chains. There are also hydrocarbons arranged in rings. Some examples are listed in Table 15-4.

When rings are combined with chains, the number of hydrocarbons is virtually infinite. And we are still using only two types of atoms (carbon and hydrogen). We will now add a third.

Table 15-4   Some examples of hydrocarbons with carbon rings

Formula	Structure	Name / Uses
C_6H_{12}	(hexagon)	**cyclohexane** - a saturated hydrocarbon with the atoms arranged in a **hexagonal** ring.
C_6H_6	(hexagon with circle)	benzene - an industrial solvent. Its alternate double and single bonds are "spread around" the ring so that the molecule is **symmetrical**. This structure is represented by a hexagon with a circle.
$C_{10}H_8$	(two fused hexagons)	**naphthalene** - used in moth balls. This can be depicted as two fused benzene rings.

Carbon, Hydrogen and Oxygen

When oxygen atoms are added, the variety of compounds grows enormously. In Table 15-5, each row discusses a series of compounds. In these examples, each molecule has a single functional group.

It is possible to have two or more functional groups on a molecule. These can be the same group (as in **oxalic acid**—a poison found in **rhubarb** leaves—which has two fatty acid groups) or different (as in **hydroxymethanoic acid**—which has a hydroxyl group and a fatty acid group):

22. carbohydrate
[ˌkɑrbo'haɪˌdret]
碳水化合物，糖类
23. sucrose
['suˌkros]
蔗糖
24. alcohols 醇类
25. ether ['iθɚ] 醚
26. anaesthetic
[ˌænɪs'θɛtɪk]
麻醉剂
27. ketone
['kiˌton]
酮
28. aldehyde
['ældəˌhaɪd]
醛
29. formaldehyde
[fɔr'mældəˌhaɪd]
甲醛
30. acetaldehyde
[ˌæsɪ'tældəˌhaɪd]
乙醛
31. formic acid
甲酸
32. acetic acid
乙酸
33. methyl acetate
乙酸甲酯
34. trinitro toluene
三硝基甲苯
35. amine
[ə'min, 'æmin]
胺
36. methylamine
甲胺
37. pungent
['pʌndʒənt]
刺激性的
38. cyanide
['saɪəˌnaɪd]
氰化物
39. methyl cyanide
甲基氰，乙腈
40. amino acid
氨基酸
41. glycine
['glaɪˌsin, -sɪn]
甘氨酸

HOOC—COOH   HOH₂C—COOH
oxalic acid   hydroxymethanoic acid

The most famous compounds containing carbon, hydrogen and oxygen are the **carbohydrates**. An example is the common sugar, **sucrose** ($C_{12}H_{22}O_{11}$).

Sucrose

Table 15-5 A series of compounds containing oxygen

General formula	Series name	Details	Examples
$C_nH_{2n+1}OH$	alcohols	Alcohols have the OH (hydroxyl) group in the molecule.	CH_3OH: methanol; wood alcohol. C_2H_5OH: ethanol; drinking alcohol.
$(C_nH_{2n+1})_2O$	ethers	Ethers have an O atom attached to two hydrocarbon chains (or rings).	$(C_2H_5)_2O$: diethyl ether; a liquid used as an **anaesthetic**.
$(C_nH_{2n+1})_2CO$	ketones	Ketones have a CO group attached to two hydrocarbon chains (or rings).	CH_3COCH_3: acetone; nail-varnish remover.
$C_nH_{2n+1}CHO$	aldehydes	Aldehydes have a CHO group attached to a hydrocarbon chain (or ring).	HCHO: **formaldehyde**; preservative in labs. CH_3CHO: **acetaldehyde.**
$C_nH_{2n+1}COOH$	fatty acids	Fatty acids contain the CO_2H (or COOH) group attached to a hydrocarbon chain or ring.	HCO_2H: **formic acid**; in ant bites and stinging nettles. CH_3CO_2H: **acetic acid**.
RCOOR'	esters	Esters are similar to fatty acids except that the H in the COOH group is another hydrocarbon chain.	$CH_3CO_2CH_3$: **methyl acetate**; essence of pear drops.

This shows how varied and complex even simple organic compounds can be. Sucrose has a pair of rings: one hexagonal, the other pentagonal. Each ring contains an oxygen atom. The rings are joined by an oxygen (ether) link. The entire compound contains several hydroxyl (OH) groups.

Adding Nitrogen

Many very important organic compounds contain nitrogen. This produces more series of compounds (Table 15-6). A famous compound containing nitrogen is **trinitro toluene** [$C_6H_2CH_3(NO_2)_3$ —usually abbreviated to TNT]. This is an artificially made explosive.

Table 15-6 A series of compounds containing nitrogen

General formula	Series name	Details	Examples
$C_nH_{2n+1}NH_2$	amines	Amines have one or more of the hydrogen atoms in ammonia (NH_3) replaced by a hydrocarbon chain or ring.	CH_3NH_2:**methylamine**; a **pungent**; water soluble gas.
$C_nH_{2n+1}CN$	cyanides	Cyanides have the CN group.	CH_3CN: **methyl cyanide**.
$C_nH_{2n}NH_2COOH$	amino acids	Amino acids have two functional groups: the amine (NH_2) group and the fatty acid (COOH) group.	CH_2NH_2COOH: **glycine**; the simplest amino acid.

42. thiamine
['θaɪəmɪn, -ˌmin]
硫胺(维生素 B₁)

43. chlorophyll
['klɔrəfɪl]
叶绿素

44. haemoglobin
血色素,血红蛋白

45. countertop
['kaʊntɚˌtɑp]
(厨房的)工作台面

46. rational
['ræʃənəl]
理性的

Other Atoms

The vast majority of organic compounds contain carbon, hydrogen, oxygen and nitrogen. Other types of atoms can be included to form even more compounds. These can contain atoms like phosphorus, sulphur (e.g. **thiamine**, Vitamin B_1), magnesium (e.g. **chlorophyll**), iron (e.g. **haemoglobin**), and chlorine (e.g. DDT). As can be imagined, these additions increase the number of compounds. Apart from the naturally occurring organic compounds, millions more can be synthesized.

I hope this introduction to organic chemistry indicates just how vast and interesting the subject is.

Hands-On Chemistry: Introduction to Countertop Chemistry

There is a lot of interesting science to investigate in this world. Not all of the science is done by men wearing white coats and working in laboratories. All over the world around us involves science. A child (or teacher) can investigate some pretty interesting stuff without requiring a laboratory or expensive laboratory equipment or dangerous chemicals.

These activities came from teacher training workshops that have been offered by the Science House since the early nineties. Many teachers have taken the workshops and have applied the activities in their own classrooms—from first grade to high school. We believe that students should be involved in active learning in which the teacher acts as a guide, not an answer machine. However, to be a good guide, the teacher has to have the road map in her/his head. So, these activities include directions for doing the activities, suggestions on finding materials and a little background on the science involved.

We realize that there are no new science demonstrations under the sun. Many of these are things that you may have seen before in another format. The point of this book is to assemble these in a **rational** format that encourages you, as a teacher or student, to try them out. A science demonstration in a book is useless until someone actually does it and uses the experience to help their understanding.

The Science House is a science and mathematics learning outreach program of the College of Physical and Mathematical Sciences at North Carolina State University. The mission of the Science House is to work in partnership with K-12 teachers to emphasize the use of hands-on learning activities in mathematics and science classes. The Science House provides a variety of in-service training and enrichment activities that reach teachers and students across North Carolina. For hints and advice on performing demonstrations and sharing activities, please read "Activities and Demonstrations" before performing the activities.

Exercises

1. Please answer the following questions.
   (1) What is organic chemistry discussing about?
   (2) What kinds of compounds are called hydrocarbons?
   (3) What is the general formula of alkane, alkene, and alkyne, respectively?
   (4) What are the structure characters of alkane?
   (5) Which hydrocarbon is most active: alkane, alkene, and alkyne?
   (6) What kind of organic compounds has oxygen atoms?
   (7) What are the differences between ketones and aldehydes?
   (8) Which elements exist in carbohydrates?
   (9) Please give the molecular formula of TNT. Why is TNT dangerous?
   (10) Besides carbon, hydrogen, oxygen, nitrogen, what other elements can be included in organic compounds?

2. Put the following words into Chinese.
   (1) formula   (2) inorganic molecules   (3) valency   (4) hydrocarbons
   (5) cyclohexane   (6) naphthalene   (7) carbohydrate   (8) trinitro toluene
   (9) phosphorous   (10) single bond

3. Put the following words into English.
   (1) 有机分子   (2) 碳环   (3) 燃料   (4) 可燃的   (5) 双键
   (6) 氢元素   (7) 化合物   (8) 乙炔   (9) 醚   (10) 甲酸

【Reading Material】

History of Organic Chemistry

(Friedrich Wöhler, 31 July 1800 ~ 23 September 1882)

At the beginning of the nineteenth century, chemists generally thought that compounds obtained from living organisms were too complex to be obtained synthetically. According to the concept of vitalism, organic matter was endowed with a "vital force". They named these compounds "organic" and directed their investigations toward inorganic materials that seemed more easily studied.

Over the course of the first half of the nineteenth century, it was realized that organic compounds could in fact be synthesized in the laboratory. Around 1816 Michel Chevreul started a study of soaps made from various fats and **alkali**(碱). He separated the different acids that, in combination with the alkali, produced the soap. Since these were all individual compounds, he demonstrated that it was possible to make a chemical change in various fats (which traditionally come from organic sources), producing new compounds, without "vital force". In 1828 Friedrich Wöhler produced the organic chemical **urea** (尿素) (carbamide), a constituent of **urine** (尿), from the inorganic **ammonium cyanate** (氰酸铵) NH$_4$OCN, in what is now called the Wöhler synthesis. Although Wöhler was, at this time as well as afterwards, cautious about claiming that he had thereby destroyed the theory of vital force, historians have looked to this event as the turning point.

A great next step was when in 1856 William Henry Perkin, while trying to manufacture

quinine (奎宁), again accidentally came to manufacture the organic dye now called Perkin's **mauve** (淡紫色), which by generating a huge amount of money greatly increased interest in organic chemistry.

The crucial breakthrough for the organic chemistry was the concept of chemical structure, developed independently and simultaneously by Friedrich August Kekule and Archibald Scott Couper in 1858. Both men suggested that tetravalent carbon atoms could link to each other to form a carbon lattice, and that the detailed patterns of atomic bonding could be discerned by skillful interpretations of appropriate chemical reactions.

Beginning in the 20th century, progress of organic chemistry allowed the synthesis of highly complex molecules via multistep procedures. Concurrently, polymers and enzymes were understood to be large organic molecules, and petroleum was shown to be of biological origin. The process of finding new synthesis routes for a given compound is called total synthesis. Total synthesis of complex natural compounds started with urea, increased in complexity to glucose and **terpineol** (萜品醇, 松油醇), and in 1907, total synthesis was commercialized the first time by Gustaf Komppa with **camphor** (樟脑, 莰酮). Pharmaceutical benefits have been substantial, for example **cholesterol** (胆固醇, 胆甾醇)-related compounds have opened ways to synthesis of complex human hormones and their modified derivatives. Since the start of the 20th century, complexity of total syntheses has been increasing, with examples such as **lysergic acid** [麦角酸(毒品麦角酸酰二乙胺原料)] and their structure and interactions **in vitro** (在试管中,在生物体外) and inside living systems, has only started in the 20th century, opening up a new chapter of organic chemistry with enormous scope. Biochemistry, like organic chemistry, primarily focuses on compounds containing carbon as well.

Lesson 16 Nomenclature of Organic Compounds

1. skeleton
['skɛlətn]
骨架, 框架
2. prefix
['priːfɪks]
前缀
3. suffix
['sʌfɪks]
后缀
4. infix
['ɪn'fɪks]
中缀
5. functional group
功能基团
6. precedence
['presɪdəns]
位次, 优先权
7. substitute
['sʌbstɪtjuːt]
取代基

Because of the great number of compounds containing carbon and the range in complexity of their **skeletons**, each specific compound requires a unique name from which an unambiguous structural formula can be created. The International Union of Pure and Applied Chemistry (IUPAC) have provided a systematic method of naming organic chemical compounds. In chemistry, a number of **prefixes, suffixes** and **infixes** are used to describe the type and position of **functional groups** in the compound. The parent functional group is selected according to the precedence, and the one with highest **precedence** should be used. The group with lower precedence is designated as a **substitute**, and shows as a prefix. The decreasing precedence of the main groups is listed in Table 16-1.

Table 16-1 Precedence order of main functional groups in organic compounds

Functional group	Structure	Prefix	Suffix
carboxylic acids	RCOOH	none	-oic acid
ester	RCOOR'	R-oxycarbonyl-	-oate
amides	$RCONH_2$	carbamoyl-	amide
aldehydes	RCHO	formyl-	-al
ketones	RCOR'	oxo-	-one
alcohols	ROH	hydroxy-	-ol
amines	$-NH_2$	amino-	-amine
ethers	ROR'	alkoxy-	-ether
alkene	$R_2C=CR'_2$	en-	-ene
alkyne	$RC\equiv CR'$	yn-	-yne
fluorine	—F	fluoro-	none
chlorine	—Cl	chloro-	none
bromine	—Br	bromo-	none
iodine	—I	iodo-	none
alkyl group	R—	alkyl	none

Straight-Chain Hydrocarbons

A hydrocarbon in which all carbon atoms are sp^3 hybridized is a member of the class of alkanes, designated by the suffix -ane. When a double bond is present, the suffix is -ene, and the compound is an alkene. When a triple bond is present, the suffix is -yne, and the compound is an alkyne. The root names indicating the number of carbon atoms in the longest continuous chain are derived from Greek or Latin, except for the first four members of the series. For a cyclic structure, the prefix cyclo -is inserted before the root. Table 16-2 gives some skeletons and names of hydrocarbons. Comparable names apply to larger systems.

Table 16-2   IUPAC nomenclature for some hydrocarbons

Alkanes		Alkanes		Alkanes	
Structure	Name	Structure	Name	Structure	Name
CH_4	methane	$n\text{-}C_{11}H_{24}$	undecane	$n\text{-}C_{21}H_{44}$	henicosane
CH_3CH_3	ethane	$n\text{-}C_{12}H_{26}$	dodecane	$n\text{-}C_{22}H_{46}$	docosane
∨	propane	$n\text{-}C_{13}H_{28}$	tridecane	$n\text{-}C_{23}H_{48}$	tricosane
∨∨$_2$	butane	$n\text{-}C_{14}H_{30}$	tetradecane	$n\text{-}C_{30}H_{62}$	triacontane
∨∨$_3$	pentane	$n\text{-}C_{15}H_{32}$	pentadecane	$n\text{-}C_{31}H_{64}$	hentriacontane
∨∨$_4$	hexane	$n\text{-}C_{16}H_{34}$	hexadecane	$n\text{-}C_{32}H_{66}$	dotriacontane
∨∨$_5$	heptane	$n\text{-}C_{17}H_{36}$	heptadecane	$n\text{-}C_{40}H_{82}$	tetracontane
∨∨$_6$	octane	$n\text{-}C_{18}H_{38}$	octadecane	$n\text{-}C_{50}H_{102}$	pentacontane
∨∨$_7$	nonane	$n\text{-}C_{19}H_{40}$	nonadecane	$n\text{-}C_{60}H_{122}$	hexacontane
∨∨$_8$	decane	$n\text{-}C_{20}H_{42}$	icosane	$n\text{-}C_{100}H_{202}$	hectane

To be continued

Alkenes		Alkynes		Cycloalkanes	
Structure	Name	Structure	Name	Structure	Name
$CH_2=CH_2$	ethene	≡	ethyne	△	cyclopropane
∨	1-propene		1-propyne	□	cyclobutane
C_2H_5∨	1-butene	≡C_2H_5	1-butyne	⬠	cyclopentane
C_3H_7∨	1-pentene	≡C_3H_7	1-pentyne	⬡	cyclohexane
C_4H_9∨	1-hexene	≡C_4H_9	1-hexyne	⬯	cycloheptane
□	Cyclobutene	≡C_5H_{11}	1-heptyne	⬰	cyclooctane

Branched Hydrocarbons

For branched hydrocarbons, IUPAC rules **dictate** that the longest continuous carbon chain be identified as the root, with branching groups named as **alkyl** substituents. An alkyl group can be considered as an alkane from which one hydrogen has been removed. The name is derived by replacing the suffix -ane with -yl. Thus, CH_3 is a **methyl** group, C_2H_5 is an **ethyl** group, and so forth. The position along the main carbon chain where an alkyl group is attached is **designated** by a number. The numbering of carbon atoms in the chain starts at the end closest to where the substituent is attached so that the lower of two possible numbers can be **assigned** to that position. The presence of more than one alkyl group along a carbon chain is indicated by a Greek prefix (di-, tri-, tetra-, penta-, for 2, 3, 4, 5 alkyl substituents). In this case, each alkyl group must be

8. dictate
   ['dɪktet]
   命令，规定
9. alkyl ['ælkɪl]
   烷基
10. methyl 甲基
11. ethyl ['εθɪl]
    乙基
12. designate
    ['dεzɪgnet]
    标明，指明，命名
13. assign [ə'saɪn]
    分配，分派

assigned a number (as low as possible) to indicate its position.

hexane   2-methylhexane   3-methylpentane   2,3-dimethylbutane   2,2-dimethylbutane

Alkyl Groups

Alkyl groups with more than two carbons are often designated by common rather than IUPAC names. The common names designate not only the structure of the alkyl group but also the point on the group at which it is attached to the main chain. Table 16-3 shows the structures and names of some alkyl groups.

Table 16-3   IUPAC nomenclature for some alkyls

Structure of alkyl	Name	Structure of alkyl	Name
—CH_3	methyl	H_3C—CH—$(CH_2)_3$— CH_3	*iso*-hexyl
—CH_2CH_3	ethyl	(benzene ring)	**phenyl**
—$CH(CH_3)_2$	*iso*-propyl	CH_2=CH—CH_2—	**allyl**
—$(CH_2)_3CH_3$	*n*-butyl	CH_2=CH—	vinyl/ethenyl
—$CH_2CH(CH_3)_2$	*iso*-butyl	CH_2=C(CH_3)—	*iso*-propenyl
CH_3CH_2—CH(CH_3)—	*s*-butyl	CH_2=	**methylene**
—$C(CH_3)_3$	*t*-butyl	CH_3—CH=	**ethylidene**
CH_3CH_2—C(CH_3)$_2$—	*t*-pentyl	—CH_2CH_2—	**ethylene**
(CH_3)$_3$C—CH_2—	*neo*-pentyl	$CH_3CH_2CH_2$—CH(CH_3)—	1-methylbutyl

The prefix *n*- (normal) refers to a straight-chain alkyl group, whose point of attachment is at a **primary carbon**——that is, one bonded to only one other carbon. The prefix *iso*- describes an alkyl group in which the point of attachment is at the end of a carbon chain that bears a methyl group at the second carbon from the opposite end. This is easier to picture than to describe, look at the isopropyl, isobutyl, and isopentyl groups in Table 16-3. The name *s*-butyl designates an alkyl group whose point of attachment is at the second carbon of a four-carbon straight chain. Here, *s*- is short for "secondary", indicating attachment at a **secondary carbon**—— that is, one attached to two other carbons. The name *t*-butyl indicates attachment at the group's central carbon. Here, *t*- means "tertiary",

14. phenyl ['fenɪl]
    苯基

15. allyl ['ælәl]
    烯丙基

16. methylene
    亚甲基

17. ethylidene
    1,1-亚乙基

18. ethylene
    ['εθɪlin]
    1,2-亚乙基

19. primary carbon
    伯碳

20. secondary carbon
    仲碳

89

21. tertiary carbon 叔碳	indicating attachment at a **tertiary carbon**——that is, one attached to three other carbons. The prefixes *s*- and *t*- are used only for butyl groups because longer alkyl groups often contain more than one type of secondary carbon, so either of these designations would not be unique. The prefixes *n*- and *iso*- are used for longer chains. The prefix neo- used almost exclusively for the **neopentyl** group, —$CH_2C(CH_3)_3$.
22. neopentyl 新戊(烷)基	## Alkenes In the name of an alkene, the position of the functional group (the double bond) is indicated by a number immediately before the root name and its designated ending (-ene). Numbering of the longest carbon chain containing the double bond starts at the end closest to that functional group, allowing it to have the lowest possible number. Use the following steps to apply the IUPAC rules for naming simple alkenes: (1) Determine the longest continuous carbon chain that contains the double bond, and name it with appropriate root and suffix -ene. (2) Assign the first carbon atom of the double bond the lowest possible number.
23. multiple ['mʌltəpl] 多的 24. geometric isomer 几何异构体	(3) Name substituent branches as alkyl groups with their positions indicated by numbers. (4) Indicate **multiple** substituents with the appropriate Greek prefix. (5) Naming **geometric isomers** of alkenes. The *cis* and *trans* **designations** for isomeric structures of alkenes are clear in simple cases such as 2-butene, in which there are two identical substituents at each of the double bond. The *cis* and *trans* designations become **ambiguous**, however, if the substituents are different. To resolve such ambiguity, IUPAC has adopted a way to **specify** such isomers uniquely. The method consists of establishing group priorities at each end of the double bond and then specifying whether the groups of higher **priority** at each end are on the same or opposite sides of the double bond. At each carbon participating in a double bond, assign priorities the two attached groups according to the atomic number of the attached atom.
25. designation [,dɛzɪg'neʃən] 标志 26. ambiguous [æm'bɪgjuəs] 模棱两可的 27. specify ['spɛsɪfaɪ] 具体指定 28. priority [praɪ'ɔrəti] 优先权, 重点	
 Before Organic Exam (I'm diene!) After Organic Exam	Next, the spatial relation between the groups having priority is indicated by using the designations *E* (from entgegen, German for "opposite") and *Z* (from zusammen, "together"). 1-butene *trans*-2-butene *cis*-2-butene *iso*-butene (*E*)-2-butene (*Z*)-butene 2-methylpropene (*Z*)-4-chloro-3- (*Z*)-4,5-dimethyloctene (2*Z*,5*E*)-3-ethylnonadiene methyl-3-heptene
29. ethyne ['ɛθaɪn] 乙炔	The same system is used to name alkynes, with the suffix -yne to indicate the triple bond, e.g. **ethyne** (acetylene), propyne (methylacetylene).

ethyne
(acetylene)

propyne
(methylacetylene)

Halogens (alkyl halides)

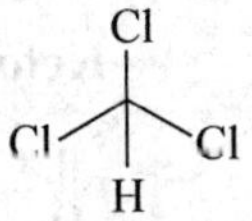

trichloromethane
(chloroform)

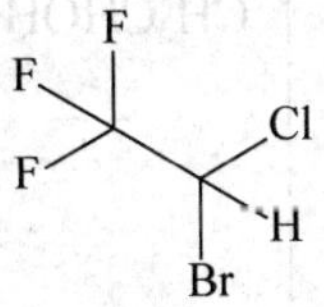

2-bromo-2-chloro-1,1,1-trifluoroethane
(halothane)

Halogen functional groups are prefixed with the bonding position and take the form fluoro-, chloro-, bromo-, iodo-, etc., depending on the halogen. For multiple groups, di-, tri-, etc., are used and different halogen groups are ordered alphabetically as before. For example, CH_2Cl_2 is dichloromethane. The **anesthetic halothane** ($CF_3CHBrCl$) is 2-bromo-2-chloro-1,1,1-trifluoroethane.

Ethers

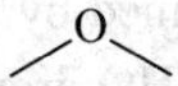

methoxymethane
(dimethyl ether)

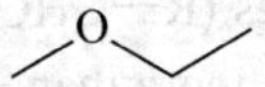

methoxyethane
(ethyl methyl ether)

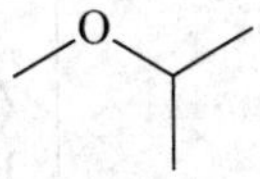

2-methoxypropane
(isopropyl methyl ether)

Ethers (R—O—R) consist of an oxygen atom between the two attached carbon chains. The shorter of the two chains becomes the first part of the name with the -ane suffix changed to -oxy, and the longer alkane chain becomes the suffix of the name of the **ether**. Thus, $CH_3OCH_2CH_3$ is **methoxyethane** (not ethoxymethane). If the oxygen is not attached to the end of the main alkane chain, then the whole shorter alkyl-plus-ether group is treated as a side chain and prefixed with its bonding position on the main chain. Thus $CH_3OCH(CH_3)_2$ is 2-methoxypropane.

Alternatively, the two alkyl groups attached to the oxygen can be put in alphabetical order with spaces between the names and they are followed by the word ether. The prefix di- is used if both alkyl groups are the same, such as dimethyl ether, ethyl methyl ether, and **isopropyl** methyl ether.

Alcohols

ethanol

propan-1-ol

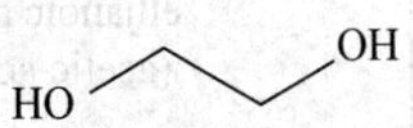

ethane-1,2-diol

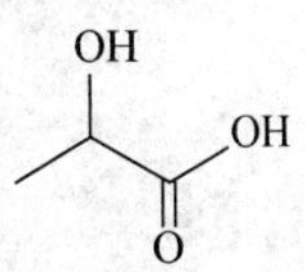

2-hydroxypropanoic acid

---

30. anesthetic
    [ˌænɪsˈθɛtɪk]
    麻醉剂，麻醉药
31. halothane
    [ˈhæləθeɪn]
    三氟溴氯乙烷，氟烷
32. ether [ˈiːθə]
    醚
33. methoxyethane
    [meˈtɒksjeθæn]
    甲基乙醚
34. isopropyl
    [ˌaɪsəʊˈprəʊpɪl]
    异丙基

35. ethylene glycol
['eθɪliːn'glaɪkɒl]
乙二醇

36. hydroxy
[haɪ'drɒksɪ]
氢氧根, 羟基

37. ketones
['ketonz]
酮类

38. aldehyde
['ældɪhaɪd]
醛, 乙醛

39. parent chain
['pɛrənt tʃein]
主链, 母链

40. carbonyl group
['kɑːbənɪl gruːp]
羰基

41. benzene
['benziːn]
苯

Alcohols (R—OH) take the suffix -ol with an infix numerical bonding position: $CH_3CH_2CH_2OH$ is propan-1-ol. For multiple —OH groups, the suffixes -diol, -triol, -tetraol, etc., are used. Thus, **ethylene glycol** CH_2OHCH_2OH is ethane-1,2-diol. If higher precedence functional groups are present, the prefix "**hydroxy**" is used with the bonding position: $CH_3CHOHCOOH$ is 2-hydroxypropanoic acid.

Ketones

propan-2-one (acetone) 2-pentanone 3-oxohexanal

In general, **ketones** (R—CO—R) take the suffix -one (pronounced own) with an infix position number: $CH_3CH_2CH_2COCH_3$ is pentan-2-one. If a higher precedence suffix is in use, the prefix oxo- is used: $CH_3CH_2CH_2COCH_2CHO$ is 3-oxohexanal.

Aldehydes

methanal (formaldehyde) cyclohexanecarbaldehyde

Aldehydes (R—CHO) are named by replacing the suffix -ane with -anal. If there is more than one —CHO group, the suffix is expanded to include a prefix that indicates the number of —CHO groups present (-anedial - there should not be more than 2 of these groups on the parent chain as they must occur at the ends). It is not necessary to indicate the position of the —CHO group because this group will be at the end of the **parent chain** and its carbon is automatically assigned as C-1. If other functional groups with higher precedence are present, a prefix form "oxo-" is used (as for ketones), with the position number indicating the end of a chain: $CHOCH_2COOH$ is 3-oxopropanoic acid. If the carbon in the **carbonyl group** cannot be included in the attached chain (for instance in the case of cyclic aldehydes), the prefix "formyl-" or the suffix "-carbaldehyde" is used: $C_6H_{11}CHO$ is cyclohexanecarbaldehyde. If an aldehyde is attached to a **benzene** and is the main functional group, the suffix becomes benzaldehyde.

Carboxylic Acids

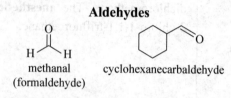

ethanoic acid (acetic acid)　　2-hydroxypropane-1,2,3-tricarboxylic acid (citric acid)　　3-oxopropanoic acid

In general, carboxylic acids are named with the suffix -oic acid (etymologically a back-formation from benzoic acid). Similar to aldehydes, they take the "1" position on the parent chain, but do not have their position number indicated. For example, $CH_3CH_2CH_2CH_2COOH$ (valeric acid) is named pentanoic acid. For common carboxylic acids, some traditional names such as acetic acid is in such widespread use, they are considered retained IUPAC names, although "systematic" names such as ethanoic acid are also acceptable. For carboxylic acids attached to a benzene ring such as Ph—COOH, these are named as benzoic acid or their derivatives.

42. carboxyl
[kɑːˈbɒksɪl]
羧基

If there are multiple **carboxyl** groups on the same parent chain, the suffix "-carboxylic acid" can be used (as -dicarboxylic acid, -tricarboxylic acid, etc.). In these cases, the carbon in the carboxyl group does not count as being part of the main alkane chain. The same is true for the prefix form, "carboxyl-". **Citric acid** is one example, it is named 2-hydroxypropane-1,2,3-tricarboxylic acid, rather than 3-carboxy-3-hydroxypentanedioic acid.

43. citric acid
[ˈsɪtrɪkˈæsɪd]
柠檬酸

Esters

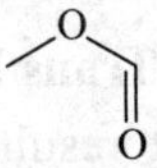

methyl methanoate (formate) 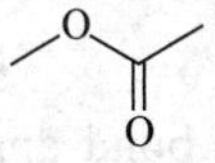 methyl ethanoate (acetate) 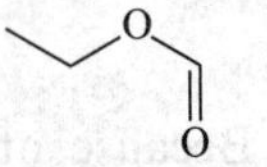 ethyl methanoate (ethyl formate) 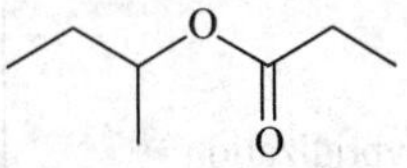 but-2-yl propanoate

Esters (R—CO—O—R′) are named as alkyl derivatives of carboxylic acids. The alkyl (R′) group is named first. The R—CO—O part is then named as a separate word based on the carboxylic acid name, with the ending changed from -oic acid to -oate. For example, $CH_3CH_2CH_2CH_2COOCH_3$ is methyl pentanoate, and $(CH_3)_2CHCH_2CH_2COOCH_2CH_3$ is ethyl 4-methylpentanoate. For esters such as ethyl acetate ($CH_3COOCH_2CH_3$), ethyl formate ($HCOOCH_2CH_3$) or dimethyl phthalate that are based on common acids, IUPAC recommends the use of these established names, called **retained** names. The -oate changes to -ate. Some simple examples, named both ways, are shown in the figure above. If the alkyl group is not attached at the end of the chain, the bond position to the ester group is infixed before "-yl": $CH_3CH_2CH(CH_3)OOCCH_2CH_3$ may be called but-2-yl propanoate or but-2-yl propionate.

44. retained
[rɪˈteɪnd]
保持，保留

Amines and Amides

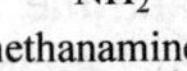

methanamine (methylamine) 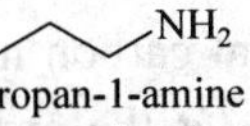   propan-1-amine (*n*-propyl amine) 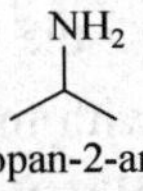   propan-2-amine isopropyl amine 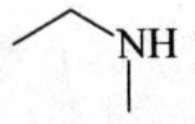   *N*-methylethanamine

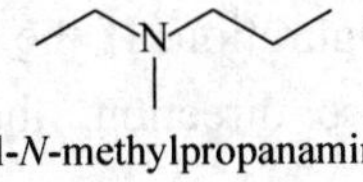

N-ethyl-*N*-methylpropanamine 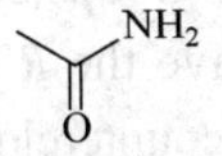   ethanamide (acetamide) 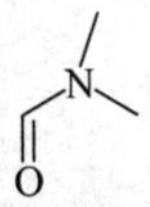   *N*,*N*-dimethylmethanamide

45. amines [əˈmiːns] 胺类，有机胺类	**Amines** (R—NH$_2$) are named for the attached alkane chain with the suffix "-amine" (e.g. CH$_3$NH$_2$ methanamine). If necessary, the bonding position is infixed: CH$_3$CH$_2$CH$_2$NH$_2$ propan-1-amine, CH$_3$CHNH$_2$CH$_3$ propan-2-amine. The prefix form is "**amino-**". For **secondary** amines (of the form R—NH—R), the longest carbon chain attached to the nitrogen atom becomes the primary name of the amine, the other chain is prefixed as an alkyl group with location prefix given as an italic *N*: CH$_3$NHCH$_2$CH$_3$ is *N*-methylethanamine. Tertiary amines (R—NR—R) are treated similarly: CH$_3$CH$_2$N(CH$_3$)CH$_2$CH$_2$CH$_3$ is *N*-ethyl-*N*-methyl- propanamine. Again, the substituent groups are ordered alphabetically.
46. amino [əˈmiːnəʊ] 氨基	
47. amides [ˈæmaɪdz] 酰胺	
48. secondary [ˈsekəndrɪ] 仲碳的	

Amides (R—CO—NH$_2$) take the suffix "-amide", or "-carboxamide" if the carbon in the amide group cannot be included in the main chain. The prefix form is both "carbamoyl-" and "amido-". Amides that have additional substituents on the nitrogen are treated similarly to the case of amines: they are ordered alphabetically with the location prefix *N*: HCON(CH$_3$)$_2$ is *N,N*-dimethylmethanamide.

Cis and Trans Isomers

Because of the bond angles that result from the sp^3 **hybridization** of carbon atoms, alkyl groups attached to the ring carbons in a cycloalkane are located either above or below the atoms that form the plane of the ring. If only one such group is present, the designations "above" and "below" are arbitrary because there is no point of reference. However, if two or more groups are present, the relative positions of the groups are fixed: two groups on the same side of the ring are said to be in a *cis* arrangement; two groups on opposite sides of the ring are said to be in a *trans* arrangement.

49. hybridization 杂化

cis-1,2-dimethylcyclohexane   *trans*-1,2-dimethylcyclohexane

Nomenclature of Enantiomers (the *R*, *S*-system)

50. enantiomer [enˈæntɪəmə] 对映体
51. chiral carbon 手性碳

First, we identify the **chiral carbon** and the four different groups bound to it. Second, we assign a priority to the four different groups according to the priority rules used to assign *E* and *Z*. Third, we view the chiral carbon along the bond from the carbon toward the group of lowest priority—that is, with the chiral carbon nearer and the low-priority group farther away. Finally, we look at the order of the remaining groups. If the priority of these groups decreases in the clockwise direction, the chiral carbon is said to have the *R* configuration. If the priority of these groups decreases in the counterclockwise direction, the compound is said to have the *S* configuration.

$$\underset{\underset{R}{Z}}{\overset{W}{\underset{|}{X\!-\!\!\!\!\!-\!\!\!\!\!-\!Y}}}$$
clockwise

$$\underset{\underset{S}{Z}}{\overset{W}{\underset{|}{Y\!-\!\!\!\!\!-\!\!\!\!\!-\!X}}}$$
anticlockwise

X>Y>Z>W

$$\begin{array}{c} CHO \\ HO\!-\!\!\!\!\!-\!\!\!\!\!-\!H \\ CH_2OH \end{array}$$

(*S*)-**Glyceraldehyde**
L-Glyceraldehyde

$$\begin{array}{c} CHO \\ H\!-\!\!\!\!\!-\!\!\!\!\!-\!OH \\ CH_2OH \end{array}$$

(*R*)-Glyceraldehyde
D-Glyceraldehyde

52. glyceraldehyde
 [ˌglɪsəˈrældəˌhaɪd]
 甘油醛

Summary

In general, the steps for naming an organic compound are:

(1) Identification of the parent hydrocarbon chain. This chain must obey the following rules, in order of precedence:

① It should have the maximum number of substituents of the suffix functional group. By suffix, it is meant that the parent functional group should have a suffix, unlike halogen substituents. If more than one functional group is present, the one with highest precedence should be used.

② It should have the maximum number of multiple bonds.

③ It should have the maximum number of single bonds.

④ It should have the maximum length.

(2) Identification of the parent functional group, if any, with the highest order of precedence.

(3) Identification of the side chains. Side chains are the carbon chains that are not in the parent chain, but are branched off from it.

(4) Identification of the remaining functional groups, if any, and naming them by their prefixes (such as hydroxy for —OH, oxy for =O, oxyalkane for O—R, etc.). Different side chains and functional groups will be grouped together in alphabetical order. (The prefixes di-, tri-, etc. are not taken into consideration for grouping alphabetically. For example, ethyl comes before dihydroxy or dimethyl, as the "e" in "ethyl" precedes the "h" in "dihydroxy" and the "m" in "dimethyl" alphabetically. The "di" is not considered in either case.) When both side chains and secondary functional groups are present, they should be written mixed together in one group rather than in two separate groups.

(5) Identification of double/triple bonds.

(6) Numbering of the chain. This is done by first numbering the chain in both directions (left to right and right to left), and then choosing the numbering which follows these rules, in order of precedence:

53. locant
 [louˈkænt]
 位次

① Has the lowest-numbered **locant** (or locants) for the suffix functional

95

group. Locants are the numbers on the carbons to which the substituent is directly attached.

② Has the lowest-numbered locants for multiple bonds. (The locant of a multiple bond is the number of the adjacent carbon with a lower number.)

③ Has the lowest-numbered locants for prefixes.

(7) Numbering of the various substituents and bonds with their locants. If there is more than one of the same types of substituent/double bond, a prefix is added showing how many there are (di-2, tri-3, tetra-4 then as for the number of carbons below with 'a' added).

The numbers for that type of side chain will be grouped in ascending order and written before the name of the side chain. If there are two side chains with the same alpha carbon, the number will be written twice. Example: 2,2,3-trimethyl- . If there are both double bonds and triple bonds, "en" (double bond) is written before "yne" (triple bond). When the main functional group is a terminal functional group (a group which can exist only at the end of a chain, like formyl and carboxyl groups), there is no need to number it.

1. Arrangement in This Form

Group of side chains and secondary functional groups with numbers made in step 3 + prefix of parent hydrocarbon chain (eth, meth) + double/triple bonds with numbers (or "ane") + primary functional group suffix with numbers. Wherever it says "with numbers", it is understood that between the word and the numbers, the prefix(di-, tri-) is used.

2. Adding of Punctuation

① Commas are put between numbers (2 5 5 becomes 2,5,5).

② Hyphens are put between a number and a letter (2 5 5 trimethylheptane becomes 2,5,5-trimethylheptane).

③ Successive words are merged into one word (trimethyl heptane becomes trimethylheptane). Note: IUPAC uses one-word names **throughout**. This is why all parts are connected.

The finalized name should look like this:

\#,\#-di<side chain>-\#-<secondary functional group>-\#-<side chain>-\#,\#,\#-tri<secondary functional group><parent chain prefix><If all bonds are single bonds, use "ane">-\#,\#-di<double bonds>-\#-<triple bonds>-\#-<primary functional group>

Note: \# is used for a number. The group secondary functional groups and side chains may not look the same as shown here, as the side chains and secondary functional groups are arranged alphabetically. The di- and tri- have been used just to show their usage. (di- after \#,\#, tri- after \#,\#,\#, etc.)

54. punctuation
['pʌŋktʃu'eʃən]
标点符号

55. throughout
[θru'aut]
始终，处处

Hands-On Chemistry: Dancing Spaghetti

56. spaghetti
[spə'gɛti]
意大利式细面条

57. sodium hydrogen carbonate
碳酸氢钠

58. vermicelli
[,vɜ·mɪ'tʃɛli]
细面条

59. tsp 茶匙

60. tbsp 汤匙

61. baking soda
小苏打，碳酸氢钠

62. raisin ['rezn]
葡萄干

63. mothball
['mɔθbɔl]
卫生球，樟脑丸

Spaghetti is denser than water and, therefore, sinks when placed in water. When spaghetti is placed in a solution of baking soda and vinegar, the spaghetti rises to the surface due to the carbon dioxide gas that adheres to it.

When the spaghetti reaches the surface, the gas is released, and the spaghetti sinks again.

Materials	Substitutions
1 1000mL beaker	1 glass mixing bowl
10g **sodium hydrogen carbonate**	3 tsp **baking soda**
45mL 3% acetic acid	4~5 **tbsp** vinegar
10 2 cm pieces of **vermicelli**	
water	

Procedure Fill a clear container 3/4 full with water. Add the sodium hydrogen carbonate (or **baking soda**) and stir to dissolve. Break the vermicelli into 2 cm, or 1 inch, pieces and add them to the container. Add the acetic acid (vinegar). If the vermicelli does not begin to "dance" after a few minutes, add more sodium hydrogen carbonate and acetic acid. If possible, substitute **raisins** or **mothballs** for the vermicelli.

Exercises

1. Draw structures that correspond to the following names.

    (1) tert-butylcyclopentane

    (2) 5-methyl-3-hexanol

    (3) 3-ethylheptan-2-one

    (4) 2,7-dimethyloctanoic acid

    (5) 2-chloropropanal

    (6) 4-methyl-1-pentanamine

    (7) N,N-dimethylpropionamide

    (8) isopropyl 2-methylpentanoate

    (9) 3-methyl-4-propylon-3-ene

    (10) 2,2-dimethylpentane

2. Give the IUPAC name for the following structure.

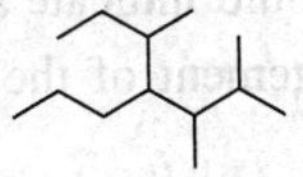

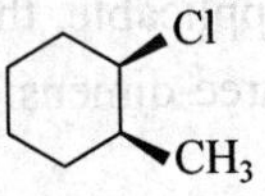

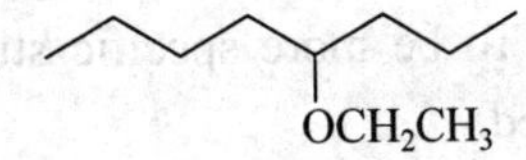

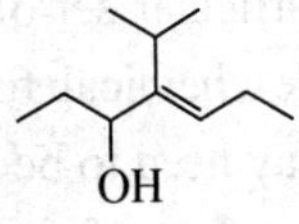

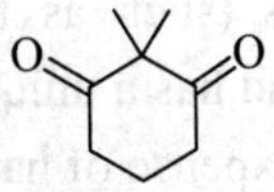

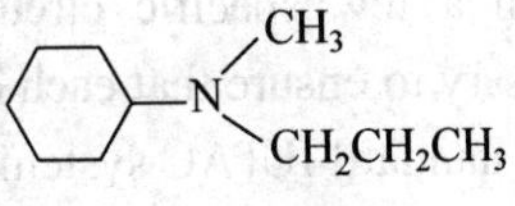

(8) [structure: 2-methylcyclopentanecarboxamide] (9) [structure: cyclohexanecarboxylic acid tert-butyl ester] (10) [structure: 2-bromo-2-ethyl... COOH]

【Reading Material】

IUPAC Nomenclature

IUPAC nomenclature is a system of naming chemical compounds and of describing the science of chemistry in general. It is developed and kept up to date under the **auspices** (赞助,支持) of the International Union of Pure and Applied Chemistry (IUPAC). The rules for naming organic and inorganic compounds are contained in two publications, known as the Blue Book and the Red Book respectively. A third publication, known as the Green Book, describes the recommendations for the use of symbols for physical quantities (in association with the IUPAP), while a fourth, the Gold Book, contains the definitions of a large number of technical terms used in chemistry.

Similar **compendium**(概要) exist for biochemistry (the White Book, in association with the IUBMB), analytical chemistry (the Orange Book), macromolecular chemistry (the Purple Book) and clinical chemistry (the Silver Book). These "colour books" are supplemented by shorter recommendations for specific circumstances which are published from time to time in the journal Pure and Applied Chemistry.

Aims of Chemical Nomenclature

The primary function of chemical nomenclature is to ensure that the person who hears or reads a chemical name is under no ambiguity as to which chemical compound it refers to: each name should refer to a single substance. It is considered less important to ensure that each substance should have a single name, although the number of acceptable names is limited.

It is also preferable that the name convey some information about the structure or chemistry of a compound. CAS numbers form an extreme example of names which do not perform this function: each refers to a single compound but none contain information about the structure.

The form of nomenclature which should be used depends on the public to which it is addressed: as such there is no single correct form, but rather different forms which are more or less appropriate in different circumstances.

A common name will often **suffice** (满足……的需要) to identify a chemical compound in a particular set of circumstances. To be more generally applicable, the name should indicate at least the chemical formula. To be more specific still, the three-dimensional arrangement of the atoms may need to be specified.

In a few specific circumstances (such as the construction of large indices), it becomes necessary to ensure that each compound has a unique name: this requires the addition of extra rules to the standard IUPAC system, at the expense of having names which are longer and less familiar to most readers. Another system gaining popularity is the International Chemical Identifier—while InChI symbols are not human-readable, they contain complete information about substance structure. That makes them more general than CAS numbers.

The IUPAC system is often criticized for the above failures when they become relevant [for example in differing reactivity of sulfur **allotropes**（同素异形体）which IUPAC doesn't distinguish]. While IUPAC has a human-readable advantage over CAS numbering, it would be difficult to claim that the IUPAC names for some larger, relevant molecules (such as **rapamycin** 雷帕霉素) are human-readable, and so most researchers simply use the informal names.

Lesson 17   Types of Organic Reactions

1. substitution reaction
   取代反应

There are four types of organic reactions:

Substitution Reactions

The reactions in which an atom or group of atoms in a molecule is replaced or substituted by different atoms or group of atoms are called substitution reaction.

$$CH_4 + Cl_2 \longrightarrow CH_3Cl + HCl$$
(H replaced by Cl)

$$CH_3CH_2I + KOH(aq) \longrightarrow CH_3CH_2OH + KI$$
(I replaced by OH)

$$C_6H_6 + HNO_3 \xrightarrow{H_2SO_4} C_6H_5NO_2$$
(H of benzene ring replaced by NO_2 group)

2. nucleophilic
   亲核的
3. nucleophile
   ['nukliə,faɪl]
   亲核试剂
4. carbocation
   碳阳离子

These reactions can be of two types:

Nucleophilic Substitution

In this type of substitution, atom or group of atoms in the molecule is replaced by a **nucleophile**. These can be either S_N1 (substitution, nucleophilic, unimolecular) or S_N2 (substitution, nucleophilic, bimolecular) type. In the S_N1 reaction only one molecule is involved which dissociates spontaneously to generate a **carbocation** intermediate. The carbocation reacts with the added nucleophile in a fast step to yield the product.

$$R-\underset{R}{\overset{R}{C^+}} \;\; + :Nu^-$$
intermediate

5. substrate
   ['sʌb,stret]
   底物

In the S_N2 reaction the nucleophile attacks the **substrate** from a position 180° away from the leaving group and that the reaction takes place in a single step without intermediates.

$$Nu^{-\delta}\cdots C \cdots X^{-\delta}$$

6. electrophilic
   亲电的

Electrophilic Substitution

In this type of substitution, an electrophile attacks the carbanion of the substrate substituting one of the hydrogens or other groups.

7. addition reaction
   加成反应

Addition Reactions

Reactions in which two molecules react to form a single product are called addition reactions. For example,

(The Art of Writing Reasonable Organic Reaction Mechanisms)

$$H-\underset{H}{\overset{H}{C}}=\underset{H}{\overset{H}{C}}-H + Br_2 \longrightarrow H-\underset{Br}{\overset{H}{C}}-\underset{Br}{\overset{H}{C}}-H$$
addition product

$$H_3C-\overset{H}{\underset{}{C}}=O + HCN \longrightarrow CH_3-\underset{CN}{\overset{H}{C}}-OH$$
addition product

Depending upon the type of attacking reagent, the substitution or addition reactions are also classified as free radical, electrophilic or nucleophilic substitution or addition reactions. For example, if the attacking reagent is nucleophile in addition reaction, the reaction is called nucleophilic addition.

Elimination Reactions

8. elimination reaction
消去反应

The reactions in which a small molecule is removed from adjacent carbon atoms resulting in the formation of additional (multiple) bond between them are called elimination reactions. For example,

$$H-\underset{H}{\overset{H}{C}}-\underset{H}{\overset{Br}{C}}-H \xrightarrow{\text{alcoholic KOH}} H_2C=CH_2$$
(H and Br are eliminated)

Rearrangement Reactions

9. rearrangement reaction
重排反应

In these reactions the migration of atoms or groups of atoms takes place from one position to another within the molecule under suitable conditions. For example,

$$H_3C-CH_2-CH_2-CH_3 \xrightarrow{\text{rearrangement}} H_3C-\underset{CH_3}{\overset{}{CH}}-CH_3$$

n-butane *iso*-butane

Hands-On Chemistry: Remove Tarnish from Silver

If you have any objects made from silver or plated with silver, you know that the bright, shiny surface of silver gradually darkens and becomes less shiny. This happens because silver undergoes a chemical reaction with sulfur-containing substances in the air. You can use chemistry to reverse the tarnishing reaction, and make the silver shiny again.

For this experiment you will need:
- ✓ a tarnished piece of silver;
- ✓ a pan or dish large enough to completely immerse the silver in;
- ✓ aluminum foil to cover the bottom of the pan;
- ✓ enough water to fill the pan;
- ✓ a vessel in which to heat the water;
- ✓ hot pads or kitchen mitts with which to handle the heated water vessel;
- ✓ baking soda, about 1 cup per gallon of water.

Line the bottom of the pan with aluminum foil. Set the silver object on top of the aluminum foil. Make sure the silver touches the aluminum. Heat the water to boiling. Remove it from the heat and place it in a sink. To the hot water, add about one cup of baking soda for each gallon of water. (If you need only half a gallon of water, use half a cup of baking soda.) The mixture will froth a bit and may spill over; this is why you put it in the sink. Pour the hot baking soda and water mixture into the pan, and completely cover the silver.

Almost immediately, the tarnish will begin to disappear. If the silver is only lightly tarnished, all of the tarnish will disappear within several minutes. If the silver is badly tarnished, you may need to reheat the baking soda and water mixture, and give the silver several treatments to remove all of the tarnish.

Hands-On Chemistry Insights: **Remove Tarnish From Silver**

When silver tarnishes, it combines with sulfur and forms silver sulfide. Silver sulfide is black. When a thin coating of silver sulfide forms on the surface of silver, it darkens the silver. The silver can be returned to its former luster by removing the silver sulfide coating from the surface. There are two ways to remove the coating of silver sulfide. One way is to remove the silver sulfide from the surface. The other is to reverse the chemical reaction and turn silver sulfide back into silver. In the first method, some silver is removed in the process of polishing. In the second, the silver remains in place. Polishes that contain an abrasive shine the silver by rubbing off the silver sulfide and some of the silver along with it. Another kind of tarnish remover dissolves the silver sulfide in a liquid. These polishes are used by dipping the silver into the liquid, or by rubbing the liquid on with a cloth and washing it off. These polishes also remove some of the silver.

The tarnish-removal method used in this experiment uses a chemical reaction to convert the silver sulfide back into silver. This does not remove any of the silver. Many metals in addition to silver form compounds with sulfur. Some of them have a greater affinity for sulfur than silver does. Aluminum is such a metal. In this experiment, the silver sulfide reacts with aluminum. In the reaction, sulfur atoms are transferred from silver to aluminum, freeing the silver metal and forming aluminum sulfide. Chemists represent this reaction with a chemical equation.

$$3 Ag_2S + 2 Al \longrightarrow 6 Ag + Al_2S_3$$

The reaction between silver sulfide and aluminum takes place when the two are in contact while they are immersed in a baking soda solution. The reaction is faster when the solution is warm. The solution carries the sulfur from the silver to the aluminum. The aluminum sulfide may adhere to the aluminum foil, or it may form tiny, pale yellow flakes in the bottom of the pan. The silver and aluminum must be in contact with each other, because a small electric current flows between them during the reaction. This type of reaction, which involves an electric current, is called an electrochemical reaction. Reactions of this type are used in batteries to produce electricity.

Exercises

1. Classify the following reactions as: (a) substitution reaction; (b) elimination reaction; (c) addition reaction; (d) rearrangement reactions.
    (1) $CH_3S^- + CH_3CH_2Br \longrightarrow CH_3CH_2SCH_3 + Br^-$ (2) $CH_3CHO + H_2 \longrightarrow CH_3CH_2OH$

(3) HO⁻ + CH₃Cl ⟶ CH₃OH + Cl⁻

(4) 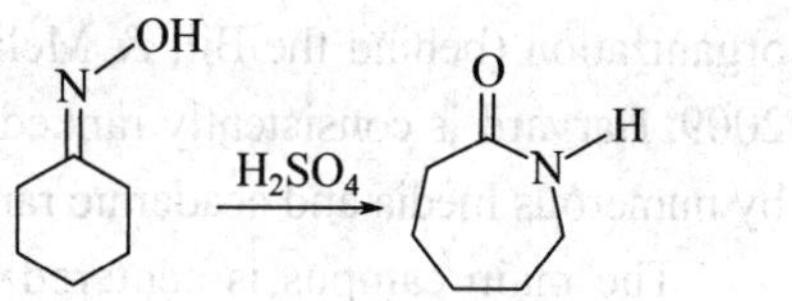 cyclohexanone =N-OH $\xrightarrow{H_2SO_4}$ caprolactam (N-H ring)

2. What is the difference between S_N1 and S_N2 reaction?
3. What catalogues can addition reactions be classified as?
4. Please give the products of the following reactions.

(1) CH₃CH₂CH₂Br + NaOH ⟶

(2) 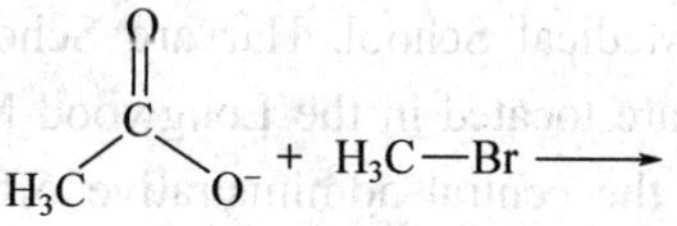 CH₃C(=O)O⁻ + H₃C—Br ⟶

(3) 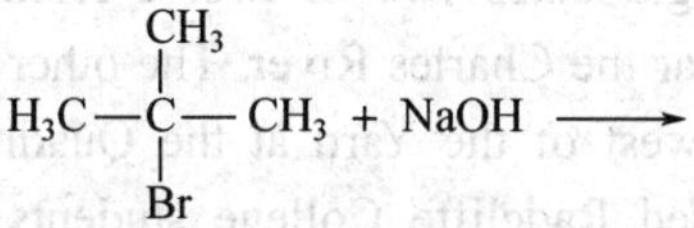 H₃C—C(CH₃)(Br)—CH₃ + NaOH ⟶

(4) 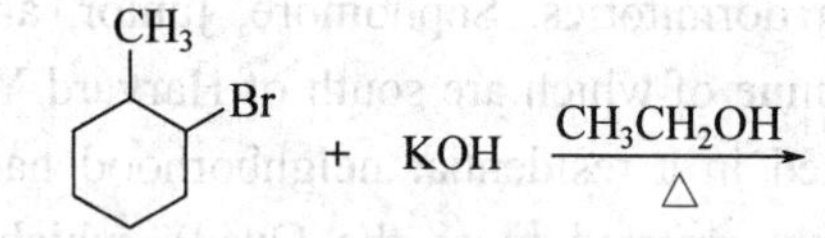 (2-methylcyclohexyl bromide) + KOH $\xrightarrow{CH_3CH_2OH, \Delta}$

(5) (CH₃)₂C=CH₂ + HCl ⟶

(6) 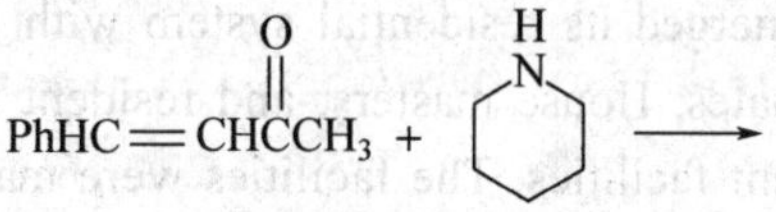 PhHC=CHCCH₃ (with C=O) + piperidine (N-H) ⟶

(7) 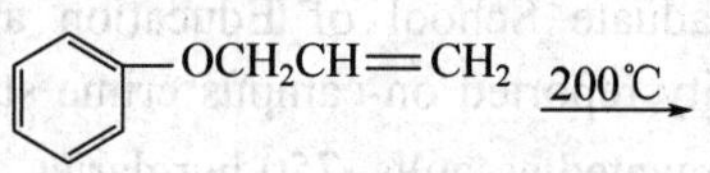 Ph—OCH₂CH=CH₂ $\xrightarrow{200℃}$

(8) 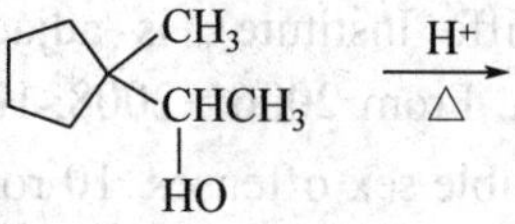 (1-methylcyclopentyl)CHCH₃ with HO on ring carbon $\xrightarrow{H^+, \Delta}$

【Reading Material】

Harvard University

Harvard University (incorporated as The President and Fellows of Harvard College) is a private university located in Cambridge, Massachusetts in the Greater Boston area and a member of the Ivy League. Established in 1636 by the colonial Massachusetts legislature, Harvard is the oldest institution of higher learning in the United States and currently comprises ten separate academic units. It was also the first corporation chartered in the United States. Initially called "New College" or "the college at New Towne", the institution was renamed Harvard College on March 13, 1639. It was named after John Harvard, a young clergyman from Southwark, Surrey, an alumnus of the University of Cambridge (after which Cambridge, Massachusetts is named), who bequeathed the College his library of four hundred books and £779 (which was half of his estate), assuring its continued operation. The earliest known official reference to Harvard as a "university" occurs in the new Massachusetts Constitution of 1780.

During his 40-year tenure as Harvard president (1869~1909), Charles William Eliot radically transformed Harvard into the pattern of the modern research university. Eliot's reforms included elective courses, small classes, and entrance examinations. The Harvard model influenced American education nationally, at both college and secondary levels. Harvard has the second-largest financial endowment of any non-profit organization (behind the Bill & Melinda Gates Foundation), standing at $26 billion as of September 2009. Harvard is consistently ranked at or near the top as a leading academic institution in the world by numerous media and academic rankings.

The main campus is centered on Harvard Yard in central Cam- bridge and extends into the surrounding Harvard Square neighborhood. The Harvard Business School and many of the university's athletics facilities, including Harvard Stadium, are located in the city of Boston's Allston neighborhood, which is situated on the other side of the Charles River from Harvard Square. The Harvard Medical School, Harvard School of Dental Medicine, and the Harvard School of Public Health are located in the Longwood Medical and Academic Area of Boston.Harvard Yard itself contains the central administrative offices and main libraries of the university, academic buildings including Sever Hall and University Hall, Memorial Church, and the majority of the freshman dormitories. Sophomore, junior, and senior undergraduates live in twelve residential Houses, nine of which are south of Harvard Yard along or near the Charles River. The other three are located in a residential neighborhood half a mile northwest of the Yard at the Quadrangle (commonly referred to as the Quad), which formerly housed Radcliffe College students until Radcliffe merged its residential system with Harvard. Each residential house contains rooms for undergraduates, House masters, and resident tutors, as well as a dining hall, library, and various other student facilities. The facilities were made possible by a gift from Yale University alumnus Edward Harkness. Radcliffe Yard, formerly the center of the campus of Radcliffe College (and now home of the Radcliffe Institute), is adjacent to the Graduate School of Education and the Cambridge Common. From 2006~2008, Harvard University reported on-campus crime statistics that included 48 forcible sex offenses, 10 robberies, 15 aggravated assaults, 750 burglaries, and 12 cases of motor vehicle theft.

Key Terms to Organic Chemistry

1. acid anhydride 酸酐
2. acid halide 酸性卤化物
3. alcohol 乙醇
4. aldehyde 醛，乙醛
5. aliphatic 脂肪族的,脂肪质的
6. alkane 烷烃
7. alkene 烯烃
8. alkoxide 醇盐
9. alkyl 烷基
10. alkyne 炔烃
11. allyl 烯丙基
12. amide 氨基化合物
13. amine 胺
14. amino acid 氨基酸
15. ammine 氨合物；氨络物
16. arene 芳烃
17. aromatic ring 芳香环，芳族环
18. aromatic compound 芳烃化合物
19. aryl 芳基(的)
20. azo 含氮的
21. carbonyl 碳酰基,羰基
22. carboxylic acid 羧酸
23. carotene 胡萝卜素
24. chelate 螯合物
25. chiral 手性，手征性
26. chiral center 手性中心
27. conformer 构象的异构体
28. coordination number 配位数
29. crystal field splitting energy 晶体场稳定化能
30. crystal field theory 晶体场理论
31. dextrorotatory 右旋性的
32. diazonium salt 重氮盐
33. diazotization 重氮化作用
34. dichloromethane 二氯甲烷
35. ester 酯
36. fatty acid 脂肪酸
37. free radical 自由基
38. functional group 官能团
39. glycerol 甘油，丙三醇
40. heterocyclic 杂环的，不同环式的
41. high spin complex 高自旋配合物
42. homolog 同系物，同系化合物
43. hydrocarbon 碳氢化合物, 烃
44. inductive effect 诱导效应
45. inorganic chemistry 无机化学
46. ketone 酮
47. levorotatory 左旋的
48. ligand 配体
49. low spin complex 低自旋配合物
50. methyl 甲基
51. mixed glyceride 甘油复酸酯
52. molecular sieve 分子筛
53. monodentate 单齿配位体
54. octane 辛烷
55. optical activity 旋光性
56. organic chemistry 有机化学
57. paraffin 石蜡
58. phenol （苯）酚, 石炭酸
59. phenyl 苯基
60. polydentate 多配位基的
61. polymer 聚合物
62. propane 丙烷
63. racemic 外消旋的
64. resonance effect 共振效应
65. superoxide 过氧化物, 超氧化物
66. tautomer 互变(异构)体
67. thin layer chromatography 薄层分析法
68. triglyceride 甘油三酸酯
69. unsaturated compound 不饱和化合物
70. water gas 水煤气
71. zwitterion 两性离子

Unit 3

Physical Chemistry & Environmental Pollution

Lesson 18 Introduction of Physical Chemistry

1. microscopic
   [ˌmaɪkrəˈskɑpɪk]
   微观的
2. macroscopic
   [ˌmækrəˈskɑpɪk]
   宏观的
3. quantum
   [ˈkwɑ:ntəm]
   量子
4. kinetics
   [kɪˈnɛtɪks]
   动力学

Physical chemistry applies the methods of physics to the study of chemical systems. A chemical system can be studied from either a **microscopic** or a **macroscopic** view-point. The microscopic viewpoint makes explicit use of the concept of molecules. The macroscopic viewpoint studies large-scale properties of matter (e.g., volume, pressure, composition) without explicit use of the molecule concept.

We can divide physical chemistry into four main areas: thermodynamics, **quantum** chemistry, statistical mechanics, and **kinetics**. Thermodynamics is a macroscopic science that studies the interrelationships between the various equilibrium properties of a system.

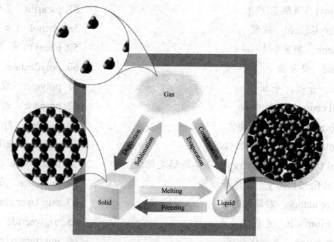

Molecules and the particles (electrons and nuclei) that compose them do not obey classical Newtonia mechanics; instead their motions are governed by the laws of quantum mechanics. Application of quantum mechanics to atomic structure, molecular bonding, and **spectroscopy** gives us quantum chemistry.

The macroscopic science of thermodynamics is a consequence of

5. spectroscopy
   [spɛkˈtrɑskəpi]
   光谱学

what is happening at a molecular (microscopic) level. The molecular and macroscopic levels are related to each other by the branch of science called statistical mechanics. Statistical mechanics gives insight into why the laws of thermodynamics hold and allows calculation of macroscopic thermodynamics properties from molecular properties.

Kinetics is the study of rate processes such as chemical reactions, diffusion, and the flow of charge in an electrochemical cell. The theory of rate processes is not as well developed as the theories of thermodynamics, quantum mechanics, and statistical mechanics. Kinetics uses relevant portions of thermodynamics, quantum chemistry, and statistical mechanics.

Thermodynamics (from the Greek words for "heat" and "power") studies heat, work, energy, and the changes they produce in the states of systems; in a broader sense, thermodynamics studies the relationships between the macroscopic properties of a system. We shall be studying classical (or equilibrium) thermodynamics, which deals with systems in equilibrium. (**Irreversible** thermodynamics deals with nonequilibrium systems and rate processes.) Classical thermodynamics is a macroscopic science and is independent of any theories of molecular structure. Strictly speaking, the word "molecule" is not part of the vocabulary of thermodynamics; however, we won't adopt a **purist** attitude but will often refer to molecular concepts to aid in understanding thermodynamics. Thermodynamics is not applicable to systems of molecular size; a system must consist of a large number of molecules for it to be treated thermodynamically. When we use the term thermodynamics, we shall always mean equilibrium thermodynamics.

The part of the universe under study in thermodynamics is called the system; the parts of the universe that can interact with the system are called **surroundings**. A thermodynamic system must be of macroscopic size.

An open system is one where transfer of matter between system and surroundings can occur. A closed system is one where no transfer of matter can occur between system and surroundings. An isolated system is one that does not interact in any way at all with its surroundings. An isolated system is obviously a closed system, but not every closed system is isolated. For an isolated system, neither matter nor energy can be transferred between system and surroundings. For a closed system, energy but not matter can be transferred between system and surroundings. For an open system, both matter and energy can be transferred between system and surroundings. Most commonly, we shall deal with closed system. It is important to note the kind of system under study, since thermodynamic statements valid for one kind of system may

6. irreversible
[ˌɪrɪ'vəsəbəl]
不可逆的

7. purist
['pjʊrɪst]
纯粹主义者，
正统主义者

8. surrounding
[sə'raʊndɪŋ]
环境

107

be invalid for other kinds.

A system surrounded by a rigid, impermeable, and adiabatic wall cannot interact with the surroundings and is isolated.

Classical thermodynamics deals with systems in equilibrium. An isolated system is in equilibrium when its properties are observed to remain constant with time. A nonisolated system is in equilibrium when the following two conditions hold: (a) the system's properties remain constant with time; (b) removal of the system from contact with its surroundings causes no change in the properties of the system. [If condition (a) holds but (b) does not hold, the system is in a steady state. An example of a steady state is a metal rod in contact at one end with a large body at 50℃ and in contact at the other end with a large body at 40℃. After sufficient time has elapsed, the metal rod satisfies condition (a); a uniform temperature gradient is set up along the rod. However, if we remove the rod from contact with its surroundings, the temperature of its parts change until the whole rod is at 45℃.]

The equilibrium concept can be divided into the following three kinds of equilibrium. For mechanical equilibrium, there is no unbalanced force acting on or within the system; hence the system undergoes no acceleration, and there is no **turbulence** within the system. For material equilibrium, no net chemical reactions are occurring in the system, nor is there any net transfer of matter from one part of the system to another. The concentration of the chemical species in the various parts of the system is constant in time. For thermal equilibrium between a system and its surroundings, there must be no change in the properties of the system or surroundings when they are separated by a thermally conducting wall. Likewise, we can insert a thermally conducting wall between two parts of a system to test whether the parts are in thermal equilibrium with each other. For thermodynamic equilibrium, all three kinds of equilibrium must be present.

Hands-On Chemistry: Atomic Weight Calculation

The atomic weight of an element depends on the abundance of its **isotopes**. If you know the mass of the isotopes and the **fractional abundance** of the isotopes, you can calculate the element's atomic weight. The atomic weight is calculated by adding the mass of each isotope multiplied by its fractional abundance. For example, for an element with 2 isotopes:

Calculating atomic weight of an element with isotopes:
atomic weight = $\text{mass}_A \times \text{fract}_A + \text{mass}_B \times \text{fract}_B$

If there were three isotopes, you would add a '*C*' entry. If there were four isotopes, you'd add a '*D*', etc.

9. turbulence
   ['təbjələns]
   涡流

10. isotope
    ['aɪsə,top]
    同位素

11. fractional abundance
    丰度分数

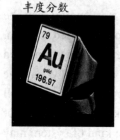

If chlorine has two naturally-occurring isotopes where:

^{35}Cl mass is 34.968852 and fract is 0.7577

^{37}Cl mass is 36.965303 and fract is 0.2423

$$\begin{aligned}\text{atomic weight} &= \text{mass}_A \times \text{fract}_A + \text{mass}_B \times \text{fract}_B \\ &= 34.968852 \times 0.7577 + 36.965303 \times 0.2423 \\ &= 26.496 \text{ amu} + 8.9566 \text{ amu} \\ &= 35.45 \text{ amu}\end{aligned}$$

Tips

The sum of the fractional abundance values must equal 1. Be sure to use the mass or weight of each isotope and not its mass number.

Exercises

1. Translate the following from Chinese into English.
   (1) 量子化学   (2) 热力学系统   (3) 封闭系统   (4) 物料平衡
2. Indicate whether the following ideas are stated or not stated (S/NS) in the passage.
   (   ) (1) Thermodynamics, quantum chemistry, statistical mechanics, and kinetics are the four main areas of physical chemistry.
   (   ) (2) Molecule is not a term in thermodynamics.
   (   ) (3) Chemical thermodynamics involves not only laboratory measurements of various thermodynamic properties, but also the application of mathematical methods to the study of chemical questions.
   (   ) (4) A closed system, is evidently an isolated system.
   (   ) (5) For an open system, both matter and energy can be transferred between system and surroundings.
   (   ) (6) Inserting a thermally conducting wall between two parts of a system can be used to test whether the parts are in thermal equilibrium with each other.
3. Choose the best answer for each of the following statements based on the passage.
   (1) A chemical system can be studied from either a _____ or a(n) _____ viewpoint.
      (a) physical···chemical   (b) molecular···atomic
      (c) microscopic···macroscopic   (d) mechanic···kinetic
   (2) _____ is a macroscopic science that studies the interrelationships between the various equilibrium properties of a system.
      (a) kinetics   (b) thermodynamics
      (c) statistical mechanics   (d) quantum chemistry
   (3) In _____, the molecular and macroscopic levels are related to each other.
      (a) quantum   (b) statistical mechanics
      (c) thermodynamics   (d) kinetics
   (4) Thermodynamics studies _____.
      (a) heat, work, energy, and the changes they produce in the states of systems
      (b) the relationships between the molecules of a system
      (c) heat, work, temperature, and the energy produced in the states of system
      (d) heat, energy, and work

(5) For a (n) _____ system, neither matter nor energy can be transferred between system and surroundings.
   (a) closed   (b) open
   (c) isolated   (d) None of the above

4. Answer the following questions briefly.
   (1) According to the reading passage, a chemical system can be studied from either a macroscopic of a microscopic viewpoint. What are the key differences between these two viewpoints?
   (2) What are the three types of equilibrium? And what is the relationship between these three types of equilibrium and thermodynamic equilibrium?
   (3) What are the three types of system? And what are the differences between them?

【Reading Material】

Legal Responsibilities of Undergraduates in Chemistry Department

You are reminded that it is your responsibility, under the **Health and Safety at Work Act**(英国健康与安全法令), 1974, while working in this laboratory:

To take reasonable care for the health and safety of yourself and of other persons who may be affected by your acts or **omissions**(疏忽).

As regards any duty or requirement imposed on the Department by or under any of relevant **statutory provisions**(法定条文), to co-operate, so far as is necessary, to enable that duty or requirement to be performed or complied with.

Not to interfere with, or misuse anything provided, in the interests of health, safety or welfare.

It is also your responsibility, under the **Control of Substances Hazardous to Health (COSHH) Act** (英国控制有害物质法令), 1994, to be aware in advance of hazards associated with the particular materials and equipment used in each experiment that you undertake. These are listed immediately before the detailed instructions for each experiment. You must read and understand this cautionary information before starting an experiment; that you have done so will be verified by the demonstrator. The information provided also includes advice as to the **remedial**(补救的) and protective action to be taken in the event of an accident involving a chemical being swallowed, inhaled, **splashed**(溅泼) on the skin, or splashed in the eye. Except where otherwise indicated in the experimental instructions, these **eventualities**(不测事件) are covered by the standard treatments.

Failure to observe any of the above may result in legal action being taken against you by **Inspectors of the Health and Safety Executive**(健康与安全检查员).

Within the department, your principal responsibilities may be summarized as follows:

You must co-operate in observing regulations provided for your safety.

You must familiarize yourself with the location of the nearest EXIT routes from the laboratory.

You must also locate and know how to operate the emergency shower on the west wall of the laboratory near the service hatch, and also the eye baths. You should also locate and familiarize yourself with the firefighting equipment-**fire extinguishers** (灭火器), **sand-buckets** (沙桶) and fire blankets for tackling small fires, and the fire alarms themselves, which are located on the south wall, next to the door, and by the main north entrance to the laboratory.

You must not interfere with any safety equipment provided, expose yourself or others as a result of deliberate carelessness, remove any material from the laboratories without the permission of a demonstrator work in the laboratory at any time unless a demonstrator is present or bring visitors into the laboratory.

You must make yourself aware, through the cautionary information provided, of the potential dangers involved in the handling of chemicals used in the experiments.

The following are examples of materials which must never be put down the sink:

Mercury and its compounds (return to the mercury residue bottle).

Liquid nitrogen (allow it to evaporate).

Non-aqueous (无水的) solvents [The laboratory is equipped with waste solvent containers. Do not mix **halogenated** (卤代的) and non-halogenated solvents. Some of these mixtures are dangerous and could involve the department in considerable expense in disposing of them].

Dilute aqueous solutions and suspensions of solids may be disposed of down the sink if they are relatively non-toxic. They must be washed down with **copious** (丰富的) quantities of water. Harmless solids should be returned to the service room in bottles labeled appropriately. When you need to dispose of material not covered by the above, seek advice from the technical staff: such material may need to be stored in a separate container.

Questions

1. What is your principal responsibilities in the department?
2. Please make a list of materials which must never be put down the sink.

Lesson 19 Chemical Equilibrium and Kinetics

1. equilibrium
[ˌikwəˈlɪbriəm]
化学平衡

2. simultaneously
[saɪməlˈtenɪəslɪ]
同时地

3. offset
[ˈɔfˌsɛt,ˈɑf-]
抵消

4. reversibility
可逆性

5. reversible
[rɪˈvɚsəbəl]
可逆的

6. outset
[ˈaʊtˌsɛt]
开始，开头

7. forward
[ˈfɔrwəd]
正向的

8. dynamic
[daiˈnæmik]
动态的

Whenever the products of a chemical reaction are capable of reacting to form the reactants, two reactions are taking place **simultaneously**, the one tending to **offset** the other. As a consequence, such reactions do not go to completion and a state of equilibrium is attained. Most chemical reactions possess the quality of **reversibility** and do not go to completion.

(The gas-phase equilibrium for the ecotbermic reaction)

Suppose we apply the law of mass action to a **reversible** reaction at a fixed temperature, such as $A+B \rightleftharpoons C+D$. This equation states that substances A and B react to form C and D, and that C and D react under the same conditions to form the original reactants A and B. Let us assume that the system at the **outset** contains only the original reactants A and B. The rate of reaction of A and B is expressed in terms of concentration in an equation referred to as the rate equation.

$$rate_f = k_1[A][B]$$

At first, no molecules of C and D are present, there is no reverse reaction, and $rate_r = 0$. However, as soon as some of the products C and D are formed they begin to react, and the rate equation of this reaction is

$$rate_r = k_2[C][D]$$

Because the concentrations of C and D at first will be small, the rate of the reverse reaction $rate_r$ increase. Meanwhile, the concentrations of A and B are becoming less and less, so that the rate of the **forward** reaction, $rate_f$, falls off. Consequently, the two reaction rates approach each other and finally become equal —a condition of **dynamic** equilibrium, which means that the opposing reactions are in full operation but at the same rate. Thus, at equilibrium $rate_r = rate_f$, and we may write

$$k_2[C][D] = k_1[A][B]$$

or, by rearranging,

$$\frac{[C][D]}{[A][B]} = \frac{k_1}{k_2}$$

Because k_1 and k_2 are constants, the ratio k_1/k_2 is also constant, and the expression may be written

$$\frac{[C][D]}{[A][B]} = K$$

K is called the equilibrium constant for the reaction. Just as k_1 and k_2 are **proportionality constants** specific for each reaction at a definite temperature, K is likewise a constant specific to this system in equilibrium at a given temperature. The values for the molar concentrations used in the above mathematical expression must always be the concentrations present after the reaction has reached a state of equilibrium.

It is important to note that at equilibrium the rates of reaction, $rate_r$ and $rate_f$ are equal but the molar concentration of the reactants and products in the equilibrium mixture are usually not equal. It is true, however, that the concentration of each reactant and product at equilibrium remains constant because the rate at which it is being used up in one reaction is equal to the rate at which it is being formed by the opposite reaction. As pointed out before, such a system is referred to as being in a state of dynamic equilibrium.

(Effects of changes in concentration on the equilibrium)

A general equation for chemical equilibria may be written

$$mA + nB + \cdots \rightleftharpoons xC + yD + \cdots$$

By applying the law of mass action to this reversible reaction, we may write

$$\frac{[C]^x[D]^y \cdots}{[A]^m[B]^n \cdots} = K$$

This is a mathematical expression of the law of chemical equilibrium which may be stated as follows: When a reversible reaction has attained equilibrium at a given temperature, the product of the molar concentrations of the substances to the right of the **arrow** in the equation

11. power
   ['pauɚ]
   乘方，幂

12. coefficient
   [ˌkoʊəˈfɪʃənt]
   系数

13. consequence
   [ˈkɑnsɪˌkwɛns,
   -kwəns]
   推论，结果

14. rigorous
   [ˈrɪgərəs]
   精确的

15. derivation
   [ˌdɛrəˈveʃən]
   推导

16. system
   [ˈsɪstəm]
   系统

17. quotient
   [ˈkwoʊʃnt]
   商

18. completeness
   完全（度）

19. convert
   [kənˈvɝt]
   转变

20. cosmic radiation
   宇宙辐射

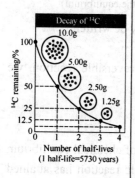

21. half-life
    半衰期

divided by the product of the molar concentrations of the substances to the left, each concentration raised to the **power** equal to the number of molecules of each substance appearing in the equation, is a constant.

The equilibrium constant expression is valid even if the reactions take place in steps, with rates involving powers of the reactant concentrations different from the **coefficients** appearing in the balanced equation. In fact, the validity of the equilibrium constant expression is a **consequence** of the laws of thermodynamics (of physical chemistry), and a more **rigorous derivation** can be made at a later stage in the student's work when a background in thermodynamics and chemical kinetics has been attained.

The meaning of the mathematical expression of the law of chemical equilibrium is that, regardless of how the individual concentrations might be varied, thereby upsetting the equilibrium temporarily, the composition of the **system** with regard to its various components will always adjust itself to a new condition of equilibrium for which the **quotient** $[C]^x[D]^y\cdots/[A]^m[B]^n\cdots$ will again have the value K.

The value of an equilibrium constant is a measure of the **completeness** of a reversible reaction. A large value K indicates that equilibrium is attained only after the reactants A and B have been largely **converted** into the products C and D. When K is very small —much less than unity— equilibrium is attained when only a small proportion of A and B has been converted to C and D.

Hands-On Chemistry: ^{14}C Dating of Organic Material

In the 1950s W.F. Libby and others (University of Chicago) devised a method of estimating the age of organic material based on the decay rate of ^{14}C. ^{14}C dating can be used on objects ranging from a few hundred years old to 50000 years old. ^{14}C is produced in the atmosphere when neutrons from **cosmic radiation** react with nitrogen atoms:

$$^{14}_{7}N + ^{1}_{0}n \longrightarrow ^{14}_{6}C + ^{1}_{1}H$$

Free carbon, including the ^{14}C produced in this reaction, can react to form carbon dioxide, a component of air. Atmospheric carbon dioxide, CO_2, has a steady-state concentration of about one atom of ^{14}C per every 10^{12} atoms of ^{12}C. Living plants and animals that eat plants (like people) take in carbon dioxide and have the same $^{14}C/^{12}C$ ratio as the atmosphere.

However, when a plant or animal dies, it stops intaking carbon as food or air. The radioactive decay of the carbon that is already present starts to change the ratio of $^{14}C/^{12}C$. By measuring how much the ratio is lowered, it is possible to make an estimate of how much time has passed since the plant or animal lived. The decay of ^{14}C is:

$$^{14}_{6}C \longrightarrow ^{14}_{7}N + ^{0}_{-1}e \quad \textbf{(half-life is 5730 years)}$$

Exercises

1. Translate the following from English to Chinese.
   (1) chemical equilibrium
   (2) equilibrium constant
   (3) forward and reverse reaction rate
   (4) chemical kinetics

2. Complete the sentences with the suitable words or phrases from the following words. Change the form where necessary.

   reaction chamber; chemical kinetics; concentration

   (1) This passage establishes the principles of _____ by showing how reaction rates may be measured and interpreted in simple cases.
   (2) Several techniques may be used to follow the changing _____.
   (3) Behind the _____ there is an observation cell fitted with a plunger.

3. Complete the following statements with prepositions.
   (1) The method used to monitor the concentrations depends on the substances involved and the rapidity _____ which they change.
   (2) The disadvantage _____ conventional flow techniques is that a large volume of reactant solution is necessary.
   (3) The study of reaction rates leads _____ an understanding of the mechanisms of reactions, their analysis _____ a sequence of elementary reactions.
   (4) _____ special laser techniques, it is now possible to observe processes occurring _____ a few femtoseconds.

4. Writing equilibrium expressions.
   (1) One of the steps in the production of sulfuric acid involves the catalytic oxidation of sulfur dioxide.
   $$2SO_2(g)+O_2(g) \longrightarrow 2SO_3(g)$$
   (2) The reaction between hydrogen gas and oxygen gas to form water vapour.
   $$2H_2(g)+O_2(g) \longrightarrow 2H_2O(g)$$
   (3) The reduction-oxidation equilibrium of iron and iodine ions in aqueous solution.
   $$2Fe^{3+}(aq)+2I^-(aq) \longrightarrow 2Fe^{2+}(aq)+I_2(aq)$$

5. Consider the reaction $A+B \rightleftharpoons C+D$. Assume that both the forward reaction and the reverse reaction are elementary processes and that the value of the equilibrium constant is very large.
   (1) Which species predominate at equilibrium, reactants or products?
   (2) Which reaction has the larger rate constant, the forward or the reverse? Explain.

6. Consider the following equilibrium (endothermic as written) $CaCO_3(s) \rightleftharpoons CaO(s)+CO_2(g)$, and predict what changes, would occur if the following stresses were applied after equilibrium was established.
   (1) add CO_2
   (2) remove CO_2
   (3) add CaO
   (4) increase T
   (5) decrease V
   (6) add a catalyst

7. Consider the following equilibrium, for which $\Delta H<0$, $2SO_2(g)+O_2(g) \longrightarrow 2SO_3(g)$
   How will each of the following changes affect an equilibrium mixture of the three gases?
   (1) $O_2(g)$ is added to the system

(2) The reaction mixture is heated
(3) A catalyst is add to the mixture
(4) The total pressure of the system is increased
(5) SO₂(g) is removed from the system
(6) adding a noble gas

【Reading Material】

Green Plastics

Driving down a dusty **gravel** (石子) road in central **Iowa** (爱荷华州), a farmer gazes toward the horizon at rows of tall, leafy corn plants shuddering in the breeze as far as the eye can see. The farmer smiles to himself, because he knows something about his crop that few people realize. Not only are **kernels** (谷粒) of corn growing in the ears, but **granules** (粒料) of plastic are **sprouting** (长出) in the stalks and leaves. This **idyllic** (田园风光的) notion of growing plastic, achievable in the foreseeable future, seems vastly more appealing than manufacturing plastic in petrochemical factories, which consume about 270 million tons of oil and gas every year worldwide. Fossil fuels provide both the power and the raw materials that transform crude oil into common plastics such as polystyrene, polyethylene and polypropylene. From milk jugs and soda bottles to clothing and car parts, it is difficult to imagine everyday life without plastics, but the sustainability of their production has increasingly been called into question. Known global reserves of oil are expected to run dry in approximately 80 years, natural gas in 70 years and coal in 700 years, but the economic impact of their **depletion** (消耗) could hit much sooner. As the resources diminish, prices will go up—a reality that has not escaped the attention of policymakers. President Bill Clinton issued an executive order in August 1999 insisting that researchers work toward replacing fossil resources with plant material both as fuel and as raw

material. With those concerns in mind, biochemical engineers were delighted by the discovery of how to grow plastic in plants. On the surface, this technological breakthrough seemed to be the final answer to the sustainability question, because this plant-based plastic would be "green" in two ways: it would be made from a renewable resource, and it would eventually break down, or biodegrade, upon disposal. Other types of plastics, also made from plants, hold similar appeal. Recent research, however, has raised doubts about the utility of these approaches. For one, biodegradability has a hidden cost: the biological breakdown of plastics releases carbon dioxide and methane, **heat-trapping** (吸热的) greenhouse gases that international efforts currently aim to reduce. What is more, fossil fuels would still be needed to power the process that extracts the plastic from the plants, an energy requirement that we discovered is much greater than anyone had thought. Successfully making green plastics depends on whether researchers can overcome these energy-consumption obstacles economically — and without creating additional environmental burdens. Traditional manufacturing of plastics uses a surprisingly large amount of fossil fuel. Automobiles, trucks, jets and power plants account for more than 90 percent of the output from crude-oil refineries, but plastics consume the bulk of the remainder, around 80 million tons a year in the U S alone. To date, the efforts of the biotechnology and agricultural industries to replace conventional plastics with plant-derived alternatives have embraced three main approaches: converting plant sugars into plastic, producing plastic inside **microorganisms** (微生物), and growing plastic in corn and other crops. **Cargill** (Cargill 农用品公司), an agricultural business giant, and **Dow Chemical** (Dow 化学公司), a top chemical firm, joined forces three years ago to develop the first approach, which turns sugar from corn and other plants into a plastic called **polylactide (PLA)**(聚丙交酯). Microorganisms transform the sugar into lactic acid, and another step chemically links the molecules of lactic acid into chains of plastic with attributes similar to **polyethylene terephthalate (PET)** (聚乙烯对苯二甲酸酯), a petrochemical plastic used in soda bottles and clothing fibers. Looking for new products based on corn sugar was a natural extension of Cargill's activities within the existing corn-wet-milling industry, which converts corn grain to products such as **high-fructose** (高果糖) corn **syrup** (糖浆), **citric acid** (柠檬酸), vegetable oil, bioethanol and animal feed. In 1999 this industry processed almost 39 million tons of corn—roughly 15 percent of the entire U S harvest for that year. Indeed, Cargill and Dow earlier this year launched a $300-million effort to begin mass-producing its new plastic, NatureWorks TM PLA, by the end of 2001. Many corporate and academic groups, including Monsanto, have channeled their efforts to produce PHA into the third approach: growing the plastic in plants. Modifying the genetic makeup of an agricultural crop so that it could synthesize plastic as it grew would eliminate the fermentation process altogether. Instead of growing the crop, harvesting it, processing the plants to yield sugar and fermenting the sugar to convert it to plastic, one could produce the plastic directly in the plant. Many researchers viewed this approach as the most efficient—and most elegant—solution for making plastic from a renewable resource. Numerous groups were (and still are) in hot pursuit of this goal.

Question

What is green plastic?

Lesson 20 Environmental Pollution

"Well, at least the EPA is doing something"

1. degradable
   [dɪ'gredəbəl]
   可降解

2. iodine
   ['aɪə,daɪn]
   碘

3. plutonium
   [plu'toniəm]
   钚

4. fertiliser
   肥料

5. manure
   [mə'nur, -'njur]
   有机肥

6. sudsy
   ['sʌdzi]
   起着泡沫的

7. sediment
   ['sɛdəmənt]
   沉积物

8. globule
   ['glɑbjul]
   球剂

9. threshold
   ['θrɛʃhold]
   阈值

With the coming of the Industrial Revolution the environmental pollution increased alarmingly. Pollution can be defined as an undesirable change in the physical, chemical, or biological characteristics of the air, water, or land that can harmfully affect health, survival, or activities of humans or other living organisms. There are four major forms of pollution—waste on land, water pollution (both the sea and inland waters), and pollution by noise.

Land can be polluted by many materials. There are two major types of pollutants: **degradable** and nondegradable. Examples of degradable pollutants are DDT and radioactive materials. DDT can decompose slowly but eventually are either broken down completely or reduced to harmless levels. For example, it typically takes about 4 years for DDT in soil to be decomposed to 25 percent of the original level applied. Some radioactive materials that give off harmful radiation, such as **iodine**-131, decay to harmless pollutants. Others, such as **plutonium**-239 produced by nuclear power plants, remains at harmful levels for thousands to hundreds of thousands of years.

Nondegradable pollutants are not broken down by natural processes. Examples of nondegradable pollutants are mercury, lead and some of their compounds and some plastics. Nondegradable pollutants must be either prevented from entering the air, water, and soil or kept below harmful levels by removal from the environment. Water pollution is found in many forms. It is contamination of water with city sewage and factory wastes; the runoff of **fertiliser** and **manure** from farms and feed lots; **sudsy** streams; **sediment** washed from the land as a result of storms, farming, construction and mining; radioactive discharge from nuclear power plants; heated water from power and industrial plants; plastic **globules** floating in the world's oceans; and female sex hormones entering water supplies through the urine of women taking birth control pills.

Even though scientists have developed highly sensitive measuring instruments, determining water quality is very difficult. There are a large number of interacting chemicals in water, many of them only in trace amounts. About 30000 chemicals are now in commercial production, and each year about 1000 new chemicals are added. Sooner or later most chemicals end up in rivers, lakes, and oceans. In addition, different organisms have different ranges of tolerance and **threshold** levels for various pollutants. To complicate matters even further, while some pollutants are either diluted to harmless levels in water or broken down to harmless forms by decomposers and natural processes, others (such as DDT, some radioactive materials, and some mercury compounds) are biologically concentrated in various organisms.

Air pollution is normally defined as air that contains one or more chemicals in high enough concentrations to harm humans, other animals, vegetation, or materials. There are two major types of air pollutants. A primary air pollutant is a chemical added directly to the air that occurs in a harmful concentration. It can be a natural air component, such as carbon dioxide, that rises above its normal concentration, or something not usually found in the air, such as a lead compound. A **secondary** air pollutant is a harmful chemical formed in the atmosphere through a chemical reaction among air components. We normally associate air pollution with **smokestacks** and cars, but volcanoes, forest fires, dust storms, **marshes**, oceans, and plants also add to the air chemicals we consider pollutants. Since these natural inputs are usually widely dispersed throughout the world, they normally don't build up to harmful levels. And when they do, as in the case of volcanic eruptions, they are usually taken care of by natural weather and chemical cycles.

As more people live closer together, and as they use machines to produce leisure, they find that their leisure, and even their working hours, become spoilt by a byproduct of their machines—namely, noise. The technical difficulties to control noise often arise from the subjective-objective nature of the problem. You can define the excessive speed of a motor-car in terms of a pointer reading on a **speedometer**. But can you define excessive noise in the same way? You find that with any existing simple "noise-meter", vehicles which are judged to be equally noisy may show considerable difference on the meter. Though the ideal cure for noise is to stop it at its source, this may in many cases be impossible. The next remedy is to absorb it on its way to the ear. It is true that the overwhelming majority of noise problems are best resolved by effecting a reduction in the sound pressure level at the receiver. Soft taped music in restaurants tends to mask the clatter of **crockery** and the conversation at the next table. Fan noise has been used in telephone **booths** to mask speech interference from adjacent booths. Usually, the problem is how to reduce the sound pressure level, either at source or on the transmission path.

Hands-On Chemistry: Etymology of Chemistry

The word chemistry comes from the earlier study of **alchemy**, which is a set of practices that encompasses elements of chemistry, metallurgy, philosophy, astrology, astronomy, mysticism and medicine. Alchemy is commonly thought of as the quest to turn lead or another common starting

(Vending machine: 'A Breath of Fresh Air')

10. secondary
    [ˈsɛkənˌdɛri]
    次级

11. smokestack
    [ˈsmokˌstæk]
    烟囱

12. marsh
    沼泽湿地

13. speedometer
    [spɪˈdɑmɪtɚ]
    车速表

14. crockery
    [ˈkrɑkəri]
    陶器类

15. booths
    电话亭，通话室

16. alchemy
    [ˈælkəmi]
    炼金术，炼丹术

119

material into gold. As to the origin of the word "alchemy" the question is a debatable one; it certainly can be traced back to the Greeks, and some, following E. Wallis Budge, have also asserted Egyptian origins. Many believe that the word "alchemy" is derived from the word Chemi or Kimi, which is the name of Egypt in Egyptian. The word was subsequently borrowed by the Greeks, and from the Greeks by the Arabs when they occupied Alexandria (Egypt) in the 7th century. The Arabs added the Arabic definite article "al" to the word, resulting in the word "al-kīmiyā", from which is derived the old French alkemie. A tentative outline is as follows:

Egyptian alchemy [**3000** BCE～**400** BCE], formulate early "element" theories such as the Ogdoad.

Greek alchemy [**332** BCE～**642** CE], the Greek king Alexander the Great conquers Egypt and founds Alexandria, having the world's largest library, where scholars and wise men gather to study.

Arab alchemy [**642** CE～**1200**], the Arabs invade Alexandria; Jabir is the main chemist.

European alchemy [**1300**～present], Pseudo-Geber builds on Arabic chemistry.

Chemistry [**1661**], Boyle writes his classic chemistry text "The Sceptical Chymist".

Chemistry [**1787**], Lavoisier writes his classic "**Elements of Chemistry**".

Chemistry [**1803**], Dalton publishes his "**Atomic Theory**".

Thus, an alchemist was called a "chemist" in popular speech, and later the suffix "-ry" was added to this to describe the art of the chemist as "chemistry".

(Symbols for Alchemical processes)

17. Elements of Chemistry
    元素化学
18. Atomic Theory
    原子学说

Exercises

1. Translate the following from English into Chinese.
   (1) pollution of the atmosphere   (2) nondegradable pollutant   (3) harmless pollutant
   (4) interacting chemicals   (5) chemical reaction   (6) sound pressure level
   (7) threshold level   (8) subjective-objective nature (9) speech interference
   (10) transmission path
2. Choose the best answer for each of the following statements based on the passage.
   (1) _____ are not broken down by natural processes, but _____ can decompose slowly but eventually to harmless level.
       (a) Degradable pollutants, nondegradable pollutants
       (b) Nondegradable pollutants, degradable pollutants
       (c) Degradable pollutants, degradable pollutants
       (d) Nondegradable pollutants, nondegradable pollutants
   (2) About 30000 chemicals are now in commercial production, and about _____ new chemicals are added each year.
       (a) 10   (b) 50   (c) 100   (d) 500

(3) Choose the major forms of pollution list below.
   (a) waste on land   (b) water pollution   (c) noise pollution   (d) plastic materials
3. Complete the summary of the text.
   (1) pollution   (2) water pollution   (3) degradable pollutants
   (4) air pollution   (5) noise   (6) nondegradable pollutants
4. Answer the following questions.
   (1) Why is the water quality very difficult to determine?
   (2) Where does the air pollution come from?
   (3) Can the noise be stopped at its source?
   (4) What are the major forms of pollution?
   (5) What is the effective way to reduce the noise pollution?
5. Topics for writting.
   (1) Environmental pollution increased alarmingly.—Do you agree with this viewpoint? Why? Why not?
   (2) According to the text, what can we do to reduce pollution in the earth? Give your own comments and then write an essay.

【Reading Material】

The Killer Lake

Disaster struck swiftly and without warning. On August 21,1986, Lake Nyos in Cameroon, a small nation on the west coast of Africa, suddenly belched a dense cloud of carbon dioxide. Speeding down a river valley, the cloud asphyxiated over 1700 people and many livestock.

How did this tragedy happen? Lake Nyos is stratified into layers that do not mix. A boundary separates the freshwater at the surface from the deeper, denser solution containing dissolved minerals and gases, including CO_2. The CO_2 gas comes from springs of carbonated groundwater that percolate upward into the bottom of the volcanically formed lake. Given the high water pressure at the bottom of the lake, the concentration of CO_2 gradually accumulated to a dangerously

high level, in accordance with Henry's law. What triggered the release of CO_2 is not known for certain. It is believed that an earthquake, landslide, or even strong winds may have upset the delicate balance within the lake, creating waves that overturned the water layers. When the deep water rose, dissolved CO_2 came out of solution, just as a soft drink fizzes when the bottle is uncapped. Being heavier than air, the CO_2 traveled close to the ground and literally smothered an entire village 15 miles away.

Now, more than 18 years after the incident, scientists are concerned that the CO_2 concentration at the bottom of Lake Nyos is again reaching saturation level. To prevent a recurrence of the earlier tragedy, an attempt has been made to pump up the deep water, thus releasing the dissolved CO_2. In addition to being costly, this approach is controversial because it might disturb the waters near the bottom of the lake, leading to an uncontrollable release of CO_2 to the surface. In the meantime, a natural time bomb is ticking away.

Questions

1. Who killed over 1700 people and many livestock on August 21,1986, Lake Nyos in Cameroon?

2. How did this tragedy happen?

3. Researchers believe a spring deep in the lake is supplying it with a slow buildup of carbon dioxide in Lake Nyos in West Africa. How do scientists remove the subsequent outburst of CO_2 gas?

Key Terms to Physical Chemistry

1. electrophoresis 电泳
2. Dyndall effect 丁达尔效应
3. molar heat capacity under constant volume 定容摩尔热容
4. constant volume thermometer 定容温度计
5. molar heat capacity under constant pressure 定压摩尔热容
6. constant pressure thermometer 定压温度计
7. faraday constant 法拉第常数
8. Faraday's law 法拉第定律
9. reverse osmosis 反渗透
10. reaction orders 反应级数
11. extent of reaction 反应进度
12. heat of reaction 反应热
13. rate of reaction 反应速率
14. constant of reaction rate 反应速率常数
15. van der Waals constant 范德华常数
16. van der Waals equation 范德华方程
17. van der Waals force 范德华力
18. van der Waals gases 范德华气体
19. van't Hoff equation 范特霍夫方程
20. van't Hoff rule 范特霍夫规则
21. non-elementary reactions 非基元反应
22. elevation of boiling point 沸点升高
23. distribution law 分配定律
24. disperse system 分散系统
25. dispersion phase 分散相
26. partial volume law 分体积定律
27. partial pressure law 分压定律
28. intermolecular force 分子间力
29. closed system 封闭系统
30. negative pole 负极
31. composite reaction 复合反应
32. Gay-Lussac law 盖·吕萨克定律
33. Hess law 盖斯定律
34. isolated system 隔离系统
35. work 功
36. photoreaction 光反应
37. extensive property 广度性质
38. oversaturated solution 过饱和溶液
39. process 过程
40. transition state theory 过渡状态理论
41. Helmholtz free energy 亥姆霍兹自由能
42. enthalpy 焓
43. molar heat capacity at constant volume 恒容摩尔热容
44. heat at constant volume 恒容热
45. constant external pressure 恒外压
46. molar heat capacity at constant pressure 恒压摩尔热容
47. heat at constant pressure 恒压热
48. chemical kinetics 化学动力学
49. stoichiometric equation of chemical reaction 化学反应计量式
50. stoichiometric coefficient of chemical reaction 化学反应计量系数
51. extent of chemical reaction 化学反应进度
52. chemical thermodynamics 化学热力学
53. entropy of mixing 混合熵
54. activation energy 活化能
55. elementary reactions 基元反应
56. Gibbs-Helmholtz equation 吉布斯-亥姆霍兹方程
57. Gibbs function 吉布斯函数
58. Gibbs function criterion 吉布斯函数判据
59. Gibbs free energy 吉布斯自由能
60. stoichiometric equation 计量式
61. stoichiometric coefficient 计量系数
62. bond enthalpy 键焓
63. colloidal nucleus 胶核
64. micelle 胶束
65. colloid 胶体
66. colloidal particles 胶体粒子
67. Joule's law 焦耳定律

Unit 4

Analytical Chemistry

Lesson 21 Introduction to Analytical Chemistry

1. qualitative analysis
定性分析
2. analyte
(被)分析物

microprobe
in vivo assay
"lab-on-a-valve"
"lab-on-chip"
μ-TAS, μ-FIA, μ-LC...

(A modern "laboratory" for *in vivo* analysis.)

3. unravel
[ʌnˈrævəl]
阐明, 阐释, 说明
4. reaction-rate
反应速率
5. nerve-signal conduction
神经信号
6. muscle contraction
肌肉收缩
7. relaxation
[ˌriːlækˈseɪʃən]
松懈, 松弛,
放松, 放宽

Analytical chemistry involves separating, identifying, and determining the relative amounts of the components making up a sample of matter. **Qualitative analysis** reveals the chemical identity of the **analytes**. Quantitative analysis tells us the relative amounts of one or more of these analytes in numerical terms. Qualitative information is required before a quantitative analysis can be undertaken. A separation step is usually a necessary part of both quantitative and qualitative analysis.

The principal topics covered in this text are quantitative methods of analysis and methods of analytical separations. We will refer occasionally to qualitative methods, however.

The Role of Analytical Chemistry in the Sciences

Analytical chemistry, or the art of recognizing different substances and determined their constituents, takes a prominent position among the applications of science, since the questions which it enables us to answer arise whenever chemical processes are employed for scientific or technical purpose. Its supreme importance has caused it to be assiduously cultivated from a very early period in the history of chemistry, and its records comprise a large part of the quantitative work which is spread over the whole domain of science. Quantitative analytical measurements also play a vital role in many research areas in chemistry, biochemistry, biology, geology, and the other sciences. For example, chemists **unravel** the mechanisms of chemical reactions through **reaction-rate** studies, the rate at which reactants are consumed or products formed in a chemical reaction can be calculated from quantitative measurements made at equal time intervals. Quantitative analysis for potassium, calcium, and sodium ions in the body fluids of animals permit physiologists to study the role these ions play in **nerve-signal conduction** and **muscle contraction** and **relaxation**.

Materials scientists rely heavily upon quantitative analysis of crystalline germanium and silicon in their studies of the behavior of semiconductor devices. Impurities in these devices are in the concentrations range of 1×10^{-6} to 1×10^{-10} percent. Archaeologists identify the source of volcanic glasses by measuring the concentrations of minor elements in samples taken from various locations. This knowledge in turn makes it possible to trace **prehistoric** trade routes for tools and weapons fashioned from obsidian.

A Classification of Quantitative Methods of Analysis

We compute the results of a typical quantitative analysis from two productive amounts of the components making up a sample of matter. Qualitative analysis reveals the chemical measurements. One is the weight or volume of sample to be analyzed. The second is the measurement of some quantity that is proportional to the amount of analyte in that sample. This seconds step normally completes the analysis.

Chemists classify analytical methods according to the nature of this final measurement. In a **gravimetric** method, the mass of the analyte or of some compounds chemically related to it is determined. In a **volumetric** method, the volume of a solution containing sufficient reagent to react completely with the analyzer is measured. Electro-analytical methods involve the measurement of such electrical properties as potential, current, resistance, and quantity of electricity. Spectroscopic methods are based upon measurements of the interaction between **electromagnetic** radiation and analyte atoms or molecules or upon the production of such radiation by analytes. Finally, a group of **miscellaneous** methods should be mentioned. These include the measurement of such properties as mass-to-charge ratio (mass spectrometry), rate of radioactive decay, heat of reaction, rate of reaction, thermal conductivity, optical activity, and refractive index.

Steps in a Typical Quantitative Analysis

Selecting a Method

Selecting which method to use to solve an analytical problem is a vital first step in any quantitative analysis. The choice is sometimes difficult, an im- portant consideration in selection is the accuracy required. Unfortunately, high reliability nearly always required a large investment of time. The chosen method is most often a compromise between **accuracy** and **economics**.

Sampling

To produce meaningful information, an analysis must be preformed on a sample whose composition faithfully represents that of the bulk of material from which the sample was taken. Where the bulk is large and inhomogeneous, great effort is required to get a representative sample. Consider, for example, a railroad car containing 25 tons of silver **ore**. Buyer and seller must agree on the value of the shipment based primarily

16. absorption
[əbˈsɔrpʃən, -ˈzɔrp-]
吸收，吸收过程，
吸收作用

17. desorption
解吸附作用

18. balance
[ˈbæləns]
平衡，均衡

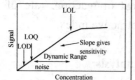

upon its silver content. The ore is inherently heterogeneous, consisting of productive amounts of the components making up a sample of matter.

Preparing a Laboratory Sample

After sampling, solid materials frequently are ground to decrease particle size, mixed to ensure homogeneity, and stored for various lengths of time before the analysis begins. During each of these steps, **absorption** or **desorption** of water may occur, depending upon the humidity of the environment. Because any loss or gain of water changes the chemical composition of a solid, it is a good idea to carefully dry samples at the start of an analysis. Alternatively, the moisture content of the sample can be determined at time of the analysis in a separate analytical procedure.

Defining Replicate Samples

Most chemical analysis are performed on replicate samples whose weight or volumes have been determined by careful measurement with an analytical **balance** or with a precise volumetric device. Obtaining replicate data on samples improves the quality of the results and provides a measure of their reliability.

Preparing Solution of the Sample

Most analysis are performed on solutions of the sample. Ideally, the solvent should dissolve the entire sample rapidly. The conditions of dissolution should be sufficient mild so that loss of the analyte cannot occur. Unfortunately, many materials that must be analyzed are insoluble in common solvents. Examples include silicate minerals, high-molecular-weight polymers, and specimens of animal tissue. Conversion of the analyte in such materials to a soluble form can be a difficult and time-consuming task.

Calibration and Measurement

Qualitative analysis reveals the chemical many lumps that vary in size as well as in silver content. The analysis to have significance, this small sample must have a composition that is representative of the 25 tons ore in the shipment. The task is to isolate 1g of material that accurately reflects the average composition of the nearly 23000000g of bulk sample. Obtaining such a representative sample is a difficult undertaking that requires a careful, systematic manipulation of the entire shipment.

Hands-On Chemistry: Atmospheric Can-Crusher

When water vapor condenses within a closed container, very low pressure is created within that container. The atmospheric pressure on the outside then has the capacity to crush the container. In this activity you will see how this works for water vapor condensing on the inside of an alumnum soda can.

What You Need water, aluminum soda can, saucepan, tongs.

Safety Note Wear safety goggles, and avoid touching the steam produced in this experiment—steam burns can be severe.

Procedure (1) Fill the saucepan with water and set aside. (2) Put about a tablespoon of water in the can and heat on a stove until steam comes out. The steam you see indicates that air has been driven out of the can and replaced by water vapor. (3) Quickly grasp the can with tongs, invert it, and dip into the water in the saucepan just enough place the can opening under water. Crunch! The can is crushed by atmospheric pressure! Why?

Hands-On Chemistry Insights: Atmospheric Can-Crusher

When the molecules of water vapor come in contact with the room-temperature water in the saucepan, they condense, leaving a very low pressure in the can, the much greater surrounding atmospheric pressure crushes the can. Here you see dramatically how pressure is reduced by condensation. This occurs because liquid water occupies much less volume than does the same mass of water vapor. As the vapor molecules come together to form the liquid, they leave a void(low pressure). This activity also shows how the atmospheric pressure surrounding us is very real and significant.

Exercises

1. Translate the following from Chinese into English.
   (1) 定性分析   (2) 分析物   (3) 准确度   (4) 反应速率
   (5) 解吸附作用   (6) 吸附   (7) 定量分析
2. What is the definition of analytical chemistry?
3. What is the definition of analyte?
4. Analytical chemistry generally involves two kinds of analytical methods, finding them from the text and listing them.
5. According to the text, identify the difference between quantitative and qualitative analysis?
6. One-choice question.
   (1) What is analytical chemistry?
      (a) The study of how chemicals react
      (b) The analysis of other chemistry studies
      (c) The identification, isolation, and measurement of chemicals
      (d) The most advanced form of chemistry
   (2) What did Robert Bunsen and Gustav Kirchhoff do to impact analytical chemistry?
      (a) Created a system of naming elements
      (b) Created a system of identifying elements
      (c) Created the first instrument to analyze compounds (flame spectrometer)
      (d) Isolated the first medical compound from a plant
   (3) What would an analytical chemist do?
      (a) Memorizes a bunch of compounds so that they know them by sight
      (b) Properly use instruments and prepare compounds to determine what an unknown substance is
      (c) Combines substances to learn how to make an explosion
      (d) Determines what chemicals are needed for our body to properly operate

【Reading Material】

Experimental Errors

Experimentalists are ware that any measurement they make is affected by errors which can seriously affect the experimental result. The actions adopted to reduce these errors as much as possible will depend on the nature of the errors themselves. A fundamental distinction is usually done between three types of errors: gross, random, and systematic.

Accuracy and precision are among the most important criteria for defining the quality of an analytical method. Usually precision is evaluated first because systematic errors (affecting accuracy) can be quantified only when the magnitude of random errors is known.

Summary of the most salient features of random and systematic errors:

"There's flaw in your experimental design. All the mice are scorpios."

(1) Random (or indeterminate) Error
① Sources include personal, instrumental and methodological uncertainties.
② Not eliminable but reducible by careful working.
③ Recognizable through the scatter around the mean.
④ Affect precision.
⑤ Quantifiable through a measure of precision (e.g., the standard deviation).

(2) Systematic (or determinate) Error
① Sources include personal, instrumental and methodological bias.
② In principle recognizable and reducible (partially or even completely).
③ Recognizable by the lack of agreement between the mean and the true value.
④ Affect accuracy.
⑤ Quantifiable by the difference between the mean and the true value.

The closeness of the replicate measurements in a set, i.e., the spreading of the data about the central value (the mean) is defined as precision.

Error is the sum of the systematic and random parts.

Bias is a constant of systematic error manifesting as a persistent positive or negative deviation of the mean from the accepted reference (true) value.

Question

The lead content in a water sample is determined by anodic stripping voltammetry. The following values (g/L) are obtained: 1.2, 1.8, 1.4, 1.6, 1.3, 1.5, 1.4, 1.3, 1.7, 1.4. Calculate the 95% and 99% confidence limits (置信区间) for the lead concentration?

Lesson 22　Volumetric Analysis and Qualitative Analysis

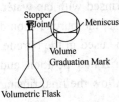

1. titration
 [taɪ'treʃən]
 滴定，滴定法

2. stoichiometric
 化学计算的，
 化学计量的

3. indicator
 ['ɪndɪˌketɚ]
 指示器，压力计，
 [化]指示剂，
 [计]指示符

4. molarity
 [mo'lærɪti]
 摩尔浓度，容模

Chemical analyses can be made by determining how much of a solution of known concentration is needed to react fully with an unknown test sample. The method is generally referred to as volumetric analysis and consists of **titrating** the unknown solution with the one of known concentration (a standard solution). By titration, you can determine exactly how much of a reagent is required to bring about complete reaction of the test solution. Usually, completion of the reaction is indicated by a sudden, visible change in the reaction system that coincides with the **stoichiometric** relationship between moles or equivalents of the reagent solution and the reactant in the test solution. A drop or two of an appropriate **indicator** solution produces a color change at the point where the reaction is complete-referred to as the endpoint.

Molarity is the number of moles (gram-molecular weights) of substance per liter of solution. The mole weight of sulfuric acid is 98.08g, and therefore, 1mol of H_2SO_4 contains 98.08g. If 49.04g are diluted to 1L then the concentration is 0.49 or $0.5 mol \cdot L^{-1}$. In the case of hydrochloric acid, HCl, a $1\ mol \cdot L^{-1}$ solution is prepared by taking 36.465g of HCl and diluting to 1L. The procedure is the same for bases.

Normality is the number of equivalent weights of substance per liter of solution. The equivalent weight of an acid is the weight of that acid capable of furnishing 1mol of protons (H^+), and the equivalent weight of a base is the weight of base capable of receiving 1mol of protons. The equivalent weight of H_2SO_4 is 98.08g/2 or 49.04g. Therefore, a normal solution (N) of H_2SO_4 contains 49.04g per liter. The normality of an acid or base of unknown concentration may be determined by titration. The advantage of using normality rather than molarity is that equal volumes of solutions of equal normalities have identical capacities for neutralization, because they contain the same number of equivalent weights.

In a titration, we compare equivalent weights of acid and base. The number of equivalents of acid is equal to the product of the volume of the acid solution and its normality:

$$V_a \times N_a = \text{equivalents of acid}$$

The number of equivalents of base is the product of the volume of the base solution and its normality:

$$V_b \times N_b = \text{equivalents of base}$$

That's true because:

$$(\text{volume})(\text{normality}) = (\text{liters})(\frac{\text{equivalents}}{\text{liter}}) = \text{equivalents}$$

Neutralization has taken place when the number of equivalents of acid is equal to the number of equivalents of base:

$$V_a \times N_a = V_b \times N_b$$

Procedure

Care must be exercised throughout the titration procedure. The **burette** should be thoroughly cleaned with soap and water, rinsed with tap water, and finally, rinsed with distilled water. Just before use, the burette should be rinsed with two 5mL portions of the solution to be used in the burette. This is done by holding the burette in a semi-horizontal position and rolling the solution around the entire inner surface. Allow the final rinsing to drain through the tip.

Fill the burette to a point above the top marking and allow the solution to run out until the bottom of the meniscus is just at the top marking of the burette. The burette tip must be completely filled to deliver the volume measured. In addition, the burette must be cleaned thoroughly after use because sodium **hydroxide** and other types of solutions will eventually frost the glass and render an expensive piece of equipment useless.

1. Titration of Vinegar

Measure 50mL of vinegar with a pipette and pour into a 250mL beaker. Add 2 drops of **phenolphthalein** indicator. Fill a burette with a 1mol/L solution of sodium hydroxide (NaOH) and draw out the excess as described above. From the burette add NaOH to the beaker of vinegar until 1 drop of NaOH produces a pale pink color in the solution. Maintain constant stirring. The appearance of pink tells you that the acid has been neutralized by the base and there is now 1 drop of excess base which has turned the indicator. Read the burette and record this reading as the volume of base used to neutralize the acid. According to the equation

$$NaOH + CH_3COOH \longrightarrow Na^+ + CH_3COO^- + H_2O$$

One molecule of NaOH neutralizes one molecule of acetic acid, or one gram-molecular weight of NaOH neutralizes one gram-molecular weight of acetic acid. Calculate the amount of acetic acid present in the vinegar. Report this amount as the percentage of acetic acid.

2. Standard Titration Curve

If a pH meter is available, repeat the above process using a pH meter for constantly determining the pH. When the endpoint is reached, continue adding the base to expand the curve further. Make a graph for this titration.

9. qualitative
   ['kwalɪˌtetɪv]
   定性的，质的

3. Equivalents of Acid

Using the 1N solution of NaOH, determine the number of equivalents in two samples of benzoic acid. Carry out the procedure for the two determinations simultaneously. From this value calculate the equivalent weight of the acid. The solid should be weighed in a beaker and should be dissolved in about 25mL of ethyl alcohol before titration with the base. Between 2.0g and 2.2g of the solid provide the best results. Record all data and make all calculations necessary to determine the equivalent weight of the solid acid. Compare your experimental value with the equivalent weight of benzoic acid (calculated from the formula) and determine the percentage of error of your work.

Qualitative Organic Analysis

Although the subject of the identification of organic compounds by application of a systematic scheme of qualitative organic analysis is covered in a full course at the senior or graduate student level in most schools, there is much to be gained by making a preliminary study of this phase of organic chemistry in the first course in the subject 1. Furthermore, of all laboratory assignments in the first year course in organic chemistry, this is the lint that students usually enjoy the most 2. You will find that the identification of an unknown poses a real challenge to you in the application of your skill in laboratory manipulations and your knowledge of the subject of organic chemistry. The analysis of a chemical substance is generally performed for one of two reasons : ① to assign unambiguously a structure to a new molecule which has not been described previously in the chemical literature; ② to characterize the properties of the compound for comparison purposes. Thus, an investigator who isolates a product as a result of some chemical reaction has two options by which he can determine the identity of this product. He can first characterize the molecule as discussed below and search the literature for a known molecule which has identical properties. Should the literature search provide such a compound, then the investigator can safely argue that he has succeeded in identifying his substance. On the other hand, if such data are not available in the literature, his remaining alternative is to carry out a full structure determination.

Hands-On Chemistry: What is the Volume of Your Lungs?

What You'll Need: clean plastic tubing, a large plastic bottle, water, kitchen sink or large water basin

Instructions: Make sure the plastic tubing is clean; Put about 10 cm of water into your kitchen sink; Fill the plastic bottle right to the top with water; Put your hand over the top of the bottle to stop water escaping when you turn it upside down; Turn the bottle upside down. Place the top of the bottle under the water in the sink before removing your hand; Push one end of the plastic tube into the bottle; Take a big breath in; Breathe out as much air as you can through the tube; Measure the volume of air your lungs had in them; Make sure you clean up the area to finish.

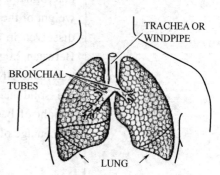

What's Happening? As you breathe out through the tube, the air from your lungs takes the place of the water in the bottle. If you made sure you took a big breath in and breathed out fully then the resulting volume of water you pushed out is equivalent to how much air your lungs can hold. Having a big air capacity in your lungs means you can distribute oxygen around your body at a faster rate. The air capacity of lungs [or $V(O_2)$ max] increases naturally as children grow up but can also be increased with regular exercise.

Exercises

1. Translate the following from Chinese into English.
   - (1) 滴定
   - (2) 指示剂
   - (3) 摩尔浓度
   - (4) 滴定管
   - (5) 酚酞
   - (6) 滴定终点
   - (7) 缓冲溶液
   - (8) 碱式滴定管
   - (9) 萃取
   - (10) 定量分析
   - (11) 滴液漏斗
   - (12) 容量分析
2. How to define the molarity, and what is the difference between the molarity and normality?
3. What is the volumetric analysis?
4. Usually the identification of organic compounds covered a systematic scheme of qualitative organic analysis, but many investigators may be seen to use the analysis of a chemical substance to identify organic compounds. Why? List two reasons.
5. Translate the following paragraphs into Chinese.
   (1) Pick the size of your separatory funnel. You will usually use 125 or 250mL, large scale reactions (1~10g) can require 500mL or 1L sizes. Remember that your separatory funnel will contain the solvent and wash liquid which must be thoroughly mixed.
   (2) Dilute the crude reaction mixture with your solvent of choice and transfer to your chosen separatory funnel. Large amounts of material require large amounts of solvent. Normal reactions (50~500mg of product) can be diluted with between 25~100mL of solvent.
   (3) Distillation is an extremely useful technique that is used to purify reagents and separate crude product mixtures. There are two varieties of distillation: atmospheric pressure and reduced pressure.

【Reading Material】

The Conjugate Acid-Base Pair and Titration

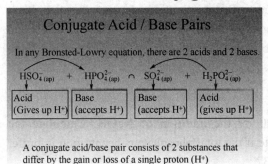

A conjugate acid/base pair consists of 2 substances that differ by the gain or loss of a single proton (H+)

In the Brösted-Lowry theory, an acid is a proton donor and a base is a proton acceptor. Each acid is related to its conjugate base and vice versa:

$$\text{acid} = \text{base} + \text{proton}$$

Therefore, the conjugate base of a strong acid must be a weak base and the conjugate base of a weak acid must be a strong base. Together they form a couple, and an acid without its conjugate base is a meaningless concept. In order to release a proton, the acid must find a base to accept it. In an aqueous solution, the proton, H^+, having an extremely small ionic radius, cannot exist as such. It is hydrated, forming the hydronium ion H_3O^+ and higher hydrates. Thus, an acid-base equilibrium is not a simple dissociation equilibrium, but the result of a proton transfer reaction in which there are at least two reagents and two products. Such a process is also called **protolysis**（质子迁移）. The overall reaction is expressed by:

$$HX + H_2O \rightleftharpoons H_3O^+ + X^-.$$

The overall equilibrium constant is

$$K = [H_3O^+][X^-] / [HX][H_2O].$$

The acid dissociation constant K_a is given by:

$$K_a = K_{[H_2O]} = [H_3O^+][X^-]/[HX]$$

K_a reflects not only the acid strength of HX, but also the base strength of water. This is why different acid dissociation constants are observed for the same acid in different solvent.

Similar proton transfer reactions exist in all solvents possessing proton donor and acceptor properties. Proton transfer reactions are extremely fast. This makes them very suitable for analytical applications and acid-base reactions have found wide use in volumetric methods and other analytical techniques.

The pH value is a measure for the acidity or basicity of a solution, aqueous or nonaqueous.

Acid-base indicators are chemical substances with acid-base properties, having different colors in their protonated and deprotonated forms.

A most important application of acid-base systems is related to the property of such a system to act as a buffer. Many chemical reactions produce protons (in aqueous solutions hydroniums) or hydroxide ions. If these products remain in the system, a corresponding pH change is observed. However, if a buffer is present in the solution it reacts with the liberated hydrogen or hydroxide ions so that only a relative small change of pH occurs. Buffer consists of a mixture of a weak acid and its conjugate base. The most efficient buffer for a given pH consists of a 1:1 ratio of the protonated and deprotonated forms of a weak acid (with $pK_a = pH$). This cannot always be achieved, but if we wish to prepare a solution of a certain pH, we select a weak acid with a pK_a value close to the desired pH. Buffer solution resists changes in pH upon adding of strong acids or strong bases. Depending on the relative concentrations of the acid and base forms of the buffer, the system can resist small or large additions of strong acid or base. This buffer capacity is defined as the number of moles of strong acid of base required to change the pH of 1 L of buffer solution by one pH unit.

Solutions with high or low pH values, formed as a result of dissolution of large quantities of a strong base or acid, are characterized by a large buffer capacity, although the electrolyte practically consists of only one of the conjugate forms (e.g., HCl or NaOH solutions).

A general requirement for all volumetric methods is that the titration process is fast and that it proceeds in a definite stoichiometric ratio, the endpoint of the reaction must by easy to detect and the reaction should be specific and not influenced by other constituents of the solution, i.e., there should be no interference.

Question

A H_3PO_4 solution is brought to pH = 7.00 by the addition of NaOH. Calculate the concentration of the various forms of orthophosphate if the total phosphate concentration in buffer is 0.200 mol/L. $pK_{a1} = 2.16$, $pK_{a2} = 7.21$, $pK_{a3} = 12.32$.

Lesson 23 Ultraviolet and Visible Molecular Spectroscopy

1. hue [hju]
   色彩，色调，颜色
2. colorimetry [ˌkʌləˈrimitri]
   比色法

Probably the first physical method used in analytical chemistry was based on the quality of the color in colored solution. The first things we observe regarding colored solutions are their **hue**, or color, and the color's depth, or intensity. These observations led to the technique historically called **colorimetry**; the color of the solution could identify the concentration of the species present (quantitative analysis). This technique was the first use of what we now understand to be absorption spectroscopy for chemical analysis. When white light passes through a solution and emerges as red light, we say that the solution is red. What has actually happened is that the solution has allowed the red component of white light to pass through, whereas it has absorbed the complementary color, yellow and blue. The more concentrated the sample solution, the more yellow and blue light is absorbed and the more intensely red the solution appears to the eye. For a long time, experimental work made use of the human eye as the detector to measure the hue and intensity of colors in solution. However, even the best analyst can have difficulty comparing the intensity of two colors with slightly different hues, and there are of course people who are color-blind and cannot see certain colors. Instruments have been developed to perform the measurements more accurately and reliably than the human eye. While the human eye can only detect visible light, this chapter will focus on both the ultraviolet(UV) and the visible(Vis) portions of the spectrum.

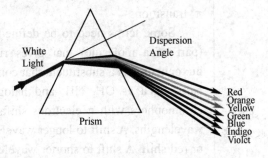

The wavelength range of UV radiation starts at the blue end of visible light(about 400nm) and ends at approximately 200nm for spectrometers operated in air. The radiation has sufficient energy to excite valence electrons in many atoms and molecules. Consequently, UV radiation is involved with electronic excitation. Visible light, considered to be light with wavelengths from 800 nm to 400nm, acts in the same way as UV light. It is also considered part of the electronic region. For this reason we

3. spectroscopic
   分光镜的,
   借助分光镜的,
   光谱(学)的

4. polyatomic
   [,pɑliə'tɑmɪk]
   多原子的

5. chromophore
   ['kromə,fɔr, -,for]
   发色团, 生色团

6. auxochrome
   助色团

7. bathochromic shift
   化合物的吸收光谱带
   的红端移动, 红移

8. hypsochromic shift
   蓝移

9. hyperchromism
   增色

10. hypochromism
    减色

find commercial **spectroscopic** instrumentation often operates with wavelengths between 800nm to 200nm. Spectrometers of this type are called UV/Visible (or UV/Vis) spectrometers. The vacuum UV region of the spectrum extends below 200nm to the X-ray region of the spectrum, at ~100Å. It has called the vacuum UV region because oxygen, water vapor, and other molecules in air absorb UV radiation below 200nm, so the spectrometer light path must be free of air to observe wavelengths <200nm. The instrument must be evacuated (kept under vacuum) or purged with an appropriate non-UV absorbing gas such as helium for this region to be used. Vacuum UV radiation is also involved in electronic excitation but the spectrometers are specialized and not commonly found in undergraduate or routine analytical laboratories. For our purposes the term UV will mean radiation between 200nm and 400nm, unless stated otherwise.

The interaction of UV and visible radiation with matter can provide qualitative identification of molecules and polyatomic species, including ions and complexes. Structural information about molecules and **polyatomic** species, especially organic molecules, can be acquired. This qualitative information is usually obtained by observing the UV/Vis spectrum, the absorption of UV and visible radiation as a function of wavelength by molecules. The shape and intensity of UV/Vis absorption bands are related to the electronic structure of the absorbing species. The molecule is often dissolved in a solvent to acquire the spectrum. We will look at how we can use the absorption maximum of a **chromophore** and a set of guidelines to predict the position of the absorption maximum in a specific molecule. We will also consider how the solvent affects the spectrum of some molecules. As a reminder, the transitions that give rise to UV/Vis absorption by organic molecules are the $n \rightarrow \sigma^*$, $\pi \rightarrow \pi^*$, and $n \rightarrow \pi^*$ transitions.

Some terms need to be defined. A chromophre is a group of atoms (part of a molecule) that gives rise to an electronic absorption. An **auxochrome** is a substituent that contains unshared (nonbonding) electron pairs, such as OH, NH, and halogens. An auxochrome attached to a chromophore with π electrons shifts the absorption maximum to longer wavelengths. A shift to longer wavelengths is called a **bathochromic shift** or red shift. A shift to shorter wavelengths is called a **hypsochromic shift** or blue shift. An increase in the intensity of absorption band (that is, an increase in ε_{max}) is called **hyperchromism**; a decrease in intensity is called **hypochromism**. These shifts in wavelength and intensity come about as a result of the structure of the entire molecule or as a result of interaction between the solute molecules and the solvent molecules. As described at the beginning of the chapter, the types of compounds that absorb UV radiation are those with nonbonded electrons (n electrons) and

conjugated double bond systems (π electrons) such as aromatic compounds and conjugated olefins. Unfortunately, such compounds absorb over similar wavelength ranges, it is necessary to purify the sample to eliminate absorption bands due to impurities. Even when pure, however, the spectra are often broad and frequently without fine structure. For these reasons. UV absorption is much less useful for the qualitative identification of functional groups or particular molecules than analytical methods such as MS, IR, and NMR, UV absorption is rarely used for organic structural methods such as MS, IR, and NMR. UV absorption is rarely used for organic structural elucidation today in modern laboratories because of the ease of use and power of NMR, IR, and MS.

UV and visible absorption spectrometry is a powerful tool for quantitative analysis. It is used in chemical research, biochemistry, chemical analysis, and industrial processing. Quantitative analysis is based on the relationship between the degree of absorption and the concentration of the absorbing material. Mathematically, it is described for many chemical systems by Beer's Law, $A = abc$. The term applied to quantitative absorption spectrometry by measuring intensity ratios is **spectrophotometry**. The use of spectrophotometry in the visible region of the spectrum used to be referred to as colorimetry. UV/Vis spectrophotometry is a widely used spectroscopic technique. It has found use everywhere in the world for research, clinical analysis, environmental analysis, and many other application. Some typical applications of UV absorption spectroscopy include the determination of ①the concentrations of phenol, **nonionic surfactants**, sulfate, sulfide, phosphates, fluoride, nitrate, a variety of metal ions; ②natural products, such as an **steroids** or **chlorophyll**; ③**dyestuff** materials; ④vitamins, proteins, DNA, and enzymes in biochemistry.

Spectrophotometry in the UV region of the spectrum is used for the direct measurement of many organic compounds, especially those with aromatic rings and conjugates multiple bonds. There are also colorless inorganic species that absorb in the UV. A good example is the nitrate ion, NO_3^-. A rapid screening method for nitrate in drinking water is performed by measuring the absorbance of the water at 220nm but not at 275nm. The measurement at 275nm is to check for interfering organic compounds that may be present.

Spectrophotometric analysis in the visible region can be used whenever the sample is colored. Many materials are inherently colored without chemical reaction (e.g. inorganic ions such as dichromate, permanganate, cupric ion, and ferric ion) and need no further chemical reaction to form colored compounds. Colored organic compounds,

11. spectrophotometry
分光光度计测定法

12. nonionic
在溶液中不分解成离子的，非离子的

13. surfactant
[sə'fæktənt]
表面活性剂（的）

14. steroid
['stɪrˌɔɪd,'stɛr-]
类固醇，甾族化合物

15. chlorophyll
['klɔrəfɪl]
叶绿素

16. dyestuff
['daɪˌstʌf]
染料

such as dyestuffs, are also naturally colored. Solution of such materials can be analyzed directly. The majority of metal and nonmetal ions, however, are colorless. The presence of these ions in a sample solution can be determined by first reacting the ion with an organic reagent to form a strongly absorbing species. If the product of the reaction is colored, absorbance can be measured in the visible region. Alternatively, the product formed may be colorless but absorb in the UV. The majority of spectrophotometric determinations result in an increase in absorbance (darker color if visible) as the concentration of the analyzer increase.

Hands-On Chemistry: **Chromatography of Leaves**

Most leaves are green due to chlorophyll. This substance is important in **photosynthesis** (the process by which plants make their food). In this experiment, the different pigments present in a leaf are separated using paper chromatography.

What to Record The chromatogram produced in this experiment can be dried and kept.

17. photosynthesis
    [,fotoˈsɪnθɪsɪs]
    光合作用

18. mortar
    [ˈmɔrtɚ]
    研钵

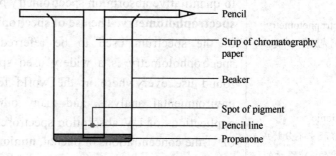

What to Do ①Finely cut up some leaves and fill a **mortar** to about 2cm depth. ②Add a pinch of sand and six drops of propanone from the teat pipette. ③Grind the mixture for at least three minutes. ④On a strip of chromatography paper, draw a pencil line 3cm from the bottom. ⑤Use a fine glass tube to put liquid from the leaf extract onto the centre of the line. Keep the spot as small as possible. ⑥Allow the spot to dry, then add another spot on top. Add five more drops of solution, letting each one dry before putting on the next. The idea is to build up a very concentrated small spot on the paper. ⑦Put a small amount of propanone in a beaker and hang the paper so it dips in the propanone. Ensure the propanone level is below the spot. ⑧Leave until the propanone has soaked near to the top. ⑨Mark how high the propanone gets on the paper with a pencil and let the chromatogram dry.

Safety Propanone is highly flammable. Wear eye protection.

Questions

1. How many substances are on the chromatogram?
2. What colours are they?
3. Which colour moved furthest?

Exercises

1. Translate the following from Chinese into English.
   (1) 紫外可见光谱   (2) 电子跃迁   (3) 光谱学   (4) 红移   (5) 蓝移
   (6) 红外光谱   (7) 标准曲线   (8) 发色团   (9) 反萃取   (10) 助色团
2. According to the text, how to identify which is a UV/Visible spectrometers?
3. What is a chromophore or an auxochrome?
4. One-choice questions.
   (1) Beer's Law states that _____.
      (a) absorbance is proportional to both the path length and concentration of the absorbing species
      (b) absorbance is proportional to the log of the concentration of the absorbing species
      (c) absorbance is equal to p_0 / p
   (2) UV/Vis spectroscopy of organic compounds is usually concerned with which electronic transition(s)?
      (a) $\sigma \rightarrow \sigma^*$ (b) $n \rightarrow \sigma^*$ (c) $n \rightarrow \pi^*$ and $\pi \rightarrow \pi^*$
   (3) Molar absorbtivity of compounds exhibiting charge transfer absorption are _____.
      (a) small   (b) moderate   (c) large
   (4) Peaks resulting from $n \rightarrow \pi^*$ transitions are shifted to shorter wavelengths (blue shift) with increasing solvent polarity.
      (a) true   (b) false   (c) unascertainable
5. Answer the questions.
   (1) An analyst is asked to determine lead in a consignment of fruit juice. The client specifies that the lead content is of the order of 100μg/kg (ppb) and that he requires an "accuracy" of 5μg/kg and accepts a 95% confidence level. Calculate the sample size necessary to satisfy this request assuming that, at the specified concentration level, the precision of the analytical method used is known to be 8μg/kg.
   (2) A student is asked to prepare 1L of a NaCl (molar mass 55.443g/mol) solution approximately 0.1mol. He weighs out, by difference, 5.8970g of the salt; he knows that standard deviations associated to the balance and volumetric flask used are 0.0001g and 0.4mL, respectively. What is the relative standard deviation on the molarity of the resulting NaCl solution? Can you tell the error originates mainly from the use of graduate flask or from the use of balance?

【Reading Material】

UV / Vis Spectroscopy

UV/Vis spectroscopy is the use of UV-visible spectrum of work equipment. Common UV/Vis spectrometer, mainly by the light source, monochromator, sample pool (absorbing light pools), detector, recording device component. UV / Vis spectroscopy is generally designed to avoid the use of the optical lens, the main use of mirrors in order to prevent the absorption of the error caused by the instrument. When the optical path can not avoid the use of transparent component should be selected on the UV-visible light are transparent materials (such as the pool and the reference pool samples are quartz glass). UV/Vis spectroscopy is the use of UV/Vis spectrometer in the wider, its main by the light source, monochromator, absorption cell, detector and data processing and records (computer) components. UV / Vis spectroscopy is mainly used for compound identification, purity checks, the determination of isomers, steric effect of determination, the determination of hydrogen bond strength, and among other relevant quantitative analysis, but usually only a supplementary analysis tool, but also requires the use of other analytical methods such as IR, NMR, EPR, comprehensive approach to analyze the test material in order to get accurate data. Here are two UV - visible spectra of the important applications. Metal complexes by UV - visible spectra are divided into three bands, first of all, there are ligands in the ultraviolet region - central ion of metal electron transfer transition bands, their intensity is usually larger; second, dd transition bands , the causes of e from the central ion d orbital transitions in the lower to the higher d orbital, usually its strength is relatively weak in the visible region, and its maximum absorption wavelength and intensity of color and depth of complex macro- corresponding; third, coordination body charge transfer, the ligand itself UV absorption. Therefore, the use of UV/Vis spectroscopy, can be of metal ions and organic ligands between the metal.

UV/Vis spectroscopy can also be used to characterize the aggregation level of metal nanoparticles. Metal surface and surface plasmon resonance absorption of free electrons of the campaign. Precious metals can be regarded as free electron system, the conduction band electron to determine its optical and electrical properties. In metal plasma theory, if the plasma inside a magnetic disturbed by its charge density is not zero in some regions, it will generate electrostatic restoring force, so that the charge distribution of the oscillation, when the electromagnetic wave frequency and the plasma oscillation frequency are same, it will produce resonance, at the macro level for the metal nanoparticles on the performance of light absorption. Metal surface plasmon resonance is to determine the optical properties of metal nanoparticles important factor. As the metal particles within the plasma resonance excitation or due to inter-band absorption, which in the UV-visible region absorption band. Different metal particles have their characteristic absorption spectra. Therefore, by UV-visible light spectrum, in particular with the Mie theory results match, can get on the particle grain size, structure, and so many of the important information. This technique is simple and convenient, liquid metal nanoparticles is characterized most commonly used technology.

Questions

1. UV-visible spectrum are used in (   ).
   (a) compound identification
   (b) purity checks
   (c) the determination of isomers
   (d) steric effect of determination, the determination of hydrogen bond strength, and among other relevant quantitative analysis

2. Which band is used by UV-visible spectrum?

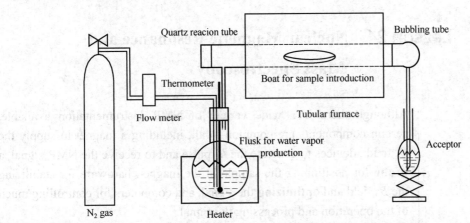

(Apparatus for the Pyrohydrolytic Determination of Fluoride in SiC powder)

Lesson 24 Nuclear Magnetic Resonance and Mass Spectroscopy

1. probe [prob]
探针，调查
2. optimizing
优化，最佳化
3. axial ['æksiəl]
轴的，轴向的
4. align [ə'laɪn]
使结盟，使成一行，匹配排列，排成一行

5. console
[kən'sol]
操纵台，控制台

6. tunable
['tunəbəl,'tju-]
可调音的，可调谐的，和谐的，音调

7. homogeneity
[,homədʒə'niːti]
同种，同质，同次性（等于homogeneousness）
8. plague [pleg]
折磨，使苦恼，使得灾祸
9. dual ['duəl]
双的，双重的

Although there is a wide variety of NMR instrumentation available, certain components are common to all, including a magnet to supply the B_0 field, devices to generate the B_1 pulse and to receive the NMR signal, a **probe** for positioning the sample in the magnet, hardware for stabilizing the B_0 field and **optimizing** the signal, and computers for controlling much of the operation and processing the signal.

Early NMR machines relied on electromagnets. Although a generation of chemists used them, the magnets had low sensitivity and poor stability. Less popular permanent magnets were simpler to maintain but still had low sensitivity. Most research grade instruments today use a superconducting magnets and operate at 3.5~18.8T(150~180MHz for protons). These magnets provide high sensitivity and stability. Possibly most important, the very high fields produced by superconducting magnets results in better separation of resonances, because chemical shifts and hence chemical shift differences increase with field strength. The superconducting magnet resembles a solid cylinder with a central **axial** hole. The direction of the B_0 field(z) is **aligned** with the axis of the cylinder.

Separate from the magnet is a **console** that contains, among other components, the transmitter and receiver of the NMR signal. The B_1 field of the transmitter in the pulse experiment is 1~40mT. In spectrometers designed to record the resonances of several nuclides (multinuclear spectrometers), the B_1 field must be **tunable** over a range of frequencies. The console in addition contains a recorder to display the signal.

The sample is placed in the most homogeneous regions of the magnetic field by means of an adjustable probe. The probe contains a holder for the sample, mechanical means for adjusting its position in the field, electronic leads for supplying the B_1 and B_2 (double resonance) fields and for receiving the signal, and devices for improving magnetic **homogeneity**.

The sample must have good solubility in a solvent, which must have no resonances in the regions of interest. For protons, $CDCl_3$, D_2O, and acetone-d_6 are traditional NMR solvents. If the spectrum is to be recorded above or below room temperature, the solvent chosen must not boil or freeze during the experiment.

The NMR experiment is **plagued** by the **dual** problems of sensitivity

10. inhomogeneity
[ɪnˈhomədʒəˈniːɪti]
不同类,不同族,
不均一(性),多相(性)

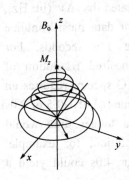

11. gyromagnetic ratio
回转磁比率
12. decoupling
退耦(装置)
13. Overhauser
奥佛豪塞（人名）

14. random
[ˈrændəm]
任意的,胡乱的,
随机的
15. amplitude
[ˈæmplɪˌtud, -ˌtjud]
振幅,广阔,
丰富,充足

and resolution. Peak separations of < 0.5Hz may need to be resolved, so the B_o field must be uniform to a very high degree(for a separation of 0.3Hz at 300MHz, field homogeneity must be better than 1 part in 10^9). Correction to field **inhomogeneity** may be made for small gradients in B_o by the use of shim coils. For example, the field along the z direction might be slightly higher at one point than at another. Such a gradient may be compensated for by applying a small current through a shim coil built into the probe. Shim coils are available for correcting gradients in all three cartesian coordinates as well as higher order gradient(x^2, xy, and so on).

The sample is spun along the axis of its cylinder at a rate of 20～50 Hz by an air flow to improve homogeneity within the tube. Spinning improves resolution because a nucleus at a particular location in the tube experiences a field that is averaged over a circular path. In the superconducting magnets, the axis of the tube is in the z direction (in electromagnets it is spun along the y direction). Spinning does not average gradients along the axis of the cylinder, so shimming is required primarily for z gradients for a superconducting magnet or y gradients for an electromagnet.

All magnet are subject to field drift, which can be minimized by electronically locking the field to the resonance of a substance contained in the sample. Because deuterated solvents are used quite commonly for this purpose, an internal lock normally is at the deuterium frequency for both ^{1}H and ^{13}C spectra. In some instruments, the field is locked to a sample contained in a separate tube locked permanently elsewhere in the probe. This external lock is usually found only in spectrometers designed for a highly specific use, such as taking only ^{1}H spectra.

Since there is an excess of only some 50 spin 1/2 nuclei per million, the NMR experiment is inherently insensitive. The sensitivity of a given experiment depends on the natural abundance and the natural sensitivity of the observed nucleus (related to the magnetic moment and **gyromagnetic ratio**). The experimentalist has no control over these factor. **Decoupling** also can bring about sensitivity enhancement through the nuclear **Overhauser** effect. Complex manipulation of pulses can raise sensitivity.

Finally, routine improvement of signal to noise is achieved. It is recorded multiple times and is stored in the same locations. Any signal present is reinforced, but noise tends to be canceled out. If n such scans are carried out and added digitally, the theory of **random** processes states that the signal **amplitude** is proportional to n and the noise is proportional to $n^{1/2}$. The signal to noise ratio (S/N) therefore increase by $n/(n^{1/2})$ or $n^{1/2}$. Thus 100 scans added together theoretically enhance S/N by a factor of about 10. Multiple scanning is routine for most nuclei and necessary for many, such as ^{13}C and ^{15}N.

16. acquisition
[ˌækwɪˈzɪʃən]
获得物,获得

(900MHz, 21.2T NMR Magnet)

In the pulsed experiment, resolution is controlled directly by the amount of time taken to acquire the signal. To distinguish two signals separated by Δv (in Hz), **acquisition** of data must continue for at lease $1/v$ seconds. For example, a desired resolution of 0.5Hz in a ^{13}C spectra requires an acquisition time (t_a) of 1/0.5 or 2.0s. Sample for a longer time would improve resolution, for example, acquisition for 4.0s could yield a resolution of 0.25Hz. Thus longer acquisition times are necessary to produce narrow lines. The tail of the FID contains mostly noise mixed with signals from narrow lines. If sensitivity is the primary concern, either shorter acquisition times should be used or the later portions of the FID may be reduced artificially by a weighting function. In this way noise is reduced, but at the possible expense of lower resolution. Decreased weighting of the data from longer time (from the bottom to the top of the figure) results in lower resolution and broader peaks.

After acquisition is complete, a delay time is necessary to the nuclei to relax before they can be examined again for the purpose of signal averaging. For optimal results, the delay time is on the order of 3~5 times the spin-lattice relaxation time. For a typical ^{13}C relaxation time of 10s, a total cycle thus might take up to 50s. Such cycle times are excessive and can be reduced by using a pulse angle θ that is less than 90°. When fewer spins are **flipped**, T_1 relaxation can occur more rapidly and the delay time can be shortened. In practice, proper balance between θ and T_a (to optimize resolution) can result in essentially no delay time, and the next pulse cycle can be initiated immediately after acquisition.

17. flip [flɪp]
轻击, 掷

18. sinusoidal
正弦曲线的

The detected range of frequencies (the spectra width) is determined by how often the detector samples the value of the FID: the sampling rate. The FID is made up of a collection of **sinusoidal** signals. A signal, specific signal must be sampled at least twice within one sinusoidal cycle to determine its frequency. For a collection of signals up to a frequency of N, the FID thus must be sampled at a rate of $2N$ Hz. For example, for a ^{13}C **spectral** width of 15000(200ppm at 75MHz), the signal must be sampled 30000 times per second.

19. spectral
[ˈspɛktrəl]
光谱的

20. signify
[ˈsɪgnəˌfaɪ]
表示……的意思,
意味

If a signal is sampled 20000 times per second, the detector spends 50μs on each point. The reciprocal of the sampling rate is called the dwell time, which **signifies** the amount of time between sampling. Reducing the dwell time means that more data points are collected in the same period of time, so that a larger computer memory is required. If the acquisition time is 4.0s(for a resolution of 0.25Hz) and the sampling rate is 20000 times per second (for a spectral width of 10000 Hz), the computer must store 80000 data points. Making do with fewer points because of computer limitations would require either lowering the resolution of decreasing the spectral width.

Hands-On Chemistry: The Lemon Battery

Materials: a lemon, a strip of copper, a strip of zinc, a voltmeter, two cables with **alligator** clips, a thermometer or clock with an LCD display.

How to Do: Roll the lemon firmly with the palm of your hand on a tabletop or other hard surface in order to break up some of the small sacks of juice within the lemon. Insert the two metal strips deeply into the lemon, being careful that the strips not touch each other. Using the voltmeter, measure the voltage produced between the two strips (figure). It should show to be about one volt. It would be nice to be able to illuminate a light bulb using your new lemon powered battery, but unfortunately it is not strong enough. If you were to try to light a bulb using this setup, the voltage across the strips would fall immediately to zero. Given this, if you want to demonstrate that the current produced by this battery is capable of powering something, try with a small device that uses an LCD display. A clock or a thermometer usually works well. An LCD display consumes an extremely small amount of current and your lemon battery is able to adequately drive this type of device. Remove any conventional battery that is in your clock or thermometer and power it with your lemon battery. You should see the device **recommence** functioning normally. If not, try **swapping** the polarity of the electricity from your lemon battery. This system allows you to demonstrate that the battery is producing energy even if you don't have a voltmeter.

21. alligator
['ælɪˌgetɚ,ˌæləˈgetɚ]
鳄口钳

The Lemon Battery

22. recommence
重新开始

23. swap [swɑp]
交换

Hands-On Chemistry Insights: The Lemon Battery

How does this battery work? The copper (Cu) atoms attract electrons more than the zinc (Zn) atoms. If you place a piece of copper and a piece of zinc in contact with each other, many electrons will pass from the zinc to the copper. As they concentrate on the copper, the electrons repel each other. When the force of repulsion between electrons and the force of attraction of electrons to the copper become equalized, the flow of electrons stops. Unfortunately there is no way to take advantage of this behavior to produce electricity because the flow of charges stops almost immediately. On the other hand, if you bathe the two strips in a conductive solution, and connect them externally with a wire, the reactions between the electrodes and the solution furnish the circuit with charges continually. In this way, the process that produces the electrical energy continues and becomes useful. As a conductive solution, you can use any electrolyte, whether it be an acid, base or salt solution. The lemon battery works well because the lemon juice is acidic. Try the same setup with other types of solutions. As you may know, other fruits and vegetables also contain juices rich in ions and are therefore good electrical conductors. You are not then, limited to using lemons in this type of battery, but can make batteries out of every type of fruit or vegetable that you wish. Like any battery, this type of battery has a limited life. The electrodes undergo chemical reactions that block the flow of electricity. The electromotive force diminishes and the battery stops working. Usually, what happens is the production of hydrogen at the copper electrode and the zinc electrode acquires deposits of oxides that act as a barrier between the metal and the electrolyte. This is referred to as the electrodes being polarized. To achieve a longer life and higher voltages and current flows, it is necessary to use electrolytes better suited for the purpose. Commercial batteries, apart from their normal electrolyte, contain chemicals with an affinity for hydrogen which combine with the hydrogen before it can polarize the electrodes.

Exercises

1. Translate the following from Chinese into English.
   (1) 探针   (2) 优化   (3) 控制台   (4) 核磁共振
   (5) 质谱   (6) 信号   (7) 灵敏性   (8) 分辨率
   (9) 检测器   (10) 分峰   (11) 化学位移   (12) 回旋磁比
2. Why do most research grade NMR machines today use a superconducting magnet in stead of electromagnets that a early generation of chemists used?
3. The NMR experiment is often plagued by a dual problem, reading the text and answer what is the problem?
4. What is mass spectrometry?
5. One choice questions.
   (1) Which of the following sub-fields of chemistry deals with using instrumentation and methods to quantify matter?
      (a) analytical chemistry   (b) organic chemistry
      (c) astrochemistry   (d) combustion chemistry
   (2) What type of instrument allows you to determine the mass of a compound?
      (a) a HPLC (High Performance Liquid Chromatography)
      (b) a mass spectrometer   (c) a thermometer   (d) a graduated cylinder
   (3) What type of instrument is effective at separating the components of a mixture based on their affinity for an adsorbent material and uses high pressure pumps?
      (a) mass spectrometer   (b) thermometer   (c) pressure tube   (d) HPLC
   (4) How does the flame method work?
      (a) Each element burns at a different color, so by burning an unknown element we can determine what it is based on the color.
      (b) Each element burns at a different temperature by burning at different temperatures we can isolate the desired compound.
      (c) Each element burns at a different temperature by burning at different temperatures we can identify the unknown compound.
      (d) By timing how long a compound burns we can determine how much of that compound is present.

【Reading Material】

Analytical Chemistry and Society

The Division of Analytical Chemistry (DAC) of the FECS (Federation of European Chemical Societies) defines: "Analytical chemistry is a scientific discipline that develops and applies methods, instruments and strategies to obtain information on the composition and nature of matter in space and time".

EINSTEIN AT THE BANK

Analytical chemistry is an in-between-science using and depending on the laws of chemistry, physics, mathematics, information science and biology. Its aim is to decipher the information hidden in the sample under investigation, not to change this intrinsic information, hence to tell the truth about the composition of the material world. This sounds trivial to a scientist, but it is not in today's complex, complicated technical and environmental matrices where analytical data frequently have to be made available in real time and in situ (in the unchanged matrix). Pressing needs of modern world trade, industry and commerce have led to the creation of national, and more importantly, international bodies for quality assurance, such as EURACHEM and CITAC. These bodies demand that even experienced laboratories prove their technical competence by passing accreditation procedures comprising both technological standards and personal skills from time to time.

Human beings are surrounded by natural and man-made environments. Air, water, soil, rocks, plants, and animals compose the natural environment. The man-made environment, on the other hand, consists of houses, clothes, foods, and all other objects such as tools, goods, books, papers, and so forth, which modern society has created for modern people to conduct their lives. These natural and artificial objects are all made out of matter, which is composed of chemical species. In order to recognize and estimate each material, man needs certain methods. Besides the shape, size, hardness, or color, etc., which are physical properties of the material, analytical chemistry is needed to obtain information on the chemical composition and properties of any of these materials. In other words, analytical chemistry is very close to our modern lives and societies.

In order to better understand the detailed relationship between analytical chemistry and society, we need to consider some of the traditional classifications of analytical chemistry.

(1) The first and most direct classification is with the material itself of which a few examples, corresponding to the material to be analyzed, or the sample material analyses, are water analysis, rock analysis, food analysis and so forth. In the industrial area, steel analysis or iron analysis is fundamentally important in order to ensure a high steel quality, which is the basis of other industries. There have been experts of analysis of steel and iron in big companies around the world from the time when wet chemical analysis was in use.

Clinical analysis is vitally important for health, and a modern hospital must be able to perform reliable blood and urine analysis in order to make a proper diagnosis of the patient. All automated physical analytical instruments are available. Pharmacological analysis and biological analysis are related fields. In recent years, environmental analytical chemistry has become very popular, and air, water, soil, and biological materials are all included for environmental investigation, which is naturally very important for our society.

(2) The classification of chemical species to be analyzed is another item. Again, giving only a few examples, total analysis is used to find all species in the sample, so that the sum of the weight of each component equals the original sample weight. The total analysis of a rock is a good example. Recently, moon rock analysis received concern from a great number of people around the world. However, often some particular constituent(s), element(s) or molecule(s), are required to be analyzed. Elemental analysis is used today to analyze all elements in both organic and inorganic compounds. Analyses for NO_x, SO_x and O_3 give information on air pollution, and PCB analysis and dioxin analysis are also needed to improve our environmental safety. These molecules, in particular, consist of numerous numbers of isomers, some of which are known to be more toxic than others. Radioactive analysis has an obvious importance for our society, as it includes the analysis of ^{90}Sr, ^{137}Cs, ^{235}U, and ^{249}Pu, which are either nuclear fission products or nuclear power materials.

(3) Today analytical chemistry encompasses qualitative analysis and quantitative analysis. The former indicates whether a particular element or compound is in the sample, whereas the latter gives the amount of the species in the sample. The relation of this concept with the society can be found in various occasions. A quantitative description is better suited to, for instance, the distance between two points, the size of a house, the amount of food, the ability of a child, the seriousness of a crime, and so forth, and such quantitative measures are often required in chemistry. When the results cannot be described in accurate figures of mg or mg/L, but may just be described by such words as fairly high, or "very little", we call them semiquantitative results. Such semiquantitative descriptions are frequently found in our daily life.

As mentioned in the previous section, physical methods of analytical chemistry such as spectrochemical analysis have been developed so widely that they are continue to grow further in the future.

Question

What is the thrust of analytical chemistry?

Key Terms to Analytical Chemistry

1. analytical chemistry 分析化学
2. qualitative analysis 定性分析
3. quantitative analysis 定量分析
4. physical analysis 物理分析
5. physico-chemical analysis 物理化学分析
6. instrumental analysis 仪器分析法
7. flow injection analysis; FIA 流动注射分析法
8. sequenctial injection analysis;SIA 顺序注射分析法
9. chemometrics 化学计量学
10. absolute error 绝对误差
11. relative error 相对误差
12. systematic error 系统误差
13. determinate error 可定误差
14. accidental error 随机误差
15. accuracy 准确度
16. precision 精确度
17. deviation 偏差
18. average deviation 平均偏差
19. relative average deviation 相对平均偏差
20. standard deviation; S 标准偏差（标准差）
21. relative standard deviation; RSD 相对标准偏差
22. coefficient of variation 变异系数
23. significant figure 有效数字
24. titrametric analysis 滴定分析法
25. volumetric analysis 容量分析法
26. stoichiometric point 化学计量点
27. equivalent point 等当点
28. charge balance 电荷平衡
29. charge balance equation 电荷平衡式
30. mass balance 质量平衡
31. material balance 物料平衡
32. spectroscopic analysis 光谱分析法
33. atomic emission spectroscopy 原子发射光谱法
34. mass spectroscopy; MS 质谱法
35. shoulder peak 肩峰
36. end absorption 末端吸收
37. blue shift 蓝（紫）移
38. molecular fluorometry 分子荧光分析法
39. mass balance equation 质量平衡式
40. acid-base titrations 酸碱滴定法
41. acid-base indicator 酸碱指示剂
42. mixed indicator 混合指示剂
43. double indicator titration 双指示剂滴定法
44. nonaqueous titrations 非水滴定法
45. protonic solvent 质子溶剂
46. acid solvent 酸性溶剂
47. basic solvent 碱性溶剂
48. compleximetry 配位滴定法
49. metallochromic indicator 金属指示剂
50. oxidation-reduction titration 氧化还原滴定法
51. precipitation titration 沉淀滴定法
52. volumetric precipitation method 容量滴定法
53. argentometric method 银量法
54. gravimetric analysis 重量分析法
55. volatilization method 挥发法
56. solvent extration 溶剂萃取法
57. counter extraction 反萃取
58. partition coefficient 分配系数
59. distribution ratio 分配比
60. ion pair 离子对（离子缔合物）
61. precipitation forms 沉淀形式
62. weighing forms 称量形式
63. hyperchromic effect 增色效应（浓色效应）
64. hypochromic effect 减色效应（淡色效应）
65. band width 谱带宽度
66. signal shot noise 散粒噪声
67. blazed grating 闪耀光栅
68. holographic grating 全息光栅
69. photodiode array detector 光二极管阵列检测器
70. electron donating group 供电子取代基
71. electron with-drawing group 吸电子取代基
72. X-ray fulorometry X射线荧光分析法
73. atomic fluorometry 原子荧光分析法

Unit 5

Biochemistry

Lesson 25 Introduction to Biochemistry

1. biochemistry ['baɪo'kɛmɪstri] 生物化学
2. organism ['ɔrgənɪzəm] 生物,有机体,(尤指)微生物
3. complexity [kəm'plɛksəti] 复杂性
4. genetics [dʒə'nɛtɪks] 遗传学
5. protein ['protin] 蛋白质
6. discipline ['dɪsəplɪn] 学科,纪律
7. molecular biology 分子生物学
8. mechanism ['mɛkənɪzəm] 机理,机制
9. encode [ɪn'kod] 给……加密,将……译成密码
10. nucleic acid [nu'kliɪk'æsɪd] 核酸
11. carbohydrate [,karbo'haɪdret] 碳水化合物,糖类
12. lipid [lɪpɪd] 脂质,类脂
13. harness ['harnɪs] 控制,利用
14. metabolism [mɪ'tæbəlɪzəm] 新陈代谢
15. nutrition [nu'trɪʃən] 营养学,营养品,营养
16. deficiency [dɪ'fɪʃənsi] 缺乏,不足
17. fertilizer ['fɜːtɪlaɪzə] 肥料,化肥
18. selenium [sə'linɪəm] 硒
19. iodine ['aɪədaɪn] 碘

Biochemistry, sometimes called biological chemistry, is the study of chemical processes within and relating to living **organisms**. Biochemical processes give rise to the **complexity** of life. Biochemistry can be divided in three fields: molecular **genetics**, **protein** science and metabolism. Over the last decades of the 20th century, biochemistry has through these three **disciplines** become successful at explaining living processes. Almost all areas of the life sciences are being uncovered and developed by biochemical methodology and research. Biochemistry focuses on understanding how biological molecules give rise to the processes that occur within living cells and between cells, which in turn relates greatly to the study and understanding of tissues, organs, organism structure and function.

Biochemistry is closely related to **molecular biology**, the study of the molecular **mechanisms** by which genetic information **encoded** in DNA is able to result in the processes of life.

Much of biochemistry deals with the structures, functions and interactions of biological macromolecules, such as proteins, **nucleic acids**, **carbohydrates** and **lipids**, which provide the structure of cells and perform many of the functions associated with life. The chemistry of the cell also depends on the reactions of smaller molecules and ions. These can be inorganic, for example water and metal ions, or organic, for example the amino acids, which are used to synthesize proteins. The mechanisms by which cells **harness** energy from their environment via chemical reactions are known as **metabolism**. The findings of biochemistry are applied primarily in medicine, **nutrition**, and agriculture. In medicine, biochemists investigate the causes and cures of diseases. In nutrition, they study how to maintain health wellness and study the effects of nutritional **deficiencies**. In agriculture, biochemists investigate soil and **fertilizers**, and try to discover ways to improve crop cultivation, crop storage and pest control.

Starting Materials: The Chemical Elements of Life

Around two dozen of the 92 naturally occurring chemical elements are essential to various kinds of biological life. Most rare elements on earth are not needed by life (exceptions being **selenium** and **iodine**), while a

Vocabulary	
20. titanium [taɪˈteniəm, tɪ-] 钛	
21. algae [ˈældʒi] 水藻，藻类	
22. boron [ˈbɔrɑn] 硼	
23. polymer [ˈpɒlɪmə] 多聚物，聚合物	
24. terminology [ˌtɜməˈnɑlədʒi] 术语	
25. monomer [ˈmɑnəmə] 单体	
26. dehydration synthesis 脱水合成	
27. aldehyde [ˈældəˌhaɪd] 醛，乙醛	
28. ketone [ˈkiːtəʊn] 酮	
29. catchall [ˈkætʃɔːl] 放各种各样物品的容器	
30. phospholipid [ˌfɒsfəˈlɪpɪd] 磷脂	
31. sphingolipid [ˌsfɪŋɡoˈlaɪpɪd] （神经）鞘脂类	
32. glycolipid [ˌɡlaɪkəˈlɪpɪd] 糖脂，醣脂类	
33. terpenoid [tɜːpinɔɪd] 萜类化合物，类萜	
34. retinoid [ˈretɪnɔɪd] 类维生素A，类视色素	
35. steroid [ˈsterɔɪd] 甾类化合物，类固醇	
36. aliphatic [ˌælɪˈfætɪk] 脂肪族的	
37. aromatic [ˌærəˈmætɪk] 芳香的，芳香族的	
38. physiologic [fɪzɪrlɒdˈʒɪk] 生理学的，生理的	
39. denote [dɪˈnot] 代表，指代，意思是	
40. conformation [ˌkɑnfɔrˈmeʃən] 构造	
41. glutamate [ˈɡlutəˌmet] 谷氨酸盐，谷氨酸酯	
42. neurotransmitter [ˈnʊroˌtrænzmɪtə] 神经传递素	

few common ones (aluminum and **titanium**) are not used.

Most organisms share element needs, but there are a few differences between plants and animals. For example, ocean **algae** use bromine, but land plants and animals seem to need none. All animals require sodium, but some plants do not. Plants need **boron** and silicon, but animals may not (or may need ultra-small amounts).

Biomolecules

The four main classes of molecules in biochemistry (often called biomolecules) are carbohydrates, lipids, proteins, and nucleic acids. Many biological molecules are **polymers**: in this **terminology**, **monomers** are relatively small micromolecules that are linked together to create large macromolecules known as polymers. When monomers are linked together to synthesize a biological polymer, they undergo a process called **dehydration synthesis**. Different macromolecules can assemble in larger complexes, often needed for biological activity.

Carbohydrates are **aldehydes** or **ketones**, with many hydroxyl groups attached, that can exist as straight chains or rings. Two of the main functions of carbohydrates are energy storage and providing structure. Sugars are carbohydrates, but not all carbohydrates are sugars. There are more carbohydrates on earth than any other known type of biomolecule. They are used to store energy and genetic information, as well as play important roles in cell to cell interactions and communications.

Lipids comprises a diverse range of molecules and to some extent is a **catchall** for relatively water-insoluble or nonpolar compounds of biological origin, including waxes, fatty acids, fatty-acid derived **phospholipids**, **sphingolipids**, **glycolipids**, and **terpenoids** (e.g., retinoids and **steroids**). Some lipids are linear **aliphatic** molecules, while others have ring structures. Some are **aromatic**, while others are not. Some are flexible, while others are rigid.

Proteins are very large molecules—macro-biopolymers—made from monomers called amino acids. An amino acid consists of a carbon atom attached to an amino group, —NH_2, a carboxylic acid group, —COOH (although these exist as —NH_3^+ and —COO^- under **physiologic** conditions), a simple hydrogen atom, and a side chain commonly **denoted** as "—R". The side chain "R" is different for each amino acid of which there are 20 standard ones. It is this "R" group that made each amino acid different, and the properties of the side chains greatly influence the overall three-dimensional **conformation** of a protein. Some amino acids have functions by themselves or in a modified form, for instance, **glutamate** functions as an important **neurotransmitter**. Amino acids can be joined via a peptide bond. In this dehydration synthesis, a water molecule is removed and the peptide bond connects the nitrogen of one amino acid's amino group to the carbon of the other's carboxylic acid group. The resulting molecule is called a **dipeptide**, and short stretches of amino acids (usually, fewer than thirty) are called peptides or **polypeptides**. Longer stretches **merit** the title proteins. As an example, the important blood **serum** protein **albumin** contains 585 amino acid residues.

Nucleic acids, so called because of their **prevalence** in cellular nuclei, is the **generic** name of the family of biopolymers. They are complex, high-molecular-weight biochemical macromolecules that can convey genetic information in all living cells and viruses. The monomers are called **nucleotides**, and each consists of three components: a **nitrogenous heterocyclic** base (either a **purine** or a **pyrimidine**), a **pentose** sugar, and a phosphate group.

Relationship to Other "Molecular-Scale" Biological Sciences

Researchers in biochemistry use specific techniques native to biochemistry, but increasingly combine these with techniques and ideas developed in the fields of genetics, molecular biology and **biophysics**. There has never been a hard-line among these disciplines in terms of content and technique. Today, the terms molecular biology and biochemistry are nearly interchangeable. The following figure is a schematic that depicts one possible view of the relationship between the fields.

Biochemistry is the study of the chemical substances and vital processes occurring in living organisms. Biochemists focus heavily on the role, function, and structure of biomolecules. The study of the chemistry behind biological processes and the synthesis of biologically active molecules are examples of biochemistry.

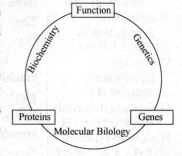

Genetics is the study of the effect of genetic differences on organisms. Often this can be inferred by the absence of a normal component (e.g. one gene), in the study of "**mutants**"——organisms with a changed gene that leads to the organism being different with respect to the so-called "wild type" or normal **phenotype**. Genetic interactions (**epistasis**) can often confound simple interpretations of such "knock-out" or "knock-in" studies.

Molecular biology is the study of molecular **underpinnings** of the process of replication, transcription and translation of the genetic material. The central **dogma** of molecular biology where genetic material is transcribed into RNA and then translated into protein, despite being an **oversimplified** picture of molecular biology, still provides a good starting point for understanding the field. This picture, however, is undergoing revision in light of emerging novel roles for RNA.

Chemical biology seeks to develop new tools based on small molecules that allow minimal **perturbation** of biological systems while

64. hybrid ['haɪbrɪd] 杂交体	
65. viral capsid 病毒衣壳	
66. margarine ['mɑrdʒərən] 人造黄油	
67. sizzle ['sɪzəl] 发咝咝声	
68. griddle ['grɪdl] 煎饼用浅锅	
69. eye dropper 吸管	
70. consistency [kən'sɪstənsi] 黏稠度	
71. saturated fats 饱和脂肪	
72. whip [hwɪp, wɪp] 搅拌	

providing detailed information about their function. Further, chemical biology employs biological systems to create non-natural **hybrids** between biomolecules and synthetic devices (for example emptied **viral capsids** that can deliver gene therapy or drug molecules).

Hands-On Chemistry: Sizzle Sources

Did you ever notice that butter and **margarine sizzle** on a hot **griddle** but vegetable oil does not? The sizzling is the sound of water boiling away rapidly as the butter or margarine hits the hotter-than-100℃ griddle. The sizzling subsides once the water is gone. Vegetable oil contains no appreciable quantities of water, and so it does not sizzle. Different brands of margarine contain different proportions of water, which is the subject of this investigation.

What You Need Several brands of margarine(be sure to include a number of "light" spreads), series of same-sized drinking glasses, microwave oven, kitchen baster or **eye dropper**.

Procedure

(1) Place each margarine sample in a separate glass. Add enough margarine so that it is at least 0.5 inch deep.

(2) Label each glass with the brand it contains.

(3) Melt all the samples in the microwave oven. (Watch carefully because this does not take long.) As the margarine melts, the water and lipid layers separate.

(4) Note the relative water content of the various brands by comparing the depths of the water layers.

(5) Use the eye dropper or kitchen baster to pull off only the water layer, which will be beneath the lipid layer. Cool the lipid layers in the refrigerator, and then look for differences in **consistencies**.

Based on the consistencies you noted in step 5, which sample do you suppose contains the greatest proportion of saturated fats?

Hands-On Chemistry Insights: Sizzle Sources

With the chilled lipid layers, you can assume that, in general, the more solid the sample, the higher its proportion of **saturated fats**. As you should have discovered from this activity, the "light" brands of spread contain fewer calories simply because they contain a greater proportion of water. Rather than water, some brands **whip** air into the spread. Either way, the net result is fewer lipid molecules per serving, which for saturated fats is not too bad a deal. Note that many of the "light" brands are labeled "for spread purposes only, not for cooking". Why do you suppose this is so?

Exercises

1. Translate the following from Chinese into English.
   - (1) 分子遗传学
   - (2) 分子生物学
   - (3) 营养不良
   - (4) 脱水合成
   - (5) 氨基酸
   - (6) 三维构象
   - (7) 肽键
   - (8) 核酸

2. Multiple choice.

   (1) Enzymes are usually _____.
   - (a) carbohydrates
   - (b) lipids
   - (c) proteins
   - (d) All of the above

   (2) Where do you find enzymes _____?
   - (a) digestive system
   - (b) blood clots
   - (c) liver
   - (d) All of the above

   (3) All of the following are carbohydrates except _____.
   - (a) starch
   - (b) glycogen
   - (c) chitin
   - (d) cholesterol

   (4) The structure CH_3COCH_3 contains which functional group?
   - (a) aldehyde
   - (b) ketone
   - (c) amino
   - (d) carbonyl

   (5) Fatty acids that are unsaturated have _____.
   - (a) an amino group
   - (b) a double bond
   - (c) an excess of protons
   - (d) a carboxyl group

   (6) Which of the following can have a quaternary structure?
   - (a) fatty acid
   - (b) protein
   - (c) polysaccharide
   - (d) DNA

   (7) An organic compound is one that _____.
   - (a) contains carbon
   - (b) is slightly acidic
   - (c) forms long chains
   - (d) is soluble in water

   (8) Which of the following elements is the least abundant in living organisms?
   - (a) oxygen
   - (b) nitrogen
   - (c) phosphorous
   - (d) sodium

   (9) Carbon can form _____ separate bonds with other elements?
   - (a) 1
   - (b) 2
   - (c) 3
   - (d) 4

   (10) The cohesion of water is caused by _____.
   - (a) ionic bonds
   - (b) hydrophobic compounds
   - (c) hydrogen bonds
   - (d) covalent bonds

   (11) One function of a carbohydrate is _____.
   - (a) to provide the body with immediate energy
   - (b) keep the heart functioning smoothly
   - (c) store and transport genetic material
   - (d) control the rate of reactions

   (12) The three types of carbohydrates are _____.
   - (a) monosaccharide, polysaccharide, disaccharide
   - (b) glycerol, polysaccharide, monosaccharide
   - (c) disaccharide, monosaccharide, glycerol
   - (d) glycerol, monosaccharide, polysaccharide

   (13) What elements make up a carbohydrate?
   - (a) hydrogen, calcium, oxygen
   - (b) hydrogen, oxygen, carbon
   - (c) carbon, potassium, oxygen
   - (d) calcium, potassium, oxygen

   (14) Glycogen, a polysaccharide, in your liver may be broken down to glucose by the process of _____.
   - (a) hydrolysis
   - (b) dehydration synthesis

(c) condensation (d) isomerization
(15) Large molecules that form when many monosaccharides bonded together are _____.
  (a) calcium (b) sugars (c) monosaccharides (d) polysaccharides
(16) The carbohydrate that provides support in plants is called _____.
  (a) chitin (b) cellulose (c) dextrose (d) lipids
(17) Single sugars, called monosaccharides supply _____ to cells.
  (a) energy (b) health (c) calcium (d) hydrolysis
(18) Which of the following is a carbohydrate?
  (a) DNA (b) insulin (c) wax (d) sucrose
(19) Carbohydrates and lipids have many carbon-hydrogen bonds, therefore they both _____.
  (a) store energy in these bonds (b) dissolve in water
  (c) dissolve in salts (d) are similar to water molecules
(20) Which of the following is an organic molecule?
  (a) water (b) ice (c) nitrogen (d) carbohydrates
(21) A substance needed by the body for growth, energy, repair and maintenance is called a _____.
  (a) nutrient (b) carbohydrate (c) calorie (d) fatty acid
(22) All of the following are nutrients found in food except _____.
  (a) plasma (b) proteins (c) carbohydrates (d) vitamins

3. A topic for speaking.
What does biochemistry deal with?

【Reading Material】

Ada Yonath

Ada E Yonath [ˈada joˈnat] (born 22 June 1939) is an **Israeli**(以色列的) **crystallographer** (晶体学家) best known for her pioneering work on the structure of the **ribosome**(核糖体). She is the current director of the Helen and Milton A Kimmelman Center for Biomolecular Structure and Assembly of the **Weizmann Institute of Science**(维茨曼科学研究所).

In 2009, she received the Nobel Prize in chemistry along with Venkatraman Ramakrishnan and Thomas A Steitz for her studies on the structure and function of the ribosome, becoming the first Israeli woman to win the Nobel Prize out of ten Israeli Nobel **laureates**(得主), the first woman from the middle east to win a Nobel Prize in the sciences, and the first woman in 45 years to win the Nobel Prize for chemistry.

Yonath (née Lifshitz) was born in the Geula quarter of **Jerusalem**(耶路撒冷). Her parents, Hillel and Esther Lifshitz, were **Zionist Jews**(犹太人中的复国主义者)who immigrated to **Palestine**(巴勒斯坦) from Zduńska Wola, Poland in 1933 before the establishment of Israel. Her father was a **rabbi**(犹太教祭司) and came from a rabbinical family. They settled in Jerusalem and ran a **grocery**(杂货店), but found it difficult to **make ends meet**(使收支相抵). They **lived in cramped quarters**(居处褊狭)with several other families, and Yonath remembers "books" being the only thing she had to keep her occupied. Despite their **poverty**(贫困), her parents sent her to

155

school in the **upscale**(高档的) Beit Hakerem neighborhood to assure her a good education. When her father died at the age of 42, the family moved to **Tel Aviv**(特拉维夫). Yonath was accepted to Tichon Hadash high school although her mother could not pay the tuition. She gave math lessons to students in return. As a youngster, she says she was inspired by the Polish and naturalized-French scientist **Marie Curie**(玛丽·居里). However, she stresses that Curie, whom she as a child was fascinated by after reading a well-written biography, was not her "role model". She returned to Jerusalem for college, graduating from the **Hebrew University of Jerusalem**(耶路撒冷希伯来大学) with a **bachelor's degree**(学士学位) in chemistry in 1962, and a **master's degree**(硕士学位) in biochemistry in 1964. In 1968, she obtained her PhD from the Weizmann Institute of Science for X-ray crystallographic studies on the structure of collagen, with Wolfie Traub as her PhD advisor.

She has one daughter, Hagit Yonath, a doctor at Sheba Medical Center, and a granddaughter, Noa. She is the cousin of anti-occupation activist Dr Ruchama Marton.

She has called for the unconditional release of all Hamas prisoners, saying that "holding Palestinians **captive**(俘虏) encourages and **perpetuates**(使持续) their motivation to harm Israel and its citizens… once we don't have any prisoners to release, they will have no reason to kidnap soldiers".

Questions

1. What was Ada E Yonath awarded the 2009 Nobel Prize in chemistry for?
2. What method did Yonath take in her studies on the structure and function of the ribosome?

Key Terms to Biochemistry

1. adenine ['ædn,in, -ɪn] 腺嘌呤
2. adenosine diphosphate 二磷酸腺苷
3. adenosine triphosphate 三磷酸腺苷
4. albumin [æl'bjumɪn] 清蛋白，白蛋白
5. alanine ['ælə,nin] 丙氨酸
6. amino acid 氨基酸
7. anabolism [ə'næbə,lɪzəm] 合成代谢
8. antibody ['æntɪ,bɑdi,æntaɪ-] 抗体
9. arginine ['ɑrʒə,nin] 精氨酸
10. aspartic acid 天冬氨酸
11. vitamin ['vaɪtəmɪn] 维生素
12. carbohydrate [,kɑrbo'haɪ,dret] 碳水化合物
13. catabolism [kə'tæbə,lɪzəm] 分解代谢
14. cholesterol [kə'lɛstə,rɔl, -,rol] 胆固醇
15. chromosome ['kromə,som] 染色体
16. clone [klon] 克隆
17. collagen ['kɑlədʒən] 胶原蛋白
18. triglyceride [traɪ'glɪsəraɪd] 甘油三酸酯
19. thymine ['θaɪmɪn] 胸腺嘧啶
20. cysteine ['sɪstə,in, -ɪn, -ti-] 半胱氨酸
21. cytoplasm ['saɪtə,plæzəm] 细胞质
22. cytosine ['saɪtə,sin] 胞核嘧啶，氧嘧啶
23. deoxyribonucleic acid 脱氧核糖核酸
24. deoxyribose [di,ɑksɪ'raɪ,bos] 去氧核糖
25. endocrine ['ɛndəkrɪn] 内分泌(腺)的，激素的
26. enzyme ['ɛnzaɪm] 酶
27. epistasis [ɪ'pɪstəsɪs] 上位，异位显性
28. fibrinogen [faɪ'brɪnədʒən] 纤维蛋白原
29. fructose ['frʌk,tos,'fruk-] 果糖
30. fungus ['fʌŋɡəs] 真菌
31. gene [dʒin] 基因
32. genetics [dʒə'nɛtɪks] 遗传学
33. gibberellin [,dʒɪbə'rɛlɪn] 赤霉素
34. threonine ['θriənin] 苏氨酸
35. glutamic acid 谷氨酸
36. glycine ['glaɪ,sin, -sɪn] 甘氨酸，氨基醋酸
37. glycolysis [glaɪ'kɑləsɪs] 糖酵解
38. guanine ['gwɑ,nin] 鸟嘌呤 (核酸的基本成分)
39. hormone ['hɔr,mon] 荷尔蒙，激素
40. hybrid ['haɪbrɪd] 杂交植物(或动物)
41. E.coli(Escherichia coli) 大肠杆菌
42. uracil ['jurəsəl] 尿嘧啶
43. inheritance [ɪn'hɛrɪtəns] 遗传
44. immunity [ɪ'mjunɪti] 免疫
45. isoleucine [,aɪsə'lu,sin] 异亮氨酸
46. keratin ['kɛrətɪn] 角蛋白
47. leucine ['lu,sin] 白氨酸，亮氨酸
48. lipid ['lɪpɪd,'laɪpɪd] 脂质
49. liposome ['lɪpə,som,'laɪpə-] 脂质体
50. lysine ['laɪ,sin, -sɪn] 赖氨酸
51. macromolecule [,mækro'mɑlɪ,kjul] 大分子
52. methionine 蛋氨酸，甲硫氨酸
53. metabolism [mɪ'tæbə,lɪzəm] 新陈代谢
54. monomer ['mɑnəmɚ] 单体
55. mutant ['mjutnt] 突变体
56. mutation [mju'teʃən] 突变
57. nucleic acid 核酸
58. nucleotide ['nukliə,taɪd,'nju-] 核苷酸
59. oligomer [ə'lɪɡəmɚ] 低聚物，低聚体
60. peptide bond 肽键
61. variation [,vɛri'eʃən] 变异
62. phenotype ['finə,taɪp] 显型
63. phenylalanine [,fɛnəl'ælə,nin,,finəl-] 苯基丙氨酸
64. phospholipid [,fɑsfo'lɪpɪd] 磷脂
65. serine ['sɛrin] 丝氨酸
66. polypeptide [,pɑli'pɛp,taɪd] 多肽
67. polysaccharide [,pɑli'sækə,raɪd] 多糖，多聚糖
68. polynucleotide [,pɑli'nukliə,taɪd, -'nju-] 多核苷酸
69. primary structure 一级结构，主(要)结构
70. protease ['proti,es, -,ez] 蛋白酶
71. protein ['pro,tin, -tiɪn] 蛋白质
72. transcription [træn'skrɪpʃən] 转录
73. replication [,rɛplɪ'keʃən] 复制
74. ribonucleic acid 核糖核酸(略作 RNA)
75. ribose ['raɪ,bos] 核糖
76. ribosome ['raɪbə,som] 核糖体

Unit 6

Polymers

Lesson 26 Introduction to Polymers

What Are Polymers?

Polymers are a very important class of materials. Polymers occur naturally in the form of proteins, **cellulose** (plants), **starch** (food) and natural rubber. Engineering polymers, however, are usually synthetic polymers. The field of synthetic polymers or plastics is currently one of the fastest growing materials industries. The interest in engineering polymers is driven by their manufacturability, recyclability, mechanical properties, and lower cost as compared to many alloys and **ceramics**. Also the macromolecular structure of synthetic polymers provides good biocompatibility and allows them to perform many **biomimetic** tasks that cannot be performed by other synthetic materials, which include drug delivery, use as grafts for **arteries** and **veins** and use in artificial **tendons**, **ligaments** and joints.

A polymer is a material whose molecules contain a very large number of atoms linked by covalent bonds, which makes polymers macromolecules. Polymers consist mainly of identical or similar units joined together. The unit forming the repetitive pattern is called a "mer" or "monomer". Usually the biggest differences in polymer properties result from how the atoms and chains are linked together in space. Polymers that have a 1-D structure will have different properties than those that have either a 2-D or 3-D structure.

Polymer Structures

One-Dimensional Polymers

One dimensional polymers are most common. They can occur whenever 2 reacting chains join to make a chain. If the long-chains pack regularly, side-by-side, they tend to form **crystalline** polymers. If the long chain molecules are irregularly **tangled**, the polymer is **amorphous** since there is no long range order. Sometimes this type of polymer is called glassy.

Word list:

1. cellulose ['sɛljə,los] 纤维素
2. starch [stɑrtʃ] 淀粉
3. ceramics 陶瓷
4. biomimetic 仿生化(技术)的
5. artery ['ɑrtəri] 动脉
6. vein [ven] 静脉
7. tendon ['tɛndən] 筋,腱
8. ligament ['lɪgəmənt] 韧带
9. crystalline ['krɪstəlɪn] 结晶(体)的
10. tangle ['tæŋgəl] 缠结
11. amorphous [ə'mɔrfəs] 无定形的

Vocabulary	
12. graphite ['græf,aɪt] 石墨	
13. lubricate ['lubrɪ,ket] 使润滑	
14. shear strength 切变强度	
15. lattice ['lætɪs] 晶[栅]格	
16. thermoplastic [,θɚmə'plæstɪk] 热塑性的	
17. remold 改造, 改铸	
18. thermosetting ['θɚmo,sɛtɪŋ] 热固(的), 热硬性(的)	
19. crosslink ['krɔs'lɪŋk, 'krɑs-] 交联	
20. long-range order 长程有序	
21. epoxy [ɪ'pɑksi] 环氧树脂的	
22. viscous ['vɪskəs] 黏性的, 黏滞的	
23. elasticity [ɪlæ'stɪsɪti] 弹性, 弹力	
24. glass transition temperature 玻璃化转变温度	

Two-Dimensional Polymers

Two dimensional polymers are rare, the best example of one would be **graphite**. It is the structure of graphite which provides its great **lubricating** capability. The condition to form this planar structure is to have 3 or more active groups all directed in the same plane and capable of forming a planar network. This structure offers low **shear strength** and good lubricating properties.

Three-Dimensional Polymers

Crystalline diamond is an example of the 3-dimensional crystalline polymer in which carbon is linked to four corners of the tetrahedra and these are packed with long range order in space to form a **lattice**. Diamond has properties which are much more like ceramics than polymers in terms of mechanical behavior (high melting point, modulus, hardness, strength, and fracture behavior) because of this.

Classification of Polymers

Thermoplastic Polymers

These are linear, one-dimensional polymers which have strong intramolecular covalent bonds and weak intermolecular Van Der Waals bonds. At elevated temperature, it is easy to "melt" these bonds and have molecular chains readily slide past one another. These polymers are capable of flow at elevated temperatures, can be **remolded** into different forms, and in general, are dissolvable. A thermoplastic, under the application of appropriate heat, can be melted into a "liquid" state.

Thermosetting Polymers

These are three-dimensional amorphous polymers which are highly **crosslinked** (strong, covalent intermolecular bonds) networks with no **long-range order**. Thermosetting polymers are those resins which are "set" or "polymerized" through a chemical reaction resulting in crosslinking of the structure into one large 3-dimensional molecular network. Once the chemical reaction or polymerization is complete, the polymer becomes a hard, infusible, insoluble material which cannot be softened, melted or molded non-destructively. A good example of a thermosetting plastic is a two-part **epoxy** systems in which a resin and hardener (both in a **viscous** state) are mixed and within several minutes, the polymerization is complete resulting in a hard epoxy plastic.

Rubbers and Elastomers

In general, a rubber material is one which can be stretched to at least twice its original length and rapidly contract to its original length. Rubber must be a high polymer (polymers with very long chains) as rubber **elasticity**, from a molecular standpoint, is due to the coiling and uncoiling of very long chains. To have "rubber-elastic properties" a rubber materials' use temperature must be above its **glass transition temperature** and it

25. aspect ratio 纵横比
26. denier 旦尼尔（用于测量丝绸、人造丝或尼龙丝等纤度的重量单位，长9000m的丝重1g为1旦尼尔，常用于表示袜子的厚度）
27. tenacity [ti'næsɪti] 韧性，韧度
28. spinneret [ˌspɪnə'rɛt] 喷丝头
29. spherulite ['sfɪrjəˌlaɪt] 球粒
30. composite [kəm'pazɪt] 复合材料
31. reinforcement [ˌriɪn'fɔrsmənt] 强化
32. parallel ['pærəˌlɛl] 平行的
33. entanglement 缠结

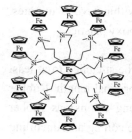

(A 54-ferrocene dendrimer)

must be amorphous in its unstretched state since crystallinity hinders coiling and uncoiling. Rubbers are lightly crosslinked in order to prevent chains from slipping past one another under stress without complete recovery. "Natural rubber" is a thermoplastic, and in its natural form it becomes "soft" and "sticky" on hot days (not a good property for an automobile tire). In fact, until Mr Goodyear discovered a curing reaction with sulfur in 1839, rubbers were not crosslinked and did not have unique mechanical, rubber-elastic properties.

Fibers

Many of the polymers used for synthetic fibers are identical to those used in plastics but the two industries developed separately and employ different testing methods and terminology. A fiber is often defined as having an **aspect ratio** (length/diameter) of at least 100. Synthetic fibers are spun into continuous filaments, or chopped in shorter staple which are then twisted into thread before weaving. The thickness of the fiber is expressed in terms of **denier** which is the weight in grams of a 9000-m length of fiber. Stresses and strength of fibers are reported in terms of **tenacity** in units of grams/denier. In melt spinning, polymer pellets are gravity fed into an extruder and subjected to shear loading at elevated temperatures. The softened polymer is delivered to the **spinneret** which has up to 1000 shaped holes for fiber formation. A molten stream of polymer is forced by pressure through shaped holes and stretched into a solid state. Then the polymer is stretched to have molecular alignment along the axial direction and crystallized in a preferred direction so that no **spherulites** form. Synthetic fibers include Kevlar, carbon, PE, PTFE, and nylon while natural fibers include silk, cotton, wool and wood pulp.

Liquid Crystals

The structure of liquid crystals such as Spectra 1000 is unique. It is a near-ideal in structure with most of its molecules virtually stretched out. While this is not useful for textiles, it is excellent for **composite reinforcement**. Continuous crystals are readily attained in liquid crystalline polymers, as the molecules are already aligned in **parallel** positions in the melt whereas the continuous morphology of PE would require elaborate processing to avoid chain **entanglement** and chain folding.

Polymer Synthesis

Step Growth

In step growth polymerization or condensation polymerization, chains of any length x and y combine to form long chains:

$$x-\text{mer} + y-\text{mer} = (x+y)-\text{mer}$$

Synthetic step growth polymers have been around since the late 1920s. The term condensation polymerization comes from this time period, because the early reaction yielded water. More recently it has been found that several other polymer reactions will condense out products such as HCl or will condense out no products.

Step growth polymerization is used to yield branched as well as networked polymers. The true initiator of a network structure lines in a multifunctional monomer. As the length and frequency of branches on the polymer chain increases so does the probability that the branches will reach from chain to chain. When all the chains are connected together the entire polymer mass becomes one giant molecule. Consider a **bowling ball**, it has a molecular weight on the order of 10^{27} g/mol, and it is one giant molecule. Crosslinked or network polymers form in one of two ways:

(1) Starting with tri functional (or higher) monomers
(2) Chemically creating crosslinks from previous linear polymers

Chain Growth

Chain growth polymerization involves an active chain site which reacts with an unsaturated (or **heterocyclic**) monomer such that the active site is recovered at the chain end. Vinyl polymers were probably the first to be synthesized in this manner when in 1839 Simon reported the conversion of styrene to some **gelatinous** mass. In general, high molecular weight species are made by the successive addition of new monomers to the chain end. There are several different techniques of synthesis depending on the active site. These are:

Active site	Synthesis technique
Free radicals	Free radical polymerization
Carbanions	Anionic polymerization
Carbenium ions	Cationic polymerization
Coordination bonds with transition metals	Ziegler-Natta polymerization

Transitions in Polymers

Amorphous polymers exhibit two distinctly different types of mechanical behavior. Some, like **polymethyl methacrylate** and **polystyrene** are hard, rigid, glassy plastics at room temperature while others like polybutadiene and polyethyl acrylate are soft, flexible, rubbery materials at room temperature.

There is a temperature, or range of temperatures, below which an amorphous polymer is in a glassy state and above which it is rubbery. This temperature is called the glass transition temperature, T_g, and it characterizes the amorphous phase. It is especially useful since all polymers are amorphous to some degree, they all have a T_g.

Molecular Motions in Amorphous Polymers

The molecular motions occurring inside an amorphous polymer influence the glass transition temperature. The important motions are:

(1) **Translation motion** of entire molecules (permits flow).

(2) Cooperative **wriggling** and jumping of segments of molecules (permits flexing and uncoiling leading to elasticity).

(3) Motions of a few atoms along the main chain or side groups on the main chain.

(4) Vibrations of atoms about equilibrium position.

The glass transition temperature is the temperature at which there is only enough energy for motions (3) and (4) to occur. Below the glass transition temperature processes (1) and (2) are frozen out. This makes the material "glassy" below T_g and "rubbery" above T_g.

Factors Influencing Glass Transition Temperature

In general the glass transition temperature depends on five other factors which are:

(1) **Free volume** of the polymer v_f, which is the volume of the polymer mass not actually occupied by the molecules themselves. The higher v_f is, the more room the molecules have to move around and the lower T_g is. For all polymers the ratio of the free volume vs. the total volume (v_f/v) is about 0.025 at T_g.

(2) The attractive forces between the molecules. The more strongly the molecules are bound together, the more thermal energy must be applied to produce motion.

(3) The internal mobility of the chains, or their freedom to rotate about the bonds.

(4) The stiffness of the chains. Stiff chains cannot easily coil and fold, causing T_g to be higher for polymers with stiff chains. Polymers with parallel bonds in the backbone, like **polyimide**, and polymers with highly aromatic backbones have extremely stiff chains and thus high T_g's.

(5) The chain length. The glass transition temperature varies according to the relation:

$$T_g = T_{inf} - C/x$$

Where C is a polymer specific constant, T_{inf} is the **asymptotic** value of the glass transition temperature for a chain of length infinity and x is the length of the chain. This relationship shows that shorter chains can move easier than longer chains. For most commercial polymers $T_g \sim T_{inf}$, since x is quite long.

Determination of the Glass Transition Temperature

The most common method used to determine T_g is to observe the variation of a **thermodynamic** property with T. T_g determined in this

---

40. translation motion
平移

41. wriggling 扭曲

42. free volume
自由体积

43. polyimide
[ˌpɑliˈɪmˌaɪd]
聚酰亚胺

44. asymptotic
渐近线的

45. thermodynamic
[ˌθɚmodaɪˈnæmɪk]
热力学的

manner will vary somewhat depending on the rate of cooling or heating, which reflects the fact that long entangled polymer chains cannot respond instantaneously to changes in temperature.

At the glass transition temperature a thermodynamic property will exhibit a discontinuity with temperature, thus it is classified as a second order thermodynamic transition.

Interesting Tidbits: About Polymers

One of the first developments of plastics was as a replacement to **ivory billiard balls**, due to the **dwindling** supply of ivory. As far back as 1866, elephants were being slaughtered at an alarming rate to keep up with the demand for ivory billiard balls, billiards having become America's favorite pastime. John Wesly Hyatt, invented the replacement, one of the first plastics, called **celluloid**, which was used in movies for a short time.

An important plastics inventor was Charles Macintosh, who developed a plastic coated waterproof fabric in the mid 19th century and gaves is named to the coat. In Britain, people still refer to their raincoats as "Macs".

The inventor of the first synthetic plastic, **bakelite**, was Leo Bakeland. Time magazine called him "The King of Plastics" and put him on the cover of its September 22, 1924 issue. It dubbed Bakelite "The material with a thousand uses".

In the 1940's a vinyl based material commonly known as Saran Wrap started as a furniture protector. During World War II defense contractors found more important uses for the plastic. Saran film "sprayed" on plane being shipped overseas protected them from the salty sea spray, which meant that the process of disassembling, greasing at the point of origin and then cleaning and reassembling at the destination were avoided. After the war there was an oversupply of the film and resulted in it being marketed as the self cling food wrap we find in kitchens everywhere.

Also during the 1940's started the craze for nylon stockings, which were rumored to be so strong that only an acetylene torch could but a run in them. Manufactures claimed that the nylon stockings were as "strong as steel" and lead many women to believe that the hose were **impervious** to razor blades and nail files. Nylons were so popular that women stampeded to get even one pair.

One of the most well known plastics of today, GORE-TEX, which is sued to make extremely warm jackets and shoes, is used extensively in **vascular** grafts and patches for heart defects and **hernias**.

46. tidbit ['tɪd,bɪt]
    趣闻
47. ivory billiard balls
    象牙台球
48. dwindling
    减少
49. celluloid ['sɛljə,lɔɪd]
    赛璐珞 (明胶), 假象牙
50. bakelite ['bekə,laɪt]
    酚醛塑料
51. impervious [ɪm'pɜvɪəs]
    不能渗透的, 透不过的
52. vascular ['væskjələ]
    血管的
53. hernias
    疝, 脱肠, (内脏) 突出

Exercises

1. Please answer the following questions.
   (1) What kind of material is defined as polymer?
   (2) What kind of forms can polymers exist naturally?
   (3) Why does graphite have low shear strength and god lubricating properties?
   (4) What is the main difference between thermoplastic polymers and thermosetting polymers?
   (5) What kind of material is defined as rubber material? Why does it have rubber elasticity?
   (6) Why were rubbers not cross-linked and did not have unique mechanical, rubber-elastic properties before 1839?
   (7) What kind of material is defined as fiber?
   (8) How to prepare a giant molecule through polymerization?
   (9) What is glassy transition temperature T_g?
   (10) What factors affect the glass transition temperature?

2. Please put the following words into Chinese.
   (1) thermoplastic polymers   (2) crosslink   (3) mechanical properties
   (4) synthetic polymers   (5) macromolecules   (6) crystallinity
   (7) hardness   (8) textile   (9) morphology
   (10) molecular weight   (11) cationic polymerization

3. Please put the following words into English.
   (1) 蛋白质   (2) 单体   (3) 无定形的   (4) 软化   (5) 环氧树脂   (6) 橡胶
   (7) 高分子链   (8) 自由基聚合反应   (9) 线性高分子   (10) 熔融纺丝   (11) 液晶

【Reading Material】

The History of Polymers

Long-Chain Molecules Which Have Become Indispensable in Modern Life

How did we come to depend on plastic, Teflon, Nylon and Lycra? Modern life would be incomparably different without synthetic chemicals called polymers. Man-made fibres, as used in clothing, carpets and curtains, plastics used in innumerable domestic and industrial applications and artificial joints, and paints and cleaning materials, are all different forms of this important discovery. What's often forgotten is that at the beginning of the 20th century the chemistry of large molecules was unknown and their synthesis unthinkable.

Large Molecules

When a German scientist named Hermann Staudinger proposed in the 1920s that it was possible to have large molecules made up of many thousands of atoms, he was ridiculed by many other scientists. The common wisdom was that the structures of such materials as rubber and bakelite were actually many small molecules held together by an unknown force.

Organic Synthesis

Staudinger **stuck to his guns**（固执己见）and, with his colleagues, he synthesized a series of organic molecules called **poly(methanal)s**（聚甲醛）. These compounds were long chains of repeating units, the units being —CH_2O—. They are made by joining lots of methanal molecules together. The German scientists made chains of different lengths and showed that their properties changed depending on the length of the chains.

The next question to answer was: Are polymers any use? Chemists working for Imperial Chemical Industries (ICI), soon discovered a polymer which put the answer to this question beyond any doubt. They were attempting to react to organic molecules, ethylene (now known as ethene) and **benzaldehyde**(苯甲醛), at very high temperature and pressure. The reaction failed to impress, but there was a small amount of a white, waxy substance on the wall of the reaction vessel. This was poly(ethene) or polythene, and soon ICI realised they had a potentially useful compound.

World War II

The new material had many properties which made it stand out. It was easy to form into different items, was tough and hard wearing, was impermeable to water and insulating to electricity. It was discovered in the 1930s and was soon being used in the Second World War to insulate the many metres of cables needed for the vital radar equipment used by the British.

HDPE

The development of the petrochemical industry after the war supplied the raw materials for this product. Then a catalysed method of producing high-density poly(ethene) (HDPE) was discovered by Karl Zeigler in Mülheim, Germany in 1950. Both these developments meant that the world would never be the same again as more and more uses for this plastic were discovered.

Nylon and Lycra [(商)莱克拉(聚氨基甲酸酯纤维)]

Soon other types of polymers were being developed. Some of the more important were poly(propene) by Giulio Natta in Milan, Nylon by Wallace Carothers in the US, poly(urethane)s, used in Lycra, and polyesters, a very important raw material for the clothing industry.

Future

The history of the development of polymers is nowhere near its end. New developments and advances are constantly being made and every year there are new patents for novel molecules which promise much. Only time will tell how we will come to depend even more on these compounds.

Lesson 27 Discovery of Polyethylene and Nylon

1. serendipity
[ˌsɛrən'dɪpɪti]
意外的运气

2. adduct
[ə'dʌkt, æ'dʌkt]
加合物

3. decomposition
[ˌdikɑmpə'zɪʃən]
分解

Polyethylene, or "polyethene", as its British inventors called it, and the nylons are two types of polymers that have particularly interesting histories of discovery that bear repeating here. Both stories illustrate the role of **serendipity** in scientific achievements of great importance.

Polyethylene was discovered accidentally by British chemists at Imperial Chemicals Industries (ICI) as an unexpected result of experiments on chemical reactions at very high pressures. In 1933, a reaction of benzaldehyde and ethylene at 170℃ and 1400 atmospheres gave no **adducts** involving the two reagents and was considered a complete failure. However, an observant chemist noticed a thin layer of "white waxy solid" on the walls of the reaction vessel used for the experiment. This was recognized as a polymer of ethylene, but additional experiments with ethylene alone to produce the same polymer only resulted in violent **decomposition** that destroyed the equipment.

Two years elapsed before better and stronger equipment was available for further experimentation. When ethylene was heat to 180℃ in this new equipment, the pressure in the apparatus dropped unexpectedly, so more ethylene was pumped in. Then, when the reaction vessel was opened, the ICI chemists found a large amount of white powdery solid, which was the long-sought polyethylene. Because they knew that the polymerization could not account for all the pressure drop that has been observed, they suspected a leak in one of the joints of the apparatus. This idea led to the proposal that the polymerization had been catalyzed by oxygen in the air that had leaked into the apparatus, and the hypothesis was confirmed by experiments in which air was intentionally included with the ethylene. Oxygen can act as a radical initiator and catalyze the polymerization by a chain-reaction mechanism.

The polyethylene produced by the oxygen-catalyzed, high-pressure, high-temperature process developed by ICI in the mid 1930s was ideal for many applications, including insulation of radar equipment, where it was used to great advantage by the Allies in World War Ⅱ. Referring to the contribution radar made to naval operations, the British Commander-in-chief said it enable the Home Fleet to "find, fix, fight, and finish the Scharnhorst (the pride of Hitler's navy)".

The group of polymers called nylons was first produced in 1939 as a textile material for women's hose and other garments, but with the onset of World War Ⅱ and the involvement of the US by 1941, nylon was taken off the domestic market because it was found to be the best available material for military parachutes.

The first nylon to be produced industrially was Nylon-66. The remarkable fact about the discovery of this polymer is not how it was first prepared from the two monomers, but how it was first prepared in a form suitable for a textile fiber. This depended on invention of the "**cold-drawing** process", and the technique was discovered almost completely by accident, as we shall see.

Wallace Hume Carothers was brought to Du Pont to direct its new basic chemical research program because his colleagues at the University of Illinois and Harvard University recommended him as the most brilliant organic chemist they knew. Carothers initiated a program aimed at understanding the composition of high polymers of Nature such as cellulose, silk, and rubber, and of producing synthetic materials like them. By 1934 his group had contributed valuable fundamental knowledge in these areas, but Carothers had just about decided that their efforts to produce a synthetic fiber like silk were a failure. It was a shrewd observation made during some "**horseplay**" among Carothers' chemists in the laboratory that turned this failure to complete with Nature into the enormous success ultimately advertised at the 1939 New York World's Fair as "nylon, the synthetic silk made from coal, air, and water!"

The Carothers group had learned how to make Nylon-66, but even though this polyamide had a molecular structure similar to that of silk, they had "put it on the back shelf" without patent protection because the polymer did not have the **tensile strength** of silk, a necessary **criterion** for a good textile fiber. The group continued its research by investigating the polyester series, polymers which were more soluble, easier to handle, and thus simpler to work with in the laboratory. It was while working with one of these softer materials that Julian Hill noted that if he gathered a small ball of such a polymer on the end of a glass **stirring rod** and drew a thread out of the mass, the thread of polymer so produced became very silky in appearance. This attracted his attention and that of others working with him, and it is reported that one day while Carothers was in downtown Wilmington, Hill and his **cohorts** tried to see how far they could stretch one of these samples. One chemist put a little ball of the polymer on a stirring rod and a second chemist touched a glass rod to the polymer ball and then ran down the hall to see how far he could stretch the thread of polymer. While doing this they noticed not only the silky appearance of the extended strands, but also their increased strength. They soon realized that this additional strength might be the result from some special **orientation** of the polymer molecules produced by the stretching procedure.

Because the polyesters they were working with at that time had melting points too low for use in textile products, a deficiency that has since been removed, the researchers returned to the polyamides (nylons) that had earlier been put aside. They soon found that these polymers, too, could be "cold-drawn" to increase their tensile strength so much that they could be made into excellent textiles. **Filaments**, gears, and other modeled objects could also be made from the strong polymer produced by cold-drawing.

The **alignment** of the long polyamide molecules in a manner that produces extensive intermolecular hydrogen-bonding binds the individual polymer molecules together in much the same way that separate **strands** in a rope, when twisted together, from a cable. This association of linear polymer molecules through hydrogen bonding is responsible for the greatly increased strength of the nylon fibers. We believe that the same principle accounts for the strength of silk fibers; the natural polyamide molecules of silk are oriented in such a way that hydrogen bonds hold the individual molecules together. Interestingly, the silkworms accomplish the equivalent of "cold-drawing" as they extrude the viscous silk filaments to produce **cocoons**!

Hands-On Chemistry: Clear Slime Polymer

Many of the materials we use every day, like starch, are made up of molecules called POLYMERS. POLY means "many" and MER means "unit". Because the units of chains are so long, the movement of polymers is restricted. **Viscosity** is a physical property of liquids that describes their rate of flow.

Procedure (1) Pour 100mL of water into a beaker. (2) Add the sodium borate to the water and stir for approximately one minute or until the solid is completely dissolved. (3) **Label** the solution. (4) Pour 80mL of water into the other beaker. (5) Add guar gum to the water while stirring. Continue stirring until the solid is completely dissolved (approximately one minute). (6) Label the solution. (7) Add food coloring of your choice to guar gum solution and stir for one minute. (8) Add 5mL of the sodium borate solution to the guar gum solution. Stir for 1minute, and then let it sit for 2minutes.

Materials	Substitutions
2.46g **sodium borate**	1tsp **borax**
0.63g **guar gum** (1/4 tsp)	
200mL water	5/6 cup of water
100mL **graduated cylinder**	measuring cup

11. filament ['fɪləmənt] 细丝
12. alignment [ə'laɪnmənt] 列队, 成直线
13. strand [strænd] (线、纤维或电线的) 股, 缕, 绞
14. cocoon [kə'kun] 茧, 蚕茧
15. slime [slaɪm] 黏质物, 黏液
16. viscosity [vɪ'skɑsɪti] 黏度
17. sodium borate 硼砂, 硼酸钠
18. guar gum 瓜尔胶
19. graduated cylinder 量筒
20. borax ['bɔrˌæks] 硼砂, 月石

	to be continued	
Materials	Substitutions	
2 ~ 250mL beakers	2 ~ 9 oz plastic cups	
2 stirring rods	2 spoons	
balance		
paper towels		
food coloring		
4 ~ 5 **zip-lock bags** (1 per person)		

21. zip-lock bag 自封袋
22. label ['lebəl]
    给……贴标签

23. stretchy
    ['strɛtʃi]
    （尤指材料或衣服）
    可延伸的，有弹性的

24. non-Newtonian fluid
    非牛顿流体

Extensions (1) Challenge students to modify the basic recipe and demonstrate each resulting product. Can they create a slime that is **stretchy** or one that bounces? (2) Extend the activity into other disciplines by having each team name their new product and create a marketing strategy, including packaging, cost analysis, and advertising. (3) Try the other **non-Newtonian fluids** in this lab manual! This recipe was adapted from a Flinn Scientific publication and an issue of the NSTA Science Scope magazine.

25. colloidal suspension
    胶状悬浮(体)

Teacher's Notes The secret to this **colloidal suspension** is the Guar Gum. It is not available through common sources and must be ordered through a chemical company.

26. diminish
    [dɪ'mɪnɪʃ]
    减少，减小

Disposal This slime can be stored in a Ziploc™ bag so students can take it with them. Its unusual properties will **diminish** over time as it dries out.

Exercises

1. Please answer the following questions.
   (1) Why polyethylene could be synthesized in a leaked apparatus?
   (2) What happened to Nylon in World War Ⅱ?
   (3) Why "cold-drawing process" can increase the strength of Nylon?
   (4) Write the structure of polystyrene and Nylon-66.
2. Please translate the following words into Chinese.
   (1) polyethylene   (2) benzaldehyde   (3) radical initiator   (4) synthetic silk
   (5) tensile strength (6) polyester   (7) cold-drawing process (8) filament
   (9) polyamide   (10) hydrogen bond

【Reading Material】

Biopolymers

Biopolymers (also called renewable polymers) are polymers produced by living organisms, in other words, they are polymeric biomolecules. There are three main classes of biopolymers, classified according to the monomeric units used and the structure of the biopolymer formed: polynucleotides, polypeptides, and polysaccharides.

Biopolymers produced from biomass are being used in the packaging industry. Biomass comes from crops such as sugar beet, potatoes or wheat. These can be converted in the following pathways:

Sugar beet > Glyconic acid > Polyglyconic acid

Starch > (fermentation) > Lactic acid > Polylactic acid (PLA)

Biomass > (fermentation) > Bioethanol > Ethene > Polyethylene

Many types of packaging can be made from biopolymers: food trays, blown starch pellets for shipping fragile goods, thin films for wrapping.

Biopolymers can be sustainable, carbon neutral and are always renewable, because they are made from plant materials which can be grown indefinitely. Therefore, the use of biopolymers would create a sustainable industry. In contrast, the feedstocks for polymers derived from petrochemicals will eventually deplete. In addition, biopolymers have the potential to cut carbon emissions and reduce CO_2 quantities in the atmosphere: this is because the CO_2 released when they degrade can be reabsorbed by crops grown to replace them: this makes them close to carbon neutral.

Biopolymers are biodegradable, and some are also compostable. Some biopolymers are biodegradable: they are broken down into CO_2 and water by microorganisms. Some of these biodegradable biopolymers are compostable: they can be put into an industrial composting process and will break down by 90% within six months. Biopolymers that do this can be marked with a "compostable" symbol, under European Standard EN 13432 (2000). Packaging marked with this symbol can be put into industrial composting processes and will break down within six months or less. An example of a compostable polymer is PLA film under 20μm thick: films which are thicker than that do not qualify as compostable, even though they are "biodegradable". In Europe there is a home composting standard and associated logo that enables consumers to identify and dispose of packaging in their compost heap.

Cellulose is the most common organic compound and biopolymer on earth with the formula of $(C_6H_{10}O_5)_n$, polysaccharide consisting of a linear chain of several hundred to many thousands of $\beta(1\rightarrow 4)$ linked D-glucose units. Cellulose is an important structural component of the primary cell wall of green plants, many forms of algae and the oomycetes. Some species of bacteria secrete it to form biofilms. Cellulose is the most abundant organic polymer on earth. The cellulose content of cotton fiber is 90%, that of wood is 40%~50%, and that of dried hemp is approximately 57%.

Cellulose is mainly used to produce paperboard and paper. Smaller quantities are converted into a wide variety of derivative products such as cellophane and rayon. Conversion of cellulose from energy crops into biofuels such as cellulosic ethanol is under investigation as an alternative fuel source. Cellulose for industrial use is mainly obtained from wood pulp and cotton.

Questions

1. If you dump a biopolymer and a traditional plastic in a landfill, which one will still be there in 20 years?

2. Which is more efficient (gives you more product): synthesis of a chemical using enzymes, or using traditional chemistry?

Key Terms to Polymer Science

1. natural polymer   天然高分子
2. oligomer   低聚物
3. prepolymer   预聚物
4. homopolymer   均聚物
5. random coiling polymer   无规卷曲聚合物
6. atactic polymer   无规立构聚合物
7. copolymer   共聚物
8. random copolymer   无规共聚物
9. alternating copolymer   交替共聚物
10. block copolymer   嵌段共聚物
11. heteropolymer   杂聚物
12. polyblend   高分子共混物
13. single-strand polymer   单股聚合物
14. linear polymer   线型聚合物
15. living polymer   活性高分子
16. three-dimensional polymer   体型聚合物
17. dendrimer   树状高分子
18. star polymer   星形聚合物
19. branched polymer   支化聚合物
20. graft polymer   接枝聚合物
21. core shell copolymer   核-壳共聚物
22. chiral polymer   手性高分子
23. gel   凝胶
24. starch   淀粉
25. chitin   甲壳质
26. natural rubber   天然橡胶
27. functional polymer   功能高分子
28. polymer drug   高分子药物
29. conducting polymer   导电聚合物
30. liquid crystal polymer   液晶高分子
31. polymer catalyst   高分子催化剂
32. resin   树脂
33. epoxy resin   环氧树脂
34. polyolefin   聚烯烃
35. polyethylene   聚乙烯
36. ultrahigh molecular weight polyethylene   超高分子量聚乙烯
37. polypropylene   聚丙烯
38. polyisobutylene   聚异丁烯
39. polystyrene   聚苯乙烯
40. polyacetylene   聚乙炔
41. polymethacrylate   聚甲基丙烯酸酯
42. polyacrylonitrile   聚丙烯腈
43. poly(vinyl alcohol)   聚乙烯醇
44. poly(vinyl chloride)   聚氯乙烯
45. poly(vinyl fluoride)   聚氟乙烯
46. polybutadiene   聚丁二烯
47. polyisoprene   聚异戊二烯
48. azo polymer   偶氮类聚合物
49. coordination polymer   配位聚合物
50. polynorbornene   聚降冰片烯
51. aldehyde polymer   醛类聚合物
52. polycondensate   缩聚物
53. polyester   聚酯
54. polycarbonate   聚碳酸酯
55. polyamide   聚酰胺
56. polypeptide   多肽
57. polyether   聚醚
58. poly(ethylene oxide)   聚环氧乙烷
59. polytetrahydrofuran   聚四氢呋喃
60. polysulfone   聚砜
61. polyimide   聚酰亚胺
62. polyurethane   聚氨基甲酸酯
63. polyaniline   聚苯胺
64. composite   复合材料
65. in situ composite   原位复合材料
66. organic-inorganic hybrid material   有机-无机杂化材料
67. elastomer   高弹体
68. butadiene-acrylonitrile rubber   丁腈橡胶
69. thermoplastic elastomer   热塑性弹性体
70. fiber   纤维
71. adhesive   黏合剂

Unit 7

Food Chemistry

Lesson 28 Introduction to Food Chemistry: The Science of Studying the Chemical Nature of Foods

Few topics interest so many different people in so many different ways as does the subject of food. Of course, people need to eat to stay alive, grow and develop, and maintain good health. This need presents ongoing challenges for humans: finding ways of growing crops and raising animals in the most efficient way in the conditions available, inventing methods for competing successfully against microorganisms and animals that also consume the crops and animals on which humans depend, developing methods for preserving foods to make sure they will be available at all times of the year, and so on.

It should be no surprise, then, to discover that a number of chemical techniques used to obtain and process foods today have their roots in human cultures of many centuries ago.

Indeed, it has become a complex scientific industry that owes as much to the development of modern chemistry as it does to folk traditions and customs.

1. dietary
['daɪˌteri]
饮食的

2. carbohydrate
[ˌkɑrbo'haɪˌdret]
碳水化合物

3. protein
['proˌtin, -tiɪn]
蛋白质

4. nutrient
['nutriənt, 'nju-]
有营养的

When Did the Food Chemistry Bring Forth

The food industry had its origins in the late 1800s, for the chemists of the 18th and 19th centuries, the understanding of the chemical nature of our foods was a major objective. They realized that this knowledge was essential if **dietary** standards, and with them health and prosperity, were to improve. Inevitably it was the food components present in large amounts, the **carbohydrates**, fats, and **proteins**, which were first **nutrients** to be described in chemical terms. So food chemistry sets out to introduce the chemistry of our diet, after all foods "macro-components"

173

5. foodstuff ['fud,stʌf] 食品
6. gel [dʒɛl] 凝胶
7. polysaccharide [,pɑli'sækə,raɪd] 多糖
8. cocoa butter 可可油
9. crystallography [,krɪstə'lɑgrəfi] 结晶
10. triglyceride [traɪ'glɪsə,raɪd] 甘油三酸酯
11. digest [də'dʒɛst] 消化
12. starch [stɑrtʃ] 淀粉
13. dessert [dɪ'zɚt] 餐后甜点
14. shelf-life 货架期，保存限期
15. synthetic [sɪn'θɛtɪk] 合成的
16. semisynthetic [,sɛmɪsɪn'θɛtɪk, ,sɛmaɪ-] 半合成的

are more overtly chemicals in character because these are the substances whose chemical properties exert the major influence on the obvious physical characteristics of **foodstuffs.** If we are to understand the properties of food **gels** we are going to need a firm grasp of the chemical properties of **polysaccharides**. Similarly we will not understand the unique properties that **cocoa butter** gives to chocolate without getting involved in the **crystallography** of **triglycerides.**

What Were the Food Chemists Concerned

Over the past two centuries, food chemists have continued to push forward the frontiers of food design and development. By the time of World War II it appeared that most of the questions being asked of food chemists by nutritionists, agriculturalists, and others had been answered. This was certainly true as far as questions of the "what is this substance and how much is there?" varieties were concerned.

However, over the past few decades new questions have been asked and many answers are still being awaited. Apart from the question of undesirable components, both natural and man-made foods chemists nowadays are required to explain the behavior of food components. What happens when food is processed, stored, cooked, chewed, **digested** and absorbed? Much of the stimulus to this type of enquiry has come from the food-manufacturing industry. For example, the observation that the **starch** in a dessert product provides a certain amount of energy has been overtaken in importance by the need to know which type of starch will give just the right degree of thickening and what is the molecular basis for the differences between one starch and another. Furthermore that **dessert** product must have a long **shelf-life** and look as pretty as the one served in the expensive restaurant.

As food chemists extended the range of their research late in the 20th century to produce foods that differed very significantly from their natural state—and, in many cases, produced entirely new and **synthetic** food products—governmental agencies have continued to be involved in efforts to make sure that such foods are safe and efficacious, efforts that have had mixed results. By the last quarter of the 20th century, researchers had begun to take advantage of the full range of new materials and techniques that had been introduced into the field of chemistry to produce a virtually endless variety of new foods for consumers.

One of the great challenges for consumers in the 21st century is to learn more about and decide how to use the host of synthetic and **semisynthetic** foods now available to them. The involvement of chemists in food modification practices is a double-edged sword. For all the improvements it may have produced in the diet available to humans, the chemical modification of foods has raised many questions about safety and benefits. Are processed foods really equivalent or preferable to natural foods? Are the processes by which foods are modified relatively safe, or do they carry significant risks for the consumer? Are there limits to the ways in which foods can and should be modified? Questions such as these have become part of the daily dialogue of concerned consumers and food chemists.

Today, virtually every technique that is available to the industrial or research chemist is employed by the food chemist to modify the composition of natural foods or even to create new foods with no counterpart in the real world.

Conclusion

The issue of "chemicals" in food is closely linked to the pursuit of "naturalness" as a guarantee of "**healthiness**". The enormous diversity of the diets consumed by **Homo sapiens** as a colonist of this planet makes it impossible to define the ideal diet. Diet-related disease, including **starvation**, is a major cause of death. But it appears that while choice of diet can certainly influence the manner of our passing (euphemistic expressions for death), diet has no influence on its inevitability. As food chemists work together with nutritionists, doctors, **epidemiologists** and other scientists to understand what it is we are eating and what it does to us we will come to understand the essential compromises the human diet entails. After all, our success on this planet is to some extent at least owed to an extraordinary ability to adapt our eating habits to what is available in the immediate environment. Whether that environment is an Arctic waste, a tropical rain forest, or a hamburger infested inner city, humans actually cope rather well.

Hands-On Food Chemistry: Volume Changes During Freezing

In a freezing environment an unprotected bottle of milk freezes. The milk expands on freezing; the volume of milk increases. The cap sits on top of a protruding cylinder of frozen milk. If the milkbootle has a solidly held cap of metal, one that would resist being pushed from the bottle, then it would burst. The liquid is not compressible and neither is the ice. As the volume increases within the bottle during freezing, either the bottle bursts or the cap is forced free.

Hands-On Food Chemistry Insights: Volume Changes During Freezing

When freezing food in a rigid container, opportunity for such expansion must be considered. However, not all food products expand during freezing.

Strawberry jam does not increase in volume when frozen. The frozen sugar solution occupies the same height in the jar frozen as unfrozen-and it may even occupy less. Water alone can be expected to increase about 10% in volume on freezing. With high sugar concentrations in water, the expansion may be nil and actually a decrease in volume can occur. Strawberries packed in sugar as ordinarily frozen will expand slightly on freezing.

Exercises

1. Translate the following from Chinese into English.
   (1) 碳水化合物   (2) 蛋白质   (3) 营养   (4) 多糖   (5) 食品制造业
   (6) 合成食品   (7) 健康   (8) 修饰   (9) 食品化学 (10) 天然食品

2. Translate the following paragraphs into Chinese.

   Food chemistry publishes original research papers dealing with the advancement of the chemistry and biochemistry of foods or the analytical methods / approach used. All papers should focus on the novelty of the research carried out.

   Topics include:

   (1) Chemistry relating to major and minor components of food, their nutritional, physiological, sensory, flavour and microbiological aspects.

   (2) Bioactive constituents of foods, including antioxidants, phytochemicals, and botanicals. Data must accompany sufficient discussion to demonstrate their relevance to food and/or food chemistry.

   (3) Chemical and biochemical composition and structure changes in molecules induced by processing, distribution and domestic conditions.

   (4) Effects of processing on the composition, quality and safety of foods, other bio-based materials, by-products, and processing wastes.

   (5) Chemistry of food additives, contaminants, and other agro-chemicals, together with their metabolism, toxicology and food fate.

3. Topics for discussing.

   (1) Are the present processes by which foods are modified relatively safe, or do they carry significant risks for the consumer?

   (2) What is the organic and natural foods?

   (3) In your view, what is the future of the food industry in China and all over the world?

【Reading Material】

Organic and Natural Foods

The last few decades of the 20th century saw a rapidly growing interest in foods labeled as "natural", "organic", "whole", "healthful", or some similar descriptive term. The precise meaning of those terms has often been difficult to determine, and the difference among them equally as hard to distinguish. Some individuals and businesses have attempted to clarify what they mean when they use each of these terms. For example, the University of Iowa Health Care program has defined a health food as "any food that contributes to overall improved health status". The program points out that the term should be used for foods that are known to benefit human health—such as fruits, vegetables, whole grains, beans, cereals, low-fat milk and dairy products, and lean meats and **poultry**(家禽)— rather than products that are simply labeled as "health foods". The Food Marketing Institute defines natural foods as "foods that are minimally processed and free of artificial color, flavors, preservatives, and additives". But it points out that the term can be misleading since there are no governmental

Gingerols, $n=4,6,8$

controls on the use of the word natural for foods and, at least in theory, any company or individual can use the term on any food product that it offers for sale.

The only term for which a clear and specific definition exists is that of organic foods. In 1990, congress passed the Organic Foods Production Act (OFPA) to set national standards governing the marketing of so-called organically produced products, to assure consumers that organically produced products meet a consistent standard; and to facilitate interstate commerce in fresh and processed food that is organically produced. The OFPA established the National Organic Program (NOP) within the US Department of Agriculture (USDA) and ordered the department to establish standards for defining foods that could be labeled as organic in the United States. The NOP promulgated those standards on October 21, 2002. They defined organic food as follows: Organic food is produced by farmers who emphasize the use of **renewable**(可更新的) resources and the conservation of soil and water to enhance environmental quality for future generations. Organic meat, poultry, eggs, and dairy products come from animals that are given no antibiotics or growth hormones. Organic food is produced without using most conventional **pesticides**(杀虫剂); **fertilizers**(化学肥料) made with synthetic ingredients or sewage sludge; **bioengineering**(生物工程学); or **ionizing radiation**(电离辐射). A critical point to be noted about this definition is that it refers to the methods by which a food is produced; it does not describe the actual food itself.

The OFPA and its administrative rules provide an exhaustive list of materials and procedures that are permitted and prohibited in the production, storage, shipping, and sale of foods that can be legally labeled as organic. The USDA's "National List of Allowed and Prohibited Substances", for example, lists dozens of synthetic products that may be used in the production of organic foods [such as alcohols, calcium hypochlorite, chlorine dioxide, hydrogen peroxide, soap-based **herbicides**(除草剂), plastic mulches, sulfur, insecticidal soaps, copper sulfate, **ethylene**(乙烯), lignin sulfonate, and **sodium silicate**(硅酸钠)] and others that may not be used in the production of organic foods [such as ash from manure burning, **arsenic**(砷), lead salts, sodium fluoaluminate, strychnine, tobacco dust [nicotine sulfate], potassium chloride [in most cases], and **sodium nitrate**(硝酸钠)].

Food products that meet the USDA's standards may be marked (but are not required to be) with a distinctive package label. The label indicates that at least 95 percent of the food in the package has been produced by methods approved by the USDA. It can be used, however, only with single-ingredient foods, such as meats, milk, eggs, cereals, and cheese. Multiple-ingredient foods that contain at least 70 percent organic ingredients cannot carry the USDA seal, but it can carry the statement "made with organic ingredients". Finally, food products that contain less than 70 percent organic foods cannot carry either the USDA label or the "made with organic ingredients" notice, although they can list organically produced ingredients on the side panel.

Food producers have reason to use terms such as natural, healthful, whole, and organic in describing their products. Public opinion surveys show that a majority of Americans prefer to purchase foods that contain fewer pesticides, are environmentally friendly, and are more nutritious. They are likely to associate natural, healthful, whole, and organic foods with these characteristics. Yet, except for the term organic, no standards exist to define other types of "healthful" foods. Absent those

"We've engineered this cow to produce 30% more milk ⋯ and look at her work that speed bag!"

standards, consumers have no guarantee that the foods they believe to be safe and nutritious actually have those qualities. This problem becomes ever more important as interest in healthful foods among consumers grows.

Questions

1. What is the organic and natural foods?
2. In your view, what is the future of the food industry in China and other nations of the world?

Key Terms to Food Chemistry

1. flavour enhancer   增味剂
2. acidity regulations   酸味剂
3. sweeteners   甜味剂
4. starch   淀粉
5. sorbitol   糖醇
6. commercial sterilization   商业无菌
7. disinfection   消毒
8. pasteurization   巴氏消毒
9. flash evaporation   闪蒸
10. ion exchange   离子交换
11. dehydration   脱水
12. rehydration   复水
13. extraction   浸取
14. pressing   压榨
15. fermentation   发酵
16. brewing   酿造
17. gelatinization   糊化
18. retrogradation   凝沉
19. liquifying (liquefaction)   液化
20. saccharification (conversion)   糖化
21. hydrogenation   氢化
22. tenderization   嫩化
23. softening   软化
24. fortification (enrichment)   营养强化
25. smoking   熏制
26. refreshment (refreshing)   保鲜
27. cold storage, frozen storage   冷藏
28. quick-freezing   速冻
29. curing preservation   腌制保存
30. salting   盐渍
31. saucing   酱渍
32. sugaring   糖渍
33. pickling   酸渍
34. basic food   主食
35. irradiation preservation   辐照保藏
36. chemical preservation   化学保藏
37. food composition   食品成分
38. food analysis   食品分析
39. acid treatment   酸处理
40. sulphuring treatment   硫处理
41. alkali treatment   碱处理
42. mashing   打浆
43. fermentation   发酵
44. nutrient   营养素
45. crude protein   粗蛋白质
46. complementary action of protein   蛋白质互补
47. essential amino acid   必需氨基酸
48. unsaturated fatty acid   不饱和脂肪酸
49. effective carbohydrate   有效碳水化合物
50. dietary fiber   膳食纤维
51. mineral matter   矿物质
52. trace element   微量元素
53. water soluble vitamin   水溶性维生素
54. water activity   水分活度
55. moisture content   水分
56. reducing sugar   还原糖
57. iodine value   碘价
58. peroxide value   过氧化值
59. chemical analyzer   化学品分析仪器
60. constituent analyzer   食品成分分析仪器
61. ingredient analyzer   食品配料分析仪器
62. food additive   食品添加剂
63. food enrichment   食品营养强化剂
64. chemical contamination   化学性污染
65. residue of pesticide   农药残留
66. residue of veterinary drug   兽药残留
67. food poisoning   食物中毒
68. rancidity   酸败
69. antibiotic   抗生素
70. butyl fermentation   丁基丙酮发酵
71. butyrate   丁酸盐
72. bacillus anthracis   炭疽杆菌
73. bacitracin   枯草秆
74. milk acid   乳酸
75. lapper milk   酸牛奶凝乳
76. colostral milk   初乳
77. toned milk   调配牛奶
78. stick liquorice   甘草茎糖
79. jellies   果胶
80. food quality inspection   食品质量检验
81. abimentary toxicosis   食物中毒

Unit 8

Pesticide Chemistry

Lesson 29 Introduction to Pesticide Chemistry

Definition of Pesticide

According to The Food and Agriculture Organization (FAO) (1989), a **pesticide** is any substance or mixture of substances **intended for** preventing, destroying, or controlling any **pest** including **vectors** of human or animal diseases, unwanted species of plants or animals, causing harm during or otherwise interfering with the production, processing, storage, transport, or marketing of food, agricultural **commodities**, wood and wood products, or animal **feedstuffs**, or substances which may be **administered to** animals for the control of insects, **arachnids** or other pests in or on their bodies. The term includes chemicals used as **growth regulators**, **defoliants**, **desiccants**, **fruit thinning agents**, or agents for preventing the **premature** fall of fruits, and substances applied to crops either before or after harvest to prevent **deterioration** during storage or transport. The term, however excludes such chemicals used as fertilizers, plant and animal nutrients, food additives and animal drugs. The term pesticide is also defined by FAO in collaboration with UNEP (1990) as chemicals designed to **combat** the attacks of various pests and vectors on agricultural crops, **domestic animals** and human beings. The definitions above imply that, pesticides are toxic chemical agents (mainly organic compounds) that are **deliberately** released into the environment to combat crop pests and disease vectors.

Historical Background of Pesticides Use in Agriculture and Public Health

The historical background of pesticides use in agriculture is dated back to the beginning of agriculture itself and it became more **pronounced** with time due to increased pest population **paralleled** with decreasing soil fertility. However, the use of modern pesticides in agriculture and public health is dated back to the 19th century. The first generation of pesticides involved the use of highly toxic compounds, arsenic (calcium arsenate and

---

1. pesticide ['pɛstɪsaɪd] 杀虫剂，农药
2. intend for 为……而准备
3. pest [pɛst] 害虫
4. vector ['vɛktɚ] （传播疾病的）媒介昆虫
5. commodity [kə'mɑdəti] 商品
6. feedstuff ['fidstʌf] 饲料
7. administered to 给予，有助于
8. arachnid [ə'ræknɪd] 蛛形动物
9. growth regulator 生长调节剂
10. defoliant [,di'folɪənt] 脱叶剂
11. desiccant ['dɛsəkənt] 干燥剂
12. fruit thinning agents 疏果剂
13. premature [,primə'tʃur] 早熟的
14. deterioration [dɪ'tɪrɪə'reʃən] 变质
15. combat ['kɑmbæt] 防止，减轻
16. domestic animals 家畜
17. deliberately [dɪ'lɪbərətli] 故意地
18. pronounced [prə'naʊnst] 明显的，显著的
19. parallel ['pærəlɛl] 与……同时发生

lead arsenate) and a **fumigant hydrogen cyanide** in 1860's for the control of such pests like **fungi**, insects and bacteria. Other compounds included Bordeaux mixture (copper sulphate, lime and water) and sulphur. Their use was abandoned because of their toxicity and **ineffectiveness**. The second generation involved the use of synthetic organic compounds. The first important synthetic organic pesticide was dichlorodiphenyltrichloroethane (DDT) which was synthesized by a German scientist Zeidler in 1873 first. Its **insecticidal** effect was discovered by a Swiss chemist Paul Muller in 1939. In its early days, DDT was **hailed** as a miracle because of its **broad-spectrum activity**, persistence, insolubility, inexpensive and ease to apply.

$$Cl--\underset{H}{\overset{CCl_3}{C}}--Cl$$
$$p,p'\text{-DDT}$$

p, p'-DDT in particular was so effective at killing pests and thus **boosting** crop yields and was so inexpensive to make its use quickly spread over the globe. DDT was also used for many non-agricultural applications as well. For example, it was used to **delouse** soldiers in the World War II and in the public health for the control of mosquitoes which are the vectors for **malaria**. Following the success of DDT, such other chemicals were synthesized to make this era what Rachel Carson (1962) in her book "The Silent Spring" described as the era of "rain of chemicals".

The intensive use of pesticides in agriculture is also well known to **be coupled with** the "green revolution". Green revolution was a worldwide agricultural movement that began in Mexico in 1944 with a primary goal of boosting grain yields in the world that was already in trouble with food supply to meet the demand of the then rapidly growing human population. The green revolution involved three major aspects of agricultural practices, among which the use of pesticides was an **integral part**. Following its success in Mexico, green revolution spread over the world. Pest control has always been important in agriculture, but green revolution in particular needed more pesticide inputs than did traditional agricultural systems because, most of the high yielding varieties were not widely resistant to pests and diseases and partly due to **monoculture system**. Each year pests destroy about 30%~48% of world's food production. For example, in 1987 it was reported that, one third of the potential world crop harvest was lost to pests. A further illustration to the pest problem in the world is shown in Table 29-1.

33. rodent ['rodnt] 啮齿动物
34. internally feeding insects 内食性昆虫
35. endosperm ['endəspɜːm] 胚乳
36. germ [dʒɜːm] 胚芽
37. deterioration [dɪˌtɪrɪə'reʃən] 变质
38. externally feeding insects 外食性昆虫
39. mystification [ˌmɪstəfə'keʃən] 神秘化
40. excrement ['ɛkskrɪmənt] 排泄物
41. contamination with 沾染
42. larval moult 幼虫蜕皮
43. cocoon 蚕茧
44. malathion [ˌmælə'θaɪon] 马拉息昂
45. impregnate [ɪm'pregneɪt] 浸透
46. vector-borne 虫媒
47. trypanosomiasis 锥虫病
48. onchocerciasis 盘尾丝虫病
49. filariasis 丝虫病
50. fatality [fə'tæləti] 灾祸

Table 29-1 Estimated losses caused by pests in some world's major crops per year

Crop	Estimated losses/%			
	Insects	Diseases	Weeds	Total
rice	26.7	8.9	10.8	46.4
maize	12.4	9.4	13.0	34.8
wheat	5.0	9.1	9.8	23.9
millet	9.6	10.6	17.8	38.0
potatoes	6.5	21.8	4.0	32.3
cassava	7.7	16.6	9.2	33.5
soybeans	4.5	11.1	13.5	29.1
peanuts	17.1	11.3	11.8	40.4
sugarcane	9.2	10.7	25.1	45.0

Insect pests and **rodents** also account for a big loss in stored agricultural products. **Internally feeding insects** feed on grain **endosperm** and the **germ**, results in the loss in grain weight, reduction in nutritive value of the grain and **deterioration** in the end use quality of the grain. **Externally feeding insects** damage grain by physical **mystification** and by **excrement contamination with** empty eggs, **larval moults** and empty **cocoons**. A common means of pest control in stored agricultural products has always been the use of insecticides such as **malathion**, chlorpyrifos-methyl or deltamethrin **impregnated** on the surfaces of the storage containers.

On the other hand, malaria remains the major **vector-borne** infectious disease in many parts of the tropics. It is estimated that over 300 to 500 million clinical cases occur each year, with cases in tropical Africa accounting for more than 90% of these figures. Other vector-borne diseases that present a serious problem especially in the tropics include **trypanosomiasis, onchocerciasis** and **filariasis**. It is therefore quite apparent that, the discovery of pesticides was not a luxury of a technical civilization but rather was a necessity for the well being of mankind.

Impacts of Pesticides Use in Agriculture and Public Health

The use of pesticides in agriculture has led to a significant improvement in crop yield per hectare of land. Studies have established a possible correlation relationship between the quantity of pesticides used per hectare and the amount of crop yields per hectare, see Table 29-2. Pesticides like DDT and others proved their usefulness in agriculture and public health. Economies were boosted, crop yields were tremendously increased, and so were the decreases in **fatalities** from insect-borne diseases. Insecticides have saved the lives of countless millions of people from insect-borne diseases.

Table 29-2 Pesticides use and the corresponding crop yield in some countries/areas

Country/Area	Pesticide use (kg/ha)	Crop yield (ton/ha)
Japan	10.8	5.5
Europe	1.9	3.4
USA	1.5	2.6
Latin America	0.2	2.0
Oceania	0.2	1.6
Africa	0.1	1.2

Side Effects of Pesticides Use to the Environment and Public Health

Despite the good results of using pesticides in agriculture and public health described above, their use is usually accompanied with **deleterious** environmental and public health effects. Pesticides hold a unique position among environmental contaminants due to their high biological activity and toxicity (**acute** and **chronic**). Although some pesticides are described to be selective in their modes of action, their selectivity is only limited to test animals. Thus pesticides can be best described as **biocides** (capable of harming all forms of life other than the target pest). Further details on the side effects of pesticides are discussed in the following chapter (ecological effects of pesticides).

Hands-On Chemistry: Cleaning Your Insects

Perhaps the most environmentally friendly insecticides are solutions of soap or detergent. Insects are **perforated** with tiny holes, called **spiracles**, through which atmospheric oxygen migrates directly into cells. These holes are easily penetrated by liquid soap or detergent, which then blocks the exchange of atmospheric gases and **suffocates** the insect. In general, the larger the insect, the more concentrated a soap solution needs to be in order to kill efficiently. A **dilute** soap solution, for example, will quickly **annihilate aphids**, but only a relatively concentrated one will kill a **cockroach**.

What You Need: Liquid soap or detergent, measuring spoons, pump-spray bottles, household or garden insect pests such as houseflies or aphids.

Procedure: Use the measuring spoons to create various concentrations of soap solution. Stir your solutions carefully to avoid excessive foaming. Pour each solution into a pump-spray bottle, and write the concentration (in teaspoons per cup) on each bottle. Test the effectiveness of each solution on the infested plants. Follow with a spray of fresh water for the plants to remove any residual soap.

---

51. deleterious [ˌdɛləˈtɪrɪəs] 有害的
52. acute [əˈkjut] 急性的
53. chronic [ˈkrɑnɪk] 慢性的
54. biocides 生物性农药
55. perforate [ˈpɝfəˈret] 打孔
56. spiracle [ˈspɪərəkl] 呼吸孔
57. suffocate [ˈsʌfəket] 窒息
58. dilute [daɪˈlut] 稀释的
59. annihilate [əˈnaɪəlet] 消灭
60. aphid [ˈeɪfɪd] 蚜虫
61. cockroach [ˈkɑkˈrotʃ] 蟑螂

Hands-On Chemistry Insights: Cleaning Your Insects

The advantages of using soap or detergent to kill insects are that these materials are inexpensive, easily washed away, and environmentally friendly. The disadvantages are that they are not selective for harmful insect pests over beneficial ones. Also, they work only by direct contact, and the plant must be wiped clean once the pests are destroyed. Getting rid of all the soap can be difficult if you are using a concentrated solution. Lastly, they have no lasting insecticidal effect. Interestingly, the fine network of spiracles is what limits the size of insects. If an insect was any heavier, its weight would collapse all the tiny channels. In prehistoric times, higher atmospheric concentrations of oxygen permitted the evolution of much larger insects, such as a dragonfly species, often depicted in dinosaur books, that could be as much as 1 meter in length.

Exercises

1. Select the correct answer to the question.
    (1) According to text, what percentage of world's food production do pests destroy each year?
        (a) 46.4% (b) 34.8% (c) 30%~48% (d) 23.9%
    (2) According to text, which of the following statements is wrong about pesticides p, p'-DDT?
        (a) p, p'-DDT was very effective at killing pests.
        (b) p, p'-DDT boosted crop yields.
        (c) p, p'-DDT spread over the globe quickly.
        (d) p, p'-DDT was used for agricultural applications only.
    (3) Which of the following is not a benefit of pesticide use?
        (a) increased food production
        (b) increased profits for farmers
        (c) disease prevention
        (d) biological magnification
    (4) What is bioaccumulation?
        (a) When a substance builds up in the body because the body does not have the proper mechanisms to remove it.
        (b) When a substance builds up in the body because the body does not want to remove it.
        (c) When a substance is stored in the body for at least a year.
        (d) When a substance is removed from the body when consumed by a predator.
2. Translate the following terms from Chinese into English.
    (1) 农药   (2) 饲料   (3) 生长调节剂   (4) 干燥剂   (5) 疏果剂
    (6) 变质   (7) 家畜   (8) 广谱活性   (9) 霉菌   (10) 疟疾
    (11) 单作体系   (12) 内食性昆虫   (13) 急性的   (14) 慢性的

【Reading Material】

Pesticides and Children

All pesticides have some level of toxicity, and pose some risk to infants and children. The risk depends on the toxicity of the pesticide ingredients and how much of the pesticide a child is exposed to.

Infants and children are more sensitive to the toxic effects of pesticides than adults. An infant's brain, nervous system, and organs are still developing after birth. When exposed, a baby's immature liver and kidneys cannot remove pesticides from the body as well as an adult's liver and kidneys. Infants may also be exposed to more pesticide than adults because they take more breaths per minute and have more skin surface relative to their body weight. Children often spend more time closer to the ground, touching baseboards and lawns where pesticides may have been applied. Children often eat and drink more relative to their body weight than adults, which can lead to a higher dose of pesticide residue per pound of body weight. Babies that crawl on treated carpeting may have a greater potential to dislodge pesticide residue onto their skin or breathe in pesticide-laden dust. Young children are also more likely to put their fingers, toys, and other objects into their mouths. Because of this, it is important to minimize your child's exposure to pesticides. One way to minimize exposure to pesticides is to take an approach called Integrated Pest Management (IPM). IPM is a pest control strategy that uses a combination of methods to prevent and eliminate pests in the most effective and least hazardous manner.

If you choose to use a pesticide, keep these tips in mind to minimize risk to infants and children:

(1) Always be sure to read the product label first. The product must be approved for the intended use and applied according to label directions.

(2) Seek the least-toxic pesticide option available. Use the signal word to identify products that are low in toxicity.

(3) Keep children out of treated areas while pesticides are being applied, and until areas are dry. The product label may have more specific instructions.

(4) Allow plenty of time for the pesticide to dry and the home to ventilate before returning.

(5) If your lawn or carpeting has recently been treated with pesticides, consider using shoes, blankets or another barrier between the treated surface and children's skin.

(6) Be sure children wash their hands before eating, especially after playing outdoors.

(7) If you apply pesticides to your pets, be sure to keep children from touching the pet until the product has completely dried.

(8) Place ant, snail and rodent baits in locked bait stations or safely out of the reach of children.

(9) Never use mothballs outside of sealed, airtight containers. Children often mistake mothballs for food when used improperly around the home.

(10) Never use illegal pesticides, such as Miraculous, Pretty Baby or Chinese Chalk. It looks and writes like normal chalk, but the pesticide dust can be breathed in, get on kids' hands or end up in their mouths.

(11) Be sure to store pesticides in their original containers. Never use food or beverage utensils or containers to mix or store pesticides.

(12) Store all pesticides out of the reach of children.

(13) If someone in the household works with pesticides, take steps to reduce the amount of pesticide residues they bring into home. If possible, wash and dry the work clothes separate from family laundry.

Question

Why does the risk depend on the toxicity of the pesticide ingredients and how much of the pesticide a child is exposed to?

Key Terms to Pesticide Chemistry

1. organochlorine pesticides　有机氯农药
2. toxic effect　毒化作用
3. detoxification　去毒作用
4. organophosphorus pesticides　有机磷农药
5. synergistic effect　协同作用
6. geographic variation　地理差异
7. dioxin contaminant　二氧（杂）芑污染物
8. persistence bioaccumulation　生物体内残留
9. grain fumigant aluminum phosphide　谷物熏蒸磷化铝消毒剂
10. phenyloxycarboxylates herbicides　苯氧羧酸类除草剂
11. cellular restriction　细胞限制因子
12. trichloroethane　三氯乙烷（DDT）
13. acute poisoning　急性中毒
14. fat-soluble　脂溶性
15. nervous system　神经系统
16. pesticide residues　农药残留
17. ecology cycle　生态循环
18. alkyl chain　烷基链
19. mineral pesticide　矿物源农药
20. biological pesticide　生物源农药
21. plant growth regulator　植物生长调节剂
22. microbial pesticide　微生物杀虫剂
23. organic arsenic species　有机砷类
24. defoliants pesticide　脱叶剂
25. insect hormones　昆虫激素
26. nematicide pesticide　杀线虫剂
27. riskiest pesticide　高毒农药
28. acaricide　杀螨剂
29. acetyltransferase　乙酰基转移酶
30. acute toxicity　急性毒性
31. alkaloid　生物碱
32. antigen　抗原
33. aphid　蚜虫
34. less-persistent pesticide　低残留农药
35. low-toxic pesticide　低毒性农药
36. avirulence　无毒性
37. baccine　疫苗
38. bactericides　杀细菌剂
39. biocontrol　生物防御
40. biological control　生物防治
41. biosynthesis　生物合成
42. biotic pesticide　生物杀虫剂
43. botanical pesticide　植物源杀虫剂
44. broad spectrum pesticide　广谱杀虫剂
45. remnant　残留
46. causal agents　病原体
47. chlorothalonil (daconil)　百菌清
48. contact pesticide　触杀性农药
49. derosal　多菌灵
50. detoxification　脱毒
51. diamondback moth　小菜蛾
52. disease-resistant cultivar　抗病品种
53. disinfectant　消毒剂
54. repellent　忌避剂
55. resistance　抗药性
56. endotoxin　内毒素
57. entomology　昆虫学
58. entomophagous insect　食虫昆虫
59. epidemiology　流行病学
60. exotic species　外来的物种
61. fatal temperature　致死温度
62. organic pesticide　有机农药
63. organomercury pesticide　有机汞农药
64. flavonod　类黄酮
65. formulation　剂型
66. fumigant　熏剂
67. fungicides　杀真菌剂
68. galactosyltransferase　乳化转移酶
69. granular pesticide　粒状农药
70. herbicide　除草剂
71. interferon　干扰素
72. median lethal dosage (LD_{50})　半数致死量
73. median lethal concentration (LD_{50})　半数致死浓度

Unit 9

Principles of Chemical Engineering

Lesson 30 Introduction to Principles of Chemical Engineering

1. microbiology
[ˌmaɪkroʊbaɪˈɑlədʒi]
微生物学
2. convert
[kənˈvɚt]
使转化

Chemistry is one of the oldest sciences, and it has certainly been one of the most productive in improving human life. Chemical engineering is the branch of engineering that deals with the application of physical science (e.g. chemistry and physics), and life sciences (e.g. biology, **microbiology** and biochemistry) with mathematics, to the process of **converting** raw materials or chemicals into more useful or valuable forms. In addition to producing useful materials, modern chemical engineering is also concerned with pioneering valuable new materials and techniques - such

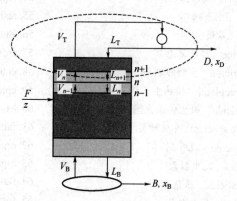

3. maintenance
[ˈmentənəns]
维护, 保持,
生活费用
4. sustainably
持续地

as nanotechnology, fuel cells and biomedical engineering. Chemical engineering largely involves the design, improvement and **maintenance** of processes involving chemical or biological transformations for large-scale manufacture. Chemical engineers ensure the processes are operated safely, **sustainably** and economically. Chemical engineers in this branch are usually employed under the title of process engineer. A related term with a wider definition is chemical technology. A person employed in this field is called a chemical engineer.

History of Chemical Engineering

In 1824, French physicist Sadi Carnot, in his "On the Motive Power of

5. thermodynamics
[,θɚ·modaɪ'næmɪks]
热力学

6. combustion
[kəm'bʌstʃən]
燃烧

7. founding
突出的，杰出的

8. timeline
时间表

9. Imperial College
帝国理工大学

10. curriculum
[kə'rɪkjələm]
课程

11. bachelor
['bætʃələ·,
'bætʃlə·]
学士学位

Fire", was the first to study the **thermodynamics** of **combustion** reactions in steam engines. In the 1850s, German physicist Rudolf Clausius began to apply the principles developed by Carnot to chemical systems at the atomic to molecular scale. During the years 1873 to 1876 at Yale University, American mathematical physicist Josiah Willard Gibbs, the first to be awarded a Ph D in engineering in the US, in a series of three papers, developed a mathematical-based, graphical methodology, for the study of chemical systems using the thermodynamics of Clausius. In 1882, German physicist Hermann von Helmholtz, published a **founding** thermodynamics paper, similar to Gibbs, but with more of an electro-chemical basis, in which he showed that measure of chemical affinity, i.e., the "force" of chemical reaction, is determined by the measure of the free energy of the reaction process. The following **timeline** shows some of the key steps in the development of the science of chemical engineering.

1805 – John Dalton published Atomic Weights, allowing chemical equations to be balanced and the basis for chemical engineering mass balances.

1882 – A course in "chemical technology" is offered at University College London.

1883 – Osborne Reynolds defines the dimensionless group for fluid flow, leading to practical scale-up and understanding of flow, heat and mass transfer.

1885 – Henry Edward Armstrong offers a course in "chemical engineering" at Central College (later **Imperial College**), London.

1888 – There is a Department of Chemical Engineering at Glasgow and West of Scotland Technical College offering day and evening classes.

1888 – Lewis M. Norton starts a new **curriculum** at Massachusetts Institute of Technology (MIT): Course X, Chemical Engineering.

1889 – Rose Polytechnic Institute awards the first **bachelor's** of science in chemical engineering in the US.

1891 – MIT awards a bachelor's of science in chemical engineering to William Page Bryant and six other candidates.

1892 – A bachelor's program in chemical engineering is established at the University of Pennsylvania.

1901 – George E. Davis produces the Handbook of Chemical Engineering.

1905 – The University of Wisconsin awards the first Ph D in chemical engineering to Oliver Patterson Watts.

1908 – The American Institute of Chemical Engineers (AIChE) is founded.

1922 – The UK Institution of Chemical Engineers (IChemE) is founded.

12. ceramic [sə'ræmɪk] 制陶制品，制陶业	
13. petrochemical [ˌpɛtro'kɛmɪkəl] 石化产品	
14. agrochemical 农用化学品	
15. insecticide [ɪn'sɛktɪˌsaɪd] 杀虫剂	
16. herbicide [ˈhəbɪˌsaɪd, 'ə-] 除草剂	
17. oleochemical 人造化学品	
18. dietary supplements 食品添加剂	
19. evaporator 蒸发器	
20. filtration [fɪl'treʃən] 过滤，筛选	
21. unit operations 单元操作	
22. momentum [mo'mɛntəm] 动力，动量	
23. sedimentation [ˌsɛdəmən'teʃən, -mɛn-] 沉淀，沉降	
24. intertwine [ˌɪntɚ'twaɪn] (使)纠缠，(使)缠绕	
25. underlying [ˈʌndɚˌlaɪɪŋ] 在下面的，根本的	
26. discrete [dɪ'skrit] 不连续的，离散的	

Applications

Chemical engineering is applied in the manufacture of a wide variety of products. The chemical industry scope manufactures inorganic and organic industrial chemicals, **ceramics,** fuels and **petrochemicals, agrochemicals** (fertilizers, **insecticides,** **herbicides**), plastics and elastomers, **oleochemicals**, explosives, detergents and detergent products (soap, shampoo, cleaning fluids), fragrances and flavors, additives, **dietary supplements** and pharmaceuticals. Closely allied or overlapping disciplines include wood processing, food processing, environmental technology, and the engineering of petroleum, glass, paints and other coatings, inks, sealants and adhesives.

Overview

Chemical engineers design processes to ensure the most economical operation. This means that the entire production chain must be planned and controlled for costs. A chemical engineer can both simplify and complicate "showcase" reactions for an economic advantage. Using a lower pressure or temperature makes several reactions easier; ammonia, for example, is simply produced from its component elements in a high-pressure reactor. On the other hand, reactions with a low yield can be recycled continuously, which would be complex, arduous work if done by hand in the laboratory. It is not unusual to build 6-step, or even 12-step **evaporators** to reuse the vaporization energy for an economic advantage. In contrast, laboratory chemists evaporate samples in a single step.

The individual processes used by chemical engineers (e.g. distillation or **filtration**) are called **unit operations** and consist of chemical reactions, mass-, heat- and **momentum**- transfer operations. Method of operation, phase equilibria and separating agents are used in these operations. It also provides the basis for the building of simple mathematical models to represent the operation of the mass transfer-based separation processes, such as distillation, gas absorption, liquid-liquid extraction, filtration, **sedimentation** and flow of fluids through beds of solid particles. Unit operations are grouped together in various configurations for the purpose of chemical synthesis and/or chemical separation. Some processes are a combination of **intertwined** transport and separation unit operations, (e.g. reactive distillation).

Three primary physical laws **underlying** chemical engineering design are conservation of mass, conservation of momentum and conservation of energy. The movement of mass and energy around a chemical process are evaluated using mass balances and energy balances, laws that apply to **discrete** parts of equipment, unit operations, or an entire plant. In doing so, chemical engineers must also use principles of thermodynamics, reaction kinetics, fluid mechanics and transport

27. process simulators
过程模拟（用计算机软件模拟各单元操作的软件）
28. encompass [ɛnˈkʌmpəs]
包围，环绕，包含
29. diverse [dɪˈvɝs, daɪ-, ˈdaɪ.vɝs]
不同的，变化多的
30. bio-compatible
生物适合的，不会引起排斥的
31. implant [ɪmˈplænt]
灌输，植入
32. prosthetics [prasˈθɛtɪks]
弥补术，修复术
33. pharmaceutical [ˌfɑrməˈsutɪkəl]
医药品
34. dielectric [ˌdaɪɪˈlɛktrɪk]
电介质，绝缘体
35. spectroscopic
分光镜的，借助分光镜的
36. mass [mæs]
使集合

phenomena. The task of performing these balances is now aided by **process simulators**, which are complex software models (see List of Chemical Process Simulators) that can solve mass and energy balances and usually have built-in modules to simulate a variety of common unit operations.

Modern Chemical Engineering

The modern discipline of chemical engineering **encompasses** much more than just process engineering. Chemical engineers are now engaged in the development and production of a **diverse** range of products including high performance materials needed for aerospace, automotive, biomedical, electronic, environmental, space and military applications. Examples include ultra-strong fibers, fabrics, dye-sensitized solar cells, adhesives and composites for vehicles, **bio-compatible** materials for **implants** and **prosthetics**, gels for medical applications, **pharmaceuticals**, and films with special **dielectric**, optical or **spectroscopic** properties for opto-electronic devices. Additionally, chemical engineering is often intertwined with biology and biomedical engineering. Many chemical engineers work on biological projects such as understanding biopolymers (proteins) and mapping the human genome. The line between chemists and chemical engineers is growing ever more thin as more and more chemical engineers begin to start their own innovation using their knowledge of chemistry, physics and mathematics to create, implement and **mass** produce their ideas.

Hands-On Chemistry: An Explanation to Chemical Engineering

In a wider sense, engineering may be defined as a scientific presentation of the techniques and facilities used in a particular industry. For example, mechanical engineering refers to the techniques and facilities employed to make machines. It is predominantly based on mechanical forces which are used to change the appearance and /or physical properties of the materials being worked, while their chemical properties are left unchanged. Chemical engineering encompasses the chemical processing of raw materials, based on chemical and physico-chemical phenomena of high complexity.

Exercises

1. Translate the following from Chinese into English.
   (1) 热力学   (2) 绝缘体   (3) 蒸发器   (4) 过滤
   (5) 沉淀   (6) 石化产品   (7) 过程模拟   (8) 单元操作

191

2. Chemical engineers design processes to ensure the most economical operation. According to the text what does this mean?
3. According to the text, what is the definition of unit operations? Explain the difference between unit operation and unit process?
4. What are the three primary physical laws underlying chemical engineering design?
5. A topic for speaking.
   Describe the application of modern chemical engineering.

【Reading Material】
Basic Trend in Chemical Engineering

Over the next few years, a confluence of intellectual advances, technologic challenges, and economic driving forces will shape a new model of what chemical engineering is and what chemical engineering do. The focus of chemical engineering has always been industrial processes that change the physical state or chemical composition of materials. Chemical engineers engage in the synthesis, design, testing scale-up, operation, control and optimization of these processes. The traditional level of size and complexity at which they worked on these problems might be termed the mesoscale. Examples of this scale include reactors and equipment for single processes(unit operations) and combinations of unit operations in manufacturing plants. Future research at the **mesoscale**（中间尺度）will be increasingly supplemented by dimensions—the **microscale**（微观尺度）and the dimensions of extremely complex systems—the **macroscale**（宏观尺度）.

Chemical engineers of the future will be integrating a wider range of scales than any other branch of engineering. For example, some may work to relate the macroscale of the environment to the mesoscale of combustion systems and the microscale of molecular reactions and transport. Other may work to relate the macroscale performance of a composite aircraft to the mesoscale chemical reactor in which the wing was formed, the design of the reactor perhaps having been influenced by studies of the microscale dynamics of complex liquids.

Thus, future chemical engineers will conceive and rigorously solve problems on a **continuum**（连续介质；连续光谱）of scales ranging from microscale to macroscale. They will bring new tools and insights to research science, and electrical engineering. And they will make increasing use of computers, artificial intelligence, and expert system in solving problem, in product and process design, and in manufacturing.

Two important developments will be part of this unfolding picture of the discipline.

Chemical engineers will become more heavily involved in product design as a **complement**（互补，补充）to process design. As the properties of a product in performance become increasingly linked to the way in which it is processed, the traditional distinction between product and process design will become blurred. There will be a special design challenge in established and emerging industries that produce proprietary, differentiated products tailored to exacting performance specifications. These products are characterized by the need for rapid innovatory as they are quickly superseded in the marketplace by newer products.

Chemical engineers will be frequent participants in multidisciplinary research efforts. Chemical engineering has a long history of fruitful interdisciplinary research with the chemical sciences, particular industry. The position of chemical engineering as the engineering discipline with the strongest tie to the molecular sciences is an asset, since such sciences as chemistry, molecular biology, biomedicine, and solid-state physics are providing the seeds for tomorrow's technologies. Chemical engineering has a bright future as the seeds for tomorrow's technologies. Chemical engineering has a bright future as the "**interfacial** (界面的，层间的) discipline", that will bridge science and engineering in the multidisciplinary environments where these new technologies will be brought into being.

Questions

1. What are the future trends in chemical engineering?
2. What is the focus of chemical engineering?

Lesson 31 Heat Transfer

Introduction

In the majority of chemical process heat is either given out or absorbed, and in a very wide range of chemical plant, fluids must often be either heated or cooled. Thus in furnaces, evaporators, distillation units, driers, and reaction vessels, one of the major problems is that of transferring heat at the desired rate. Alternatively, it may be necessary to prevent the loss of heat from a hot vessel or **steam pipe**. The control of the flow of heat in the desired manner forms one of the most important sections of chemical engineering. Provided that a temperature difference exists between two parts of a system, heat transfer will take place in one or more of three different ways.

Conduction

In a solid, the flow of heat by conduction is the result of the transfer of vibrational energy from one molecule to another, and in fluids it occurs in addition as a result of the transfer of kinetic energy. Heat transfer by conduction may also arise from the movement of free electrons. This process is particularly important with metals and accounts for their high thermal conductivities.

Convection

Heat transfer by convection is attributable to **macroscopic** motion of the fluid and therefore is confined to liquids and gases. In natural convection it is caused by differences in density arising from **temperature gradients** in the system. In forced convection, it is due to **eddy currents** in a fluid in **turbulent motion.**

Radiation

All materials radiate thermal energy in the form of electromagnetic waves. When this radiation falls on a second body it may be partially reflected, transmitted, or absorbed. It is only the fraction that is absorbed appears as heat in the body.

Basic Considerations

In many of the applications of heat transfer in chemical engineering, each of the mechanisms of conduction, convection, and radiation is involved. In the majority of heat exchanger units, the process is complicated in that the heat has to pass through a number of intervening layers before it reaches the material whose temperature is to be raised and the form of the equation may be complex.

As an example take the problem of transferring heat to oil in a crude oil still from a flame obtained by burning waste refinery gas. The heat from the flame is transferred by a combination of radiation and convection

---

1. steam pipe 蒸汽管

2. conduction
   [kən'dʌkʃən]
   热传导

3. convection
   [kən'vɛkʃən]
   对流

4. macroscopic
   [ˌmækrə'skɑpɪk]
   肉眼可见的, 宏观

5. temperature gradients
   温度梯度

6. eddy ['ɛdi]
   旋转, 旋涡

7. eddy currents
   涡流

8. turbulent
   ['tɝbjələnt]
   狂暴的, 吵闹的

9. turbulent motion
   湍流

to the outer surface of the pipes, then passes through the walls by conduction, and, finally, is transferred to the boiling oil by convection. After prolonged usage, solid **deposits** may form on both the inner and outer walls of the pipes, and these will then contribute additional resistance to the transfer of heat. The simplest form of equation, which represents this heat transfer operation, can be written as:

$$Q = UA\Delta T \tag{1}$$

where Q is the heat transferred per unit time, A is the area available for the flow of heat, ΔT is the difference in temperature between the flame and the boiling oil, and U is known as the **overall heat transfer coefficient** (W/m^2 · K^{-1} in SI units).

At first sight, equation (1) implies that the relationship between Q and ΔT is linear. Whereas this is **approximately** so, over limited ranges of temperature difference for which U is nearly constant, in practice U may well be influenced both by the temperature difference and by the **absolute value** of the temperatures.

If it is required to know the area needed for the transfer of heat at a specified rate, the temperature difference ΔT, and the value of the overall heat-transfer coefficient must be known, thus the calculation of the value of U is a key requirement in any design problem in which heating or cooling is involved. A large part of the study of heat transfer is therefore devoted to the **evaluation** of this coefficient.

The value of the coefficient will depend on the mechanism by which heat is transferred, on the **fluid dynamics** of both the heated and the cooled fluid, on the properties of the materials through which the heat must pass, and on the **geometry** of the fluid paths.

Suppose heat is being transferred through three **media**, each of area A, that the individual coefficients for each of the media are h_1, h_2 and h_3, and the corresponding temperature changes are ΔT_1, ΔT_2 and ΔT_3. If there is no accumulation of heat in the media, the heat transfer rate Q will be the same through each. Three equations, analogous to equation (1) can therefore be written:

$$Q = h_1 A \Delta T_1$$
$$Q = h_2 A \Delta T_2$$
$$Q = h_3 A \Delta T_3 \tag{2}$$

Rearranging: $\Delta T_1 = Q/A \times 1/h_1$, $\Delta T_2 = Q/A \times 1/h_2$, $\Delta T_3 = Q/A \times 1/h_3$

Adding: $\Delta T_1 + \Delta T_2 + \Delta T_3 = Q/A(1/h_1 + 1/h_2 + 1/h_3)$ (3)

Noting that $\Delta T_1 + \Delta T_2 + \Delta T_3 =$ total temperature difference ΔT:
$\Delta T = Q/A(1/h_1 + 1/h_2 + 1/h_3)$ (4)

But from equation (1): $\Delta T = Q/(AU)$ (5)

Comparing equations (4) and (5): $1/U = 1/h_1 + 1/h_2 + 1/h_3$ (6)

The **reciprocals** of the heat transfer coefficients are resistances, and

20. thick pipe wall 厚壁管	
21. in terms of 根据, 依据	
22. per unit time 单位时间	
23. through each of the media 通过单位介质	
24. broken down 分成, 细分	

equation (6) therefore illustrates that the resistances are additive.

In some cases, particularly for the **radial flow** of heat through a **thick pipe wall** or cylinder, the area for heat transfer is a function of position. Thus the area for heat transfer applicable to each of the three media could differ and may be A_1, A_2 and A_3. Equation (3) then becomes:

$$\Delta T_1 + \Delta T_2 + \Delta T_3 = Q(1/h_1 A_1 + 1/h_2 A_2 + 1/h_3 A_3) \qquad (7)$$

Equation (7) must then be written **in terms of** one of the area terms A_1, A_2, and A_3 or sometimes in terms of a mean area. Since Q and ΔT must be independent of the particular area considered, the value of U will vary according to which area is used as the basis. Thus equation (7) can be written, for example:

$$Q = U_1 A_1 \Delta T \quad \text{or} \quad \Delta T = Q/(U_1 A_1)$$

This will then give U_1 as:

$$1/U_1 = 1/h_1 + A_1/A_2 (1/h_2) + A_1/A_3 (1/h_3) \qquad (8)$$

In the above analysis it is assumed that the heat flowing **per unit time through each of the media** is the same. Now that the overall coefficient U has been **broken down** into its component parts, each of the individual coefficient h_1, h_2 and h_3 must be evaluated. This can be done from knowledge of the nature of the heat transfer process in each of the media. A study will therefore be made of how these individual coefficients can be calculated for conduction, convection, and radiation.

Hands-On Chemistry: **Heat Transfer**

When two objects at different temperatures are brought into thermal contact, heat flows from the object at the higher temperature to that at the lower temperature. The net flow is always in the direction of the temperature decrease. The mechanisms by which the heat may flow are three: conduction, convection and radiation.

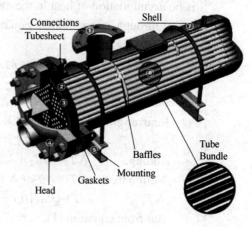

(Heat exchanger)

196

Exercises

1. Translate the following terms into English.
   (1) 对流   (2) 涡流   (3) 湍流   (4) 温度梯度   (5) 径向流动   (6) 单位介质
2. How physicists define heat energy?
3. What is the definition of heat transfer?
4. What are the forms of heat transfer?
5. One choice questions.
   (1) Higher the temperature difference, _____ will be heat transfer rate.
       (a) lower   (b) higher
       (c) heat transfer rate independent of temperature difference
   (2) The heat content of a substance is called _____.
       (a) sensible heat   (b) enthalpy   (c) latent heat   (d) total heat
   (3) The space heater does heat the air.
       (a) true   (b) false
   (4) Heat that is known to be added to or removed from a substance but no temperature change is recorded is called as _____.
       (a) entropy   (b) enthalpy   (c) latent heat   (d) specific heat
   (5) The boiling point or boiling temperature of water can be changed by changing the pressure on the water.
       (a) true   (b) false
   (6) On the Fahrenheit scale, absolute zero is _____.
       (a) −273 degrees (b) −183 degrees (c) −460 degrees (d) −360 degrees
   (7) Any temperatures on the absolute Celsius scale (Kelvin) can be found by adding _____ to the thermometer reading.
       (a) 460   (b) 273   (c) −273   (d) −460
   (8) The rate of heat consumption (BTU/h) has the unit of _____.
       (a) energy   (b) power   (c) work
   (9) Heat transfer will occur when temperature difference exist within the same medium.
       (a) true   (b) false
   (10) The greater the temperature difference between the two bodies, heat transfer rate will be _____.
       (a) less   (b) same   (c) greater
   (11) When the temperature of the substance is reduced to −273℃ or 0K, molecular movement in the substance _____.
       (a) slows down   (b) accelerates   (c) does not move
6. A topic for speaking.
   List the three ways of heat transfer and give some examples.

【Reading Material】

Applications of Thermodynamics

Before committing a great deal of time and effort to the study of a subject, it is reasonable to ask the following two questions: what is it? What is it good for? Regarding thermodynamics, the second question is more easily answered, but an answer to the first is essential to an understanding of the subject. Although it is doubtful that many experts or scholars would agree on a simple and precise definition of thermodynamics, necessity demands that a definition be attempted. However, this is best accomplished after the applications of thermodynamics have been discussed.

There are two major applications of thermodynamics, both of which are important to chemical engineers:

(1) The calculation of heat and work effects associated with processed as well as the calculation of the maximum work obtainable from a process or the minimum work required to drive a process.

(2) The establishment of relationships among the variables describing systems at equilibrium.

The first application is suggested by the name thermodynamics, which implies heat in motion. Most of these calculations can be made by the direct implementation of the first and second laws. Examples are calculating the work of compressing a gas, performing an energy balance on an entire process or a process unit, determining the minimum work of separating a mixture of ethanol and water, or evaluating the efficiency of an ammonia synthesis plant.

The application of thermodynamics to a particular system results in the definition of useful properties and the establishment of a network of relationships among the properties and other variables such as pressure, temperature, volume, and mole fraction. Actually, application 1 would not be possible unless a means existed for evaluating the necessary thermodynamic property changes required in implementing the first and second laws. The property changes are calculated from experimentally determined data via the established network of relationships. Additionally, the network of relationships among the variables of a system allows the calculation of values of variables which are either unknown or difficult to determine experimentally from variables which are either available or easier to measure. For example, the heat of vaporizing a liquid can be calculated from measurements of the vapor pressure at several temperatures and the densities of the liquid and vapor phases at several temperatures, and the maximum conversion obtainable in a chemical reaction at any temperature can be calculated from calorimetric measurements performed on the individual substances participating in the reaction.

Questions

1. Identify the application of thermodynamics.
2. State and describe briefly each of the three thermodynamic laws that you have learned.

Key Terms to Chemical Engineering

1. unbound electron 自由电子
2. conductivity 传导率，导热系数
3. diffuse 扩散，散布
4. opaque 不透明的，不传导的，无光泽的
5. enthalpy 焓
6. flux 通量
7. buoyancy 浮力，浮性
8. agitator 搅拌器，搅拌装置
9. fuse 熔融，熔化
10. unlagged 未保温的，未绝缘的
11. cross section 截面，剖面
12. adiabatically 绝热地
13. centrifugal 离心力的
14. potential 势能，位能，电势
15. shaft （传动，旋转）轴
16. reflux 回流，倒流
17. condenser 冷凝器
18. entropy 熵
19. hydraulics 水力学，液压系统
20. momentum 动量，冲量
21. potential flow 势流
22. tension 张力，弹力
23. laminar 层流的，层状的
24. cross-current 错流，正交流
25. conduit 导管，输送管
26. mainstream 干流，主流
27. critical velocity 临界速度
28. viscous 黏（性，滞，稠）的
29. dissipate 使耗散，消除，消耗
30. lixiviate 浸提，溶滤
31. humidification 增湿作用
32. leaching 浸取，浸提
33. relative volatility 相对挥发度
34. rectify 精馏
35. stripping 洗提，汽提，解吸
36. stripping section 提馏段
37. multiple-feed 多口进料
38. sidestream 侧线馏分
39. mass separating agent 质量分离剂
40. flash drum 闪蒸槽
41. rectifier 精馏器，整流器
42. cross-flow 错流
43. downcomer 降液管
44. shower plate 喷淋塔
45. sieve 筛子，滤网
46. valve plate 浮阀塔
47. flap 活盖，簧片，阀门
48. mild steel 低碳钢
49. turn-down ratio 极限负荷比，操作弹性
50. entrainment 挟带，雾沫
51. wire web 金属丝网，网体填料
52. grid 栅格
53. structured packing 结构填料
54. mesh 网，筛，目
55. void fraction 孔隙率
56. revamp 改进，整形
57. pulsed column 脉冲塔
58. HETS=height equivalent to a theoretical stage 等板高度
59. rotary-agitation column 回流搅拌塔
60. centrifugal extractor 离心萃取器
61. reformate （汽油）重整产品
62. extract 萃取相
63. raffinate 萃余液
64. activity coefficient 活度系数
65. ventilate 通风，排气

Unit 10

Fine Chemicals

Lesson 32　Introduction to Fine Chemicals

1. bulk chemicals
散装化学品，大宗化学品
2. commodity
[kə'mɑdəti]
商品
3. ethylene oxide
环氧乙烷
4. pseudo commodities
半通用商品
5. formulate
['fɔrmjə'let]
构想出，规划，阐述
6. low volume chemicals
小批量化学品
7. specification
['spɛsəfə'keʃən]
说明，规格
8. performance
性能
9. specialty chemicals
专用化学品
10. pharmaceuticals
医药品
11. pesticide
['pɛstɪsaɪd]
杀虫剂
12. flavors　调味品
13. intermediates
中间体
14. active ingredients
活性成分
15. agrochemical
农用化学品
16. bulk actives
原料药

What are Bulk, Specialty, and Fine Chemicals?

Chemicals are divided on the basis of volume and character. **Bulk chemicals**, or **commodities**, are produced in large quantities and sold on the basis of an industry specification. Typical examples would be acetone, **ethylene oxide**, and phenol. **Pseudo commodities** are also made in large quantities but are sold on the basis of their performance. In many cases the product is **formulated** and properties can differ from one supplier to another. Examples include large volume polymers, surfactants, paints, etc.

Similarly, **low volume chemicals** are classified according to whether they are sold primarily on the basis of **specification** or **performance**. Specialties are generally formulations that are sold on the basis of their performance and their prices reflect their value rather than cost of production. Producers of **specialty chemicals** often provide extensive technical service to their customers. Examples of specialty chemicals include **pharmaceuticals**, **pesticides**, **flavors** and fragrances, specialty polymers, etc. Fine chemicals, on the other hand, are produced to customer specifications and are often **intermediates** or **active ingredients** for specialty chemicals, e.g. pharmaceutical and **agrochemical** intermediates and **bulk actives**.

Characteristic Features of Fine Chemicals Manufacture

Fine chemicals are products of high and well-defined purity, which are manufactured in relatively small amounts and sold at relatively high price. Fine chemicals can be divided in two basic groups: those that are used as intermediates for other products, and those that by their nature have a specific activity and are used based on their performance characteristics. Performance chemicals are used as active ingredients or additives in

17. multifunctional 多功能的	
18. labile ['lebɪl] 易变化的	
19. elevated ['ɛlɪvetɪd] 升高的	
20. quenching 猝灭	
21. profitable ['prɑfɪtəbl] 赢利的	
22. isolation [,aɪsə'leʃən] 分离	
23. sideproducts 副产品	
24. optical isomers 光学异构体	
25. single-pass 单程	
26. hazardous ['hæzədəs] 有危险的	
27. flammable ['flæməbl] 易燃的,可燃的	
28. cyanides 氰化物	
29. phosgene ['fɑzdʒin] 光气,碳酰氯	
30. isocyanate 异氰酸酯	
31. toxic 有毒的	
32. effluent ['ɛfluənt] 污水,工业废水	
33. batch stirred-tank reactors 一批搅拌式反应器	
34. inventory ['ɪnvəntɔri, -,tori] 目录,清单	
35. thermal runaways 热的释放	

formulations, and as aids in processing.

Complex, **multifunctional** large molecules take up a significant proportion of the fine chemicals. These molecules are **labile**, unstable at **elevated** temperature, and sensitive towards (occasionally even minor) changes in their environment (e.g. pH). Therefore, processes are needed with inherent protective measures (e.g. chemical or physical **quenching**) or a precise control system to operate exactly within the allowable range. Otherwise the yield of the desired product can drop to nearly zero.

Fine chemicals are high-added-value products. In general, expensive raw materials are processed to obtain fine chemicals, and therefore, the degree of their utilization is very important. With complex reaction pathways, selectivity is the key problem to make the process **profitable**. Selectivity is significant also because of difficulties in **isolation** and purification of the desired product from many **sideproducts**, especially those with physical-chemical properties similar to those of the desired product (close boiling points, **optical isomers**, etc.). Furthermore, a low selectivity results in large streams of pollutants to be treated before they can be disposed of. Selectivity is even more an issue because in contrast with bulk chemicals production, where a limited **single-pass** conversion coupled with separation and recycling of unreacted raw materials is often applied, usually complete conversion is aimed at. Selectivity can be controlled by chemical factors such as chemical route, solvent, catalyst and operating conditions, but it is also strongly dependent on engineering solutions. Catalysis is the key to increasing the selectivity.

In the manufacture of fine chemicals many **hazardous** chemicals are used, such as highly **flammable** solvents, **cyanides**, **phosgene**, halogens, volatile amines, **isocyanates** and phosphorous compounds. The use of hazardous and **toxic** chemicals produces severe problems associated with safety and **effluent** disposal. Moreover, fine chemistry reactions are predominantly carried out in **batch stirred-tank reactors** characterized by ① a large **inventory** of dangerous chemicals; ② a limited possibility to transfer the generated heat to the surroundings. Therefore, the risk of **thermal runaways**, explosions, and emissions of pollutants to the surroundings is greater than in bulk (usually continuous) production. That is why much attention must be paid to safety, health hazards, and waste disposal during development, scale-up, and operation of the process.

Fine chemicals are often manufactured in multistep conventional syntheses, which results in a high consumption of raw materials and, consequently, large amounts of by-products and wastes. The high raw materials-to-product ratio in fine chemistry justifies extensive search for selective catalysts. Use of effective catalysts would result in a decrease of reactant consumption and waste production, and the simultaneous reduction of the number of steps in the synthesis.

36. fluctuation [ˌflʌktʃuˈeʃən] 波动	
37. dedicated [ˈdɛdəˌketɪd] 专用的	
38. enormous [ɪˈnɔrməs] 巨大的	
39. multipurpose 多目标的	
40. multiproduct 多产品	
41. versatile equipment 通用装置	
42. batch [bætʃ] 批	
43. residence time 停留时间	
44. foregoing 前面提到的	
45. profoundly 极大的	
46. constraint [kənˈstrent] 系统规定参数	
47. impurities 杂质	
48. regulations and specifications 规章规范	
49. isomer [ˈaɪsəmɚ] 同分异构体	
50. enantiomer [ɪˈnæntiəmɚ] 对映体	
51. racemic [reˈsimɪk, -ˈsɛmɪk, rə-] 外消旋的	
52. stereoselective 立体选择性的	

One of the most important features of fine chemicals manufacture is the great variety of products, with new products permanently emerging. Therefore, significant **fluctuations** in the demand exist for a variety of chemicals. If each product would be manufactured using a plant **dedicated** to the particular process, the investment and labour costs would be **enormous**. In combination with the ever changing demand and given the fact that plants are usually run below their design capacity, this would make the manufacturing costs very high. Therefore, only larger volume fine chemicals or compounds obtained in a specific way or of extremely high purity are produced in dedicated plants. Most of the fine chemicals, however, are manufactured in **multipurpose** or **multiproduct** plants (MPPs). They consist of **versatile equipment** for reaction, separation and purification, storage, effluent treatment, solvent recovery, and equipment for utilities. By changing the connections between the units and careful cleaning of the equipment to be used in the next campaign, one can adapt the plant to the intended process. The investment and labour costs are significantly lower for MPPs than for dedicated plants, while the flexibility necessary to meet changing demands is provided.

The most versatile reactors are stirred-tank reactors operated in **batch** or semibatch mode, so such reactors are mainly used in multiproduct plants. Continuous plants with reactors of small volume are sometimes used despite the small capacity required. This is the case when the **residence time** of reactants in the reaction zone must be short or when too much hazardous compounds could accumulate in the reaction zone.

From the **foregoing**, it will be clear that in fine chemicals process development and the strategy differ **profoundly** from that in the bulk chemical industry. The major steps are ① adaptation of procedures to **constraints** imposed by the existing facilities with some necessary equipment additions, or ② choice of appropriate equipment and determination of procedures for a newly built plant, in such a way that procedures in both cases guarantee the profitable, competitive, and safe operation of a plant.

The accuracy of analytical methods has increased enormously in the past decades and this has enabled detection of even almost negligible traces of **impurities**. The consequence is that both **regulations and specifications** for intermediates and final fine chemicals have become stricter. Therefore, very pure compounds must often be produced with impurities at ppm or ppb level.

The production of complex molecules in many cases results in mixtures containing **isomers**, including optical isomers. The demand for **enantiomeric** materials is growing at the expense of their **racemic** counterparts, driven primarily by the pharmaceutical industry. Therefore, both **stereoselective** and effective synthesis, often non-conventional

53. chromatography
[ˌkrɒməˈtɑːrəfi]
色谱
54. supercritical
超临界的
55. racemate
外消旋物
56. disappearing ink
消色墨水
57. acid-base indicator
酸碱指示剂
58. thymolphthalein
麝香草，酚酞，百里酚酞
59. phenolphthalein
[ˌfiːnɒlˈθæliːn]
酚酞

methods of purification [e.g. High Performance Liquid **Chromatography** (HPLC) and treatment under **supercritical** conditions] are in a wider use to meet the stricter requirements. The increasing need for optical purity, with complex and often not very effective methods of **racemate** resolution, stimulates the development of new stereoselective catalysts, including biocatalysts. In this respect, biotechnology is becoming a competitor to classical chemical technology.

Hands-On Chemistry: **How to Make Disappearing Ink**

Disappearing ink is a water-based **acid-base indicator** (pH indicator) that changes from a colored to a colorless solution upon exposure to air. The most common pH indicators for the ink are **thymolphthalein** (blue) or **phenolphthalein** (red or pink). The indicators are mixed into a basic solution that becomes more acidic upon exposure to air, causing the color change. Note that in addition to disappearing ink, you could use different indicators to make color-change inks, too.

Exercises

1. Translate the following phrase from English to Chinese.
   (1) multi-phase reactions (2) competing reactions (3) continuous vs. batch processes
   (4) continuous /semicontinuous processes (5) hygiene products and toiletries
   (6) high-added-value products (7) bulk chemicals (8) specialty chemicals
   (9) fine chemicals (10) intrinsic properties (11) pseudo commodities
   (12) low volume chemicals (13) active ingredients (14) bulk active
   (15) performance chemicals (16) batch stirred tank reactor
   (17) thermal runaways (18) regulations and specifications
   (19) existing facilities (20) scale up

2. Translate the following phrase from Chinese to English.
   (1) 原料   (2) 副产物   (3) 目标产物   (4) 保护措施   (5) 中间体
   (6) 理化性能   (7) 工程方案   (8) 危险化学品   (9) 废水处理   (10) 批量生产
   (11) 小容量   (12) 停留时间   (13) 分子模拟   (14) 立体选择性催化剂
   (15) 超临界条件

3. Translate the following passage from English to Chinese.

   (1) Chemicals are divided on the basis of volume and character. Bulk chemicals, or commodities, are produced in large quantities and sold on the basis of an industry specification. Typical examples would be acetone, ethylene oxide, and phenol. Pseudo commodities are also made in large quantities but are sold on the basis of their performance. In many cases the product is formulated and properties can differ from one supplier to another. Examples include large volume polymers, surfactants, paints, etc.

   (2) Fine chemicals are products of high and well-defined purity, which are manufactured in relatively small amounts and sold at relatively high price. Fine chemicals can be divided in two basic groups: those that are used as intermediates for other products, and those that by their nature have a specific activity and are used based on their performance

characteristics. Performance chemicals are used as active ingredients or additives in formulations, and as aids in processing.

(3) Fine chemicals are often manufactured in multistep conventional syntheses, which results in a high consumption of raw materials and, consequently, large amounts of by-products and wastes. The high raw materials-to-product ratio in fine chemistry justifies extensive search for selective catalysts. Use of effective catalysts would result in a decrease of reactant consumption and waste production, and the simultaneous reduction of the number of steps in the synthesis.

4. Translate the following sentence from Chinese to English.
   (1) 通常必须生产出杂质是 ppm 或 ppb 级的高纯产品。
   (2) 对各种化学品存在明显的需求波动。
   (3) 其他一些蒸发过程将简单地加以讨论。
5. A topic for speaking.
   What you can do if you are going to work for a fine chemistry plant?

【Reading Material】

Chemists Discover How Antiviral Drugs Bind To and Block Flu Virus

Antiviral drugs(抗病毒药物) block **influenza A viruses**(A 型流感病毒) from reproducing and spreading by attaching to a site within a **proton channel**(质子通道) necessary for the virus to **infect**(感染) healthy cells, according to a research project led by **Iowa State University's**(爱荷华州立大学) Mei Hong and published in the Feb. 4 issue of the journal Nature.

Hong's research concluded that when amantadine is present at the **pharmacologically**(药理学) relevant amount of one molecule per channel, it attaches to the lumen inside the proton channel. But the paper also reports that when there are high concentrations of amantadine in the membrane, the drug will also attach to a second site on the surface of the virus protein near the channel.

"Our study using solid-state NMR technology **unequivocally**(无疑) shows that the true binding site is in the channel lumen, while the **surface-binding site**(表面结合点) is occupied only by excess drug," Hong said. "The previous solution NMR study used 200-fold excess drug, which explains their observation of the surface-binding site. The resolution of this **controversy**(矛盾) means that medical chemists can now try to design new drugs to target the true binding site of the channel."

Here's how a flu virus uses its proton channel and how amantadine blocks that channel.

The virus begins an infection by attaching itself to a healthy cell. The healthy cell surrounds the flu virus and takes it inside the cell through a process called **endocytosis**(内吞作用). Once inside the cell, the virus uses a protein called M2 to open a channel to the healthy cell. Protons from the healthy cell flow through the channel into the virus and raise its acidity. That **triggers**(触发) the release of the virus' **genetic**(基因的) material into the healthy cell. The virus **hijacks**(劫持) the healthy cell's resources and uses them to reproduce and spread.

When amantadine binds to and blocks the M2 proton channel, the process doesn't work and a virus can't infect a cell and spread.

Hong and the research team developed powerful techniques to study the proton channel using solid-state NMR spectroscopy, the technology behind medical **magnetic resonance imaging**(磁共振成像). The techniques provided the researchers with a detailed look at the antiviral drug within the proton channel, showed them the structure of the protein at the drug-binding site and allowed them to make accurate measurements of the distances between the drug and the protein.

The researchers also found that amantadine **spins**(旋转) when it binds to the inside of the proton channel. That means it doesn't fill the channel. And Hong said that leaves room for development of other drugs that do a better job blocking the channel, stopping the flu and **evading**(躲避) development of drug **resistance**(抗药性).

Questions

1. According to this reading material, how can an antiviral drug block influenza viruses from reproducing and spreading effectively?

2. What kind of technique did Hong's team use for this research? Describe the function of this kind of technique.

Key Terms to Fine Chemistry

1. fine chemicals　精细化学品
2. heavy chemicals　大宗化学品
3. specialty chemicals　专用化学品
4. accelerating agent (accelerator)　促进剂
5. adhesive　黏合剂
6. antibiotics　抗菌素
7. antifreezing agent　抗冻剂
8. antioxidant　抗氧剂
9. azeotrope　共沸混合物
10. acaricide　杀螨剂
11. batch reactors　批量生产反应器
12. biocatalysis　生物催化
13. bubble-cap tower　泡罩塔
14. cabinet dryer　干燥箱
15. cable oil　电缆油
16. caking agent of coal　煤黏结剂
17. calcium antagonist　钙拮抗剂
18. chemical reactor　化学反应器
19. conceptual design　概念设计
20. colorant　颜料，着色剂
21. commodity　用品
22. condiment　调味品
23. container　容器
24. cooler　冷却器
25. cosmetic　化妆品
26. decolorant　脱色剂
27. denatured alcohol　变性酒精
28. binder　黏合剂
29. destructive distillation　分解蒸馏
30. detergent　洗涤剂
31. developer　显影剂
32. droplet　液滴
33. dyestuff　染料
34. emulsion　乳剂
35. enzyme　酶
36. essential oil　（香）精油
37. fermentation　发酵
38. filament　细丝，丝状体
39. filter　过滤器，滤色片
40. favoring　香剂，调味剂
41. flow reactors　连续生产反应器
42. fractional distillation　分馏
43. fumigant　熏蒸(消毒)剂
44. fungicide　杀真菌剂
45. herbicide　除草剂
46. hold-up　塔储量，容纳量
47. humectant　润湿剂
48. ingestion　吸收，吸入
49. inlet　进口，入口
50. insecticide　杀虫剂
51. insulin　胰岛素
52. jacket cooling　套管冷却
53. jasmin oil　茉莉花油
54. jet condenser　喷水凝汽器
55. lining　衬里，衬料，衬套
56. lubricating grease　润滑脂
57. overhead　塔顶馏出物
58. overheat　过热
59. oxidant　氧化剂
60. perfume　香料
61. pesticide　杀虫剂
62. pharmaceuticals　药物
63. pigment　颜料
64. preservative　防腐剂
65. propellant　推进剂
66. recover　回收
67. recrystallization　重结晶
68. rectifier　精馏器
69. scale-up ratio　放大生产比例
70. sequential scale-up　连续放大生产
71. settle　(使)沉淀，澄清
72. setup　装置，装配
73. sewage　污水
74. still pot　蒸馏釜
75. suspension　悬浮液
76. surfactant　表面活性剂
77. sweetener　增甜剂
78. stoichiometric ratio　化学计量比
79. per pass conversion　单程转化率
80. theoretical yield　理论收率
81. orthogonal experiment　正交实验法

Part 3
Chemical Reactions and Stoichiometry

Lesson 33 Chemical Reactions

What is a Chemical Reaction?

A chemical reaction occurs when substances (the reactants) **collide**（碰撞） with enough energy to rearrange to form different compounds (the products). The change in energy that occurs when a reaction take place is described by **thermodynamics**（热力学） and the rate or speed at which a reaction occurs is described by **kinetics**（动力学）. Reactions in which the reactants and products coexist are considered to be in **equilibrium**（处于平衡）. A chemical equation consists of the **chemical formula**（化学式） of the reactants, and the chemical formula of the products. The two are separated by an "⟶" usually read as "yields" and each chemical formula is separated from others by a **plus sign**（加号）. Sometimes a triangle is drawn over the arrow symbol to denote energy must be added to the substances for the reaction to begin. Each chemical formula may be preceded by a **scalar**（数量的） coefficient indicating the **proportion**（比例） of that substance necessary to finish the reaction in formula. For instance, the formula for the burning of methane ($CH_4 + 2 O_2 \longrightarrow CO_2 + 2H_2O$) indicates that twice as much O_2 as CH_4 is needed, and when they react, twice as much H_2O as CO_2 will be produced. This is because during the reaction, each atom of carbon needs exactly two atoms of oxygen to combine with, to produce the CO_2, and every two atoms of hydrogen need an atom of oxygen to combine with to produce the H_2O. If the proportions of the reactants are not respected, when they are forced to react, either not all of the substance used will participate in the reaction, or the reaction that will take place will be different from the one noted in the equation.

Some Types of Chemical Reactions

In a **combination reaction**（化合反应）, two reactants combine to give a single product, i.e., A+B ⟶ AB.

Example: $2H_2 + O_2 \longrightarrow 2H_2O$; $C + O_2 \longrightarrow CO_2$

When two elements react, a combination reaction occurs producing a **binary compound**（二元化合物）(that is, one consisting of only two types of atoms). If a metal and a nonmetal react, the product is ionic with a formula determined by the charges on the ions the elements form. If two nonmetals react, the product is a molecule with polar covalent bonds, with a formula consistent with the normal valences of the atoms involved.

In a **decomposition reaction**（分解反应）, a single compound breaks down to give two or more other substances, i. e., AB ⟶ A + B

Example: $2 H_2O \longrightarrow 2H_2 + O_2$; $2 HgO \longrightarrow 2Hg + O_2$

Reaction of a metal oxide with water produces a metal hydroxide, that is, a strong base. Reaction of a nonmetal oxide with water produces an **oxyacid**（含氧酸）in which the nonmetal is in the same oxidation state as in the oxide you started with.

Both of these are combination reactions, and both can be reversed by heating the products. Metal hydroxides decompose on heating to give the metal oxide and water. Oxyacids decompose on heating to give water and the nonmetal oxide in the appropriate oxidation state.

In a **displacement reaction**（置换反应）, atoms or ions of one substance replace other atoms or ions in a compound (A+BC ⟶ AC+B). Metals can be arranged in an activity series based on their ability to displace hydrogen from water or acids and their ability to displace each other in soluble ionic compounds. Zinc metal reacts with aqueous hydrochloric acid to happen a displacement reaction.

$$Zn(s) + HCl(aq) \longrightarrow ZnCl_2 + H_2(g)$$

Partner exchange reaction（double decomposition, double displacement, and metathesis，复分解反应）have the general form AC + BD ⟶ AD +BC. Often such reactions occur between ionic compounds in solution when one product is an insoluble solid, known as a precipitate.

Example: $AgNO_3(aq) + NaCl(s) \longrightarrow AgCl(s) + NaNO_3(aq)$

The Bronsted-Lowry definition describes acids as proton donors and bases as proton acceptors. Acids and bases can react with water and also react together. An acid in water will dissociate and a base will undergo **hydrolysis**（水解）, which means that it splits a water molecule. The acid dissociation and base hydrolysis reactions are described in the acid-base equilibrium document.

The reaction of an acid and a base is called a **neutralization**（中和） reaction. Mixing acids and bases results in neutralization because the base will accept the proton that the acid donates. What remains when an acid and a base react depends on the relative amounts of the acid and base. The following example shows neutralization of equal molar quantities of a strong acid and a strong base.

$$HNO_3 + NaOH \longrightarrow H_2O(l) + Na^+(aq) + NO_3^-(aq)$$

The spectator ions are usually left out of the reaction and the "net" reaction is:

$$H^+(aq) + OH^-(aq) \longrightarrow H_2O(l)$$

Precipitation Reactions

Many compounds have limited solubility in aqueous (water) solution. When the concentrations of the ions in a solution rise above the solubility limit, the ions combine to form solid particles that precipitate from solution. The concentrations of the ions remaining in solution are governed by the equilibrium constant, K_{sp}, which is called the solubility product.

Example: When chloride is added to a silver solution, solid silver chloride precipitates from solution. The resulting equilibrium is always written in the direction of the solid dissolving:

$$AgCl(s) \rightleftharpoons Ag^+(aq) + Cl^-(s)$$

Complexation Reactions

Metal ions in solution can bind with ligands to form soluble complexes. These reactions are always described by the equilibrium expression written in the direction of complex formation. The equilibrium constant, K_f, is called the formation constant.

Example: When ethylenediaminetetraacetic acid (EDTA) is added to a solution containing calcium ion, a calcium (EDTA) complex forms in solution:

$$Ca^{2+}(aq) + EDTA^{4-}(aq) \rightleftharpoons Ca(EDTA)^{2-}(aq)$$

Redox Reactions (氧化还原反应)

Reduction-oxidation (redox) reactions involve the transfer of electrons from one species to another. One species will be oxidized and the other will be reduced.

Example: $Zn(s) + Cu^{2+}(aq) \longrightarrow Zn^{2+}(aq) + Cu(s)$

In this reaction two electrons are transferred from each zinc atom to each copper ion. The zinc metal is oxidized to zinc ions and the copper ions are reduced to copper metal.

Hands-On Chemistry: Any Fun Things I Can Make with Ammonium Nitrate?

Leaning Towards Rockets?

Playing with ammonium nitrate (NH_4NO_3) will either land you in jail or in a hospital (possibly both)! Ammonium nitrate does not **explode**（爆炸）when **ignited**（点燃）, contrary to popular **belief**（看法）. In fact, it will not even burn. It only explodes when initiated to undergo **detonation**（爆炸声）by a sufficiently strong **hypersonic**（超音速）**shock wave**（冲击波）, such as that produced by **blasting caps**（雷管）. Regarding fun things with rockets, search for **amateur**（业余的）ammonium nitrate rocket sites. Usually they use magnesium as a fuel. However, ammonium nitrate is harder and harder to get nowadays, and there are better rocket fuels.

Exercises

1. Balance and classify each of the following chemical equations as a (a) combination reactions, (b) decomposition reaction, (c) displacement reaction, or partner-exchange reaction.

   (1) $Fe_3O_4 + H_2(g) \longrightarrow Fe(s) + H_2O(l)$

   (2) $KClO_3(s) \longrightarrow KCl(s) + O_2(g)$

   (3) Steam and hot carbon react to form gaseous hydrogen and gaseous carbon monoxide.

   (4) $Cl_2O_7(g) + H_2O(l) \longrightarrow HClO_4(aq)$

   (5) $Br_2(l) + H_2O(l) \longrightarrow HBr(aq) + HBrO(aq)$

   (6) $Ca_3(PO_4)_2(s) + H_2SO_4(aq) \longrightarrow CaSO_4(s) + H_3PO_4(aq)$

   (7) Potassium reacts with water to give aqueous potassium hydroxide and gaseous hydrogen.

   (8) Solid magnesium carbonate decomposes to form solid magnesium oxide and gaseous carbon monoxide.

2. Translate the following passages.

   (1) When two elements react, a combination reaction occurs producing a binary compound (that is, one consisting of only two types of atoms). If a metal and a nonmetal react, the product is ionic with a formula determined by the charges on the ions the elements form. If two nonmetals react, the product is a molecule with polar covalent bonds, with a formula consistent with the normal valences of the atoms involved.

   (2) Both of these are combination reactions, and both can be reversed by heating the products. Metal hydroxides decompose on heating to give the metal oxide and water. Oxyacids decompose on heating to give water and the nonmetal oxide in the appropriate oxidation state.

   (3) Many compounds have limited solubility in aqueous (water) solution. When the concentrations of the ions in a solution rise above the solubility limit, the ions combine to form solid particles that precipitate from solution. The concentrations of the ions remaining in solution are governed by the equilibrium constant, K_{sp}, which is called the solubility product.

   (4) Metal ions in solution can bind with ligands to form soluble complexes. These reactions are always described by the equilibrium expression written in the direction of complex formation. The equilibrium constant, K_f, is called the formation constant.

Lesson 34 Chemical Calculations

Stoichiometry

When chemical species (atoms, molecules, and ions) go into "action", we have chemical reactions—processes of chemical change. The calculation of the quantitative relationships in chemical change is called **stoichiometry** (化学计量). Stoichiometry problems can easily be solved when amounts of substances are converted from mass (in units of g, kg, etc.), volume (L), into moles. Amounts in moles depend on how the chemical formula or the chemical reaction equations are written. Chemical reaction equations are the basis for reaction stoichiometry, but when the reactants are not stoichiometric mixtures, some reactants will be in excess whereas others will be in limited supply.

Equation:	$2 H_2(g)$	+	$O_2(g)$	$\longrightarrow$	$2 H_2O(l)$
Molecules:	2 molecules H_2	+	1 molecule O_2	$\longrightarrow$	2 molecules H_2O
Mass/amu:	4.0 amu H_2	+	32.0 amu O_2	$\longrightarrow$	36.0 amu H_2O
Amount/mol:	2 mol H_2	+	1 mol O_2	$\longrightarrow$	2 mol H_2O
Mass/g:	4.0 g H_2	+	32.0 g O_2	$\longrightarrow$	36.0 g H_2O

Solving Stoichiometry Problems

In all stoichiometric problems, the **mole ratios**（摩尔比）from the balanced chemical equation provide the connection between the known and unknown quantities, whether these are masses, pressure-volume-temperature data for gases, or molarities for substances in solution. To solve a stoichiometry problem, first write the balanced chemical equation, convert the known information to moles, use mole ratios to find the unknown in terms of moles, and convert the answer from moles to the desired quantity. Notice that we cannot directly convert from grams of one compound to grams of another. Instead we have to go through moles. Many stoichiometry problems follow a pattern:

$$\text{grams}(x) \leftrightarrow \text{moles}(x) \leftrightarrow \text{moles}(y) \leftrightarrow \text{grams}(y)$$

Example: How many grams of water can be got when 1g of $C_6H_{12}O_6$ are oxidized completely?

$$C_6H_{12}O_6 + 6 O_2 \longrightarrow 6CO_2 + 6H_2O$$

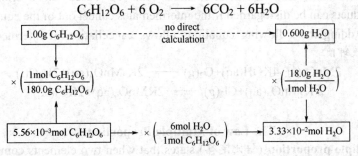

Starting with 1.00g of $C_6H_{12}O_6$, then we calculate the moles of $C_6H_{12}O_6$, use the **coefficients** (系数) to find the moles of H_2O, then turn the moles of water to grams.

Limiting Reactants

The exact amount of a substance required by a balanced chemical equation is the stoichiometric amount. When reactants are present in non-stoichiometric amounts, the one that determines the amount of product that can be formed is called the **limiting reactant** (限量反应物). At the completion of the reaction, some of the other reactant(s) will be left over.

Example: 10.0g of aluminum reacts with 35.0g of chlorine gas to produce aluminum chloride. Which reactant is limiting, which is in excess, and how much product is produced?

$$2Al + 3Cl_2 \longrightarrow 2AlCl_3$$

We get 49.4g of aluminum chloride from the given amount of aluminum, but only 43.9g of aluminum chloride from the given amount of chlorine. Therefore, chlorine is the limiting reactant, aluminum is excess. Once the 35.0g of chlorine is used up, the reaction comes to a complete stop.

By calculating the amount of the excess reactant needed to completely react with the limiting reactant, we can **subtract**（减去）that amount from the given amount to find the amount of excess.

Theoretical and Actual Yields

For known amounts of reactants, theoretical amounts of products can be calculated in a chemical reaction or process. Calculated amounts of products are called **theoretical yield**（理论产量）. In these calculations, the limiting reactant is the limiting factor for the theoretical yields of all products. However, in a reaction to prepare a compound, you may get less than the theoretical yield, because of incomplete reactions or loss. The amount recovered divided by the theoretical yield gives a **percent yield**（%, 百分产率）or **actual yield**（实际产率）.

Example: A chemist ran the following reaction with 2.5g of salicylic acid and excess acetic anhydride (over 2.0g). The actual yield was 2.43g of aspirin. What was the percent yield?

$$HOC_6H_4COOH(s) + (CH_3CO)_2O(l) \longrightarrow CH_3COOC_6H_4COOH(s) + CH_3COOH(l)$$

salicylic acid   acetic anhydride   acetylsalicylic acid   acetic acid

We get 3.26g aspirin from the given amount of salicylic acid.

percent yield = [(actual yield)/(theoretical yield)] ×100 % = (2.43g/3.26g) ×100% = 74.5%

Stoichiometry in Industrial Chemistry

Calculations can be simplified by using such units as pound-moles or ton-moles instead of moles, in order to find masses in pounds, or tons, and so on, instead of in grams. For processes that include consecutive reactions, stoichiometric calculations can be based on overall equations and intermediate products can be disregarded if the intermediates cancel out of the equations.

Example: Adding the following equations allows cancellation of intermediate, **potassium manganate** (锰酸钾).

$$2MnO_2(s) + 4KOH(aq) + O_2(g) \longrightarrow 2K_2MnO_4(aq) + 2H_2O(l)$$
$$2 K_2MnO_4(aq) + Cl_2(g) \longrightarrow 2KMnO_4(aq) + 2KCl(aq)$$

The Law of Multiple Proportions

The law of multiple proportions(倍比定律) states that when two elements combine to form two or more different compounds, the weights of one compound that can combine with a given weight of the second compound form small whole number ratios. For example, consider one experiment in which 10.0 grams of sulfur is combined with 10.0 grams of oxygen to form an oxide of sulfur, and another experiment under different conditions in which 3.21 grams of sulfur is combined with 4.82 grams of oxygen to form a different oxide. For each 10.0 grams of sulfur used in the second experiment, 15.0 grams (4.82 × 10.0/3.21) of oxygen is used. The ratios of the masses of oxygen that combine with a fixed mass of sulfur are 10.0:15.0, which is equal to the whole number ratio 2:3. This conforms to the law of multiple proportions.

Non-Stoichiometric Compounds

Most of chemistry is governed by simple whole-number ratios of molecules and atoms. Simple stoichiometry, although valid for the vast majority of mole ratios, is not universal: there are compounds with non-integral mole ratios. Substances such as **alloys**（合金） and glasses created problems for the initial acceptance of Dalton's atomic theory. There are, in addition, simple non-stoichiometric compounds that have varying ratios of **constituent**（组成的） atoms. Such compounds are generally crystalline solids with defects in their **crystal lattices**（晶格）, the lack of simple stoichiometry may give them important properties. **Wustite**（铁酸盐）, an oxide of iron, is an example of a non-stoichiometric compound. Its formula can be written Fe_nO, where n may have values varying from 0.88 to 1.00 and its physical and chemical properties will vary somewhat depending on the value of n.

Current Applications of Stoichiometry

Most chemical reactions are complex, occurring via many steps. In such cases, can an overall reaction be written that describes the stoichiometry of a system under consideration? Consider an example in which sulfur is burned in oxygen to simultaneously form sulfur dioxide (mostly) and some sulfur trioxide:

$$S + O_2 \longrightarrow SO_2 \qquad (1)$$
$$S + 1.5O_2 \longrightarrow SO_3 \qquad (2)$$

If the two reactions are added, the resulting equation is: $2S + 2.5O_2 \longrightarrow SO_2 + SO_3$. This representation of the reaction is plainly wrong because it states that one mole of SO_2 is obtained for every mole of SO_3, whereas most of the products consist of SO_2. The reason for this inconsistency is that the arrows in reactions (1) and (2) mean "becomes", they are not equivalent to equal signs because they involve time dependence. In order to obtain an overall stoichiometric description of the reaction, both equations (1) and (2) are necessary, as is knowledge about their relative importance in the overall reaction.

Hands-On Chemistry: Let Make Chocolate Chip Cookies!

What You Will Need   1 cup **butter**（黄油）, 1/2 cup white sugar, 1 cup packed **brown sugar**（红糖）, 1 teaspoon **vanilla extract**（香草精）, 2 eggs, 2 ½ cups all-purpose **flour**（面粉）, 1 teaspoon **baking soda**（发酵粉）, 1 teaspoon salt, 2 cups semisweet chocolate chips, makes 3 dozen.

Then answer the following questions.

How many eggs are needed to make 3 dozen cookies?

How much butter is needed for the amount of chocolate chips used?

How many eggs would I need to make 9 dozen cookies?

How much brown sugar would I need if I had 1 1/2 cups white sugar?

Exercises

1. Show all the following calculations.

    (1) $2C_4H_{10} + 13O_2 \longrightarrow 8CO_2 + 10H_2O$

    (a) What mass of O_2 will react with 400g C_4H_{10}?

    (b) How many moles of water are formed in (a)?

    (2) $3HCl + Al(OH)_3 \longrightarrow 3H_2O + AlCl_3$

    How many grams of aluminum hydroxide will react with 5.3moles of HCl?

    (3) $Ca(ClO_3)_2 \longrightarrow CaCl_2 + 3 O_2$

    What mass of O_2 results from the decomposition of 1.00kg of calcium chlorate?

    (4) The reaction of Ca with water can be predicted using the activity series. What mass of water is needed to completely react with 2.35g of Ca?

    (5) 15.0g of potassium reacts with 15.0g of iodine. Calculate which reactant is limiting and how much product is made.

    (6) 17.56g of ethanol (C_2H_6O) reacts with 102.5g of oxygen to produce carbon dioxide and water. How many grams of carbon dioxide are produced in this reaction? What is the limiting reactant? If only 30.00g of carbon dioxide are produced, what is the percent yield?

    (7) $4NH_3 + 5O_2 \longrightarrow 6H_2O + 4NO$

    (a) How many moles of H_2O can be made using 0.5mol NH_3?

    (b) What mass of NH_3 is needed to make 1.5mol NO?

    (c) How many grams of NO can be made from 120g of NH_3?

2. Translate the following passages.

    (1) When chemical species (atoms, molecules, and ions) go into "action", we have chemical reactions—processes of chemical change. The calculation of the quantitative relationships in chemical change is called stoichiometry. Stoichiometry problems can easily be solved when amounts of substances are converted from mass (g, kg), volume (L), into moles. Amounts in moles depend on how the chemical formula or the chemical reaction equations are written. Chemical reaction equations are the basis for reaction stoichiometry, but when the reactants are not stoichiometric mixtures, some reactants will be in excess whereas others will be in limited supply.

    (2) Calculations can be simplified by using such units as pound-moles or ton-moles instead of moles, in order to find masses in pounds, or tons, and so on, instead of in grams. For processes that include consecutive reactions, stoichiometric calculations can be based on overall equations and intermediate products can be disregarded if the intermediates cancel out of the equations.

    (3) Most chemical reactions are complex, occurring via many steps. In such cases, can an overall reaction be written that describes the stoichiometry of a system under consideration? Consider an example in which sulfur is burned in oxygen to simultaneously form sulfur dioxide (mostly) and some sulfur trioxide.

Part 4
Chemical Laboratory

Lesson 35 Common Labware

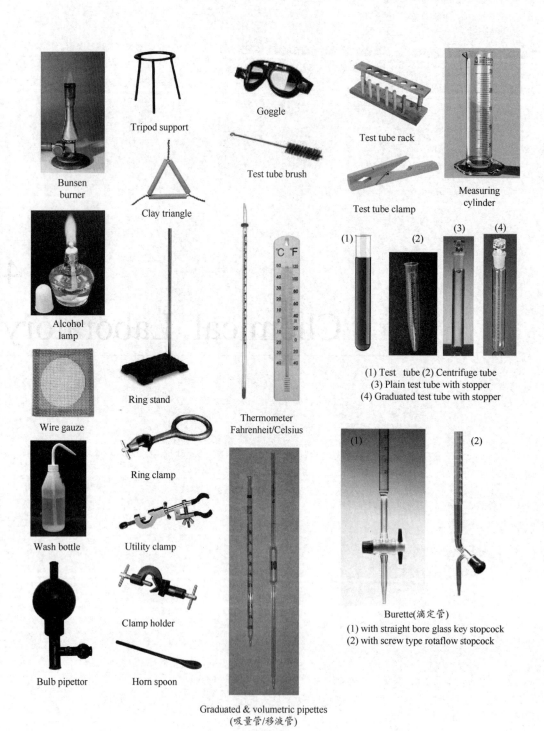

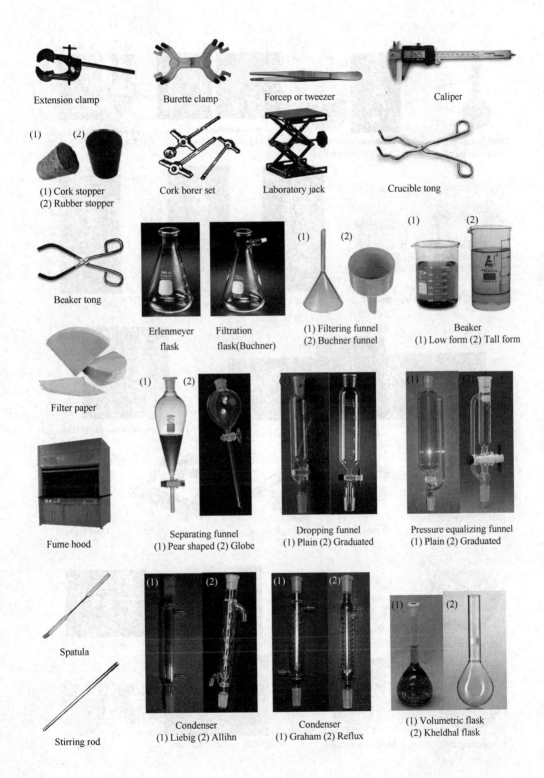

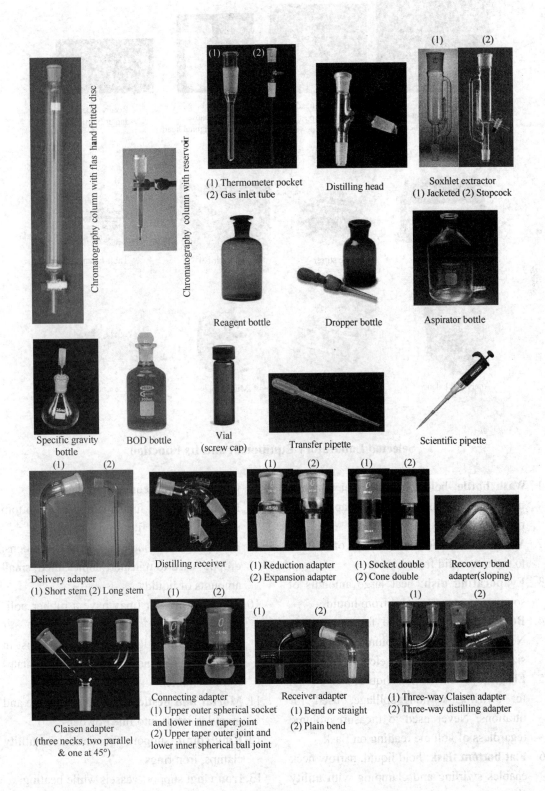

Adaptor
(1) Cone to rubber tubing
(2) Socket to rubber tubing

Splash heads
(1) Pear shape sloping
(2) Pear shape vertical

Stopper
(1) Penny head
(2) Flat hexagonal head

Receiver bend adapter (vertical)

Hot plate   Magnetic stirrer   Magnetic stir bar   Table balance

Analytical balance   Top pan balance   Microscope   pH test paper

Selected Laboratory Equipment and Its Function

1. **Wash bottle**: hold distilled water which is used to make solutions or holds tap water to rinse a precipitate from a beaker.
2. **Watch glass**: heat small amounts of solution to separate solid from liquids.
3. **Evaporating dish**: heat large amounts of solution to separate solid from liquid.
4. **Beaker**: hold liquid, use for water bath. Never used to measure volume even though sizes read 50mL, 100mL, etc.
5. **Erlenmeyer flask**: hold liquid, shape ideal for swirling without spillage, used for titrations. Never used to measure volume regardless of volume reading on flask.
6. **Flat bottom flask**: hold liquid, narrow neck enables swirling and clamping with utility clamp. Never used to measure volume.
7. **Filtration flask**: triangular flask with a side jet to assist in vacuum filtration.
8. **Buchner funnel**: funnel with porous bottom used for vacuum filtration.
9. **Dropper bottle**: bottle used to hold solutions; dropper used to transport unmeasured, small amounts of liquids.
10. **Stirring rod**: stir [may have a rubber policeman (橡胶刮铲) attached to help stir material in the edges of a beaker], assist in pouring liquids, and for removing precipitates from container.
11. **Mortar and pestle**: grind up crystalline and granular solids into fine powder.
12. **Ring stand**: support, foundation for utility clamps, iron rings.
13. **Iron ring**: support vessels while heating.
14. **Wire gauze**: spread heat evenly as it rests on an iron ring, ideal for supporting beaker

that is being heated.

15. **Thermometer**: measure temperature, typically in degrees Celsius.
16. **Thermometer pocket**: hold thermometer upright heating/cooling.
17. **Test tube**: hold chemicals, may be heated.
18. **Utility clamp**: hold vessels securely while heating, attach vessel to ring stand.
19. **Test tube clamp**: grasps hot test tubes, transport hot test tubes.
20. **Crucible**: heat solids.
21. **Clay triangle**: hold crucible on iron ring while heating.
22. **Crucible tong**: transport hot crucible.
23. **Desiccator**: store solids in a moisture-free environment.
24. **Funnel**: channel material into a narrow opening or separate insoluble matter from a liquid.
25. **Beaker tong**: transport hot beakers. Never used to transport hot ceramics as rubber will melt.
26. **Bunsen burner**: heat source using flame.
27. **Hot plate**: heat source without using flame, ideal when heating liquids with low flammability.
28. **Tubing clamp**: close rubber hose opening by pinching hose.
29. **Pinch clamp**: clamp a rubber connector.
30. **Rubber stopper**: close a vessel such as a test tube, Erlenmeyer flask or Florence flask.
31. **Spot plate**: small plate with several depressions or wells used to hold small samples of solutions.
32. **Test tube brush**: clean test tube.
33. **Balance**: measure mass.
34. **Spatula**: transport small amounts of solid in weighing.
35. **Separatory funnel**: separate liquids based on density.
36. **Graduated cylinder**: measure liquid volume accurately.
37. **Burette**: measure liquid volume accurately, ideal for titration analysis.
38. **Pipet**: measure small liquid volumes accurately (use pipettor, not ones mouth).
39. **Volumetric flask**: prepare solutions of known concentration, accurate liquid volume measurement.

Exercises

1. Give the English names of the following labware.
   (1) 烧杯　　　　　(2) 量筒　　　　　(3) 滴定管　　　　(4) 移液管
   (5) 刻度吸量管　　(6) 布氏漏斗　　　(7) 滴液漏斗　　　(8) 圆底烧瓶
   (9) 干燥塔　　　　(10) 塞子　　　　 (11) 坩埚钳　　　 (12) 铁架台
   (13) 托盘天平　　 (14) 煤气灯　　　 (15) 通风柜

2. Which piece of laboratory equipment would be most useful for each of the following tasks?
   (1) Accurately measuring liquid volume:＿＿＿＿　(2) Heating a liquid:＿＿＿＿
   (3) Heating a solid:＿＿＿＿　　　　　　　　　　(4) Holding or picking up an item:＿＿＿＿
   (5) Preparation of solutions:＿＿＿＿　　　　　　(6) Cleaning equipment:＿＿＿＿
   (7) Clamps to a ring stand:＿＿＿＿　　　　　　　(8) Grinding a solid:＿＿＿＿
   (9) Mixing 2 liquids together:＿＿＿＿　　　　　 (10) Titrating a solution:＿＿＿＿
   (11) Holding test tubes:＿＿＿＿　　　　　　　　 (12) Evaporating a liquid:＿＿＿＿
   (13) Preventing exposure to atmospheric moisture:＿＿＿＿
   (14) Transferring a chemical reagent from one container to another:＿＿＿＿

Lesson 36　Work in Chemical Laboratory

General Directions

The Object of Laboratory Work　laboratory work is a form of study. A thorough knowledge of elementary chemistry can not be acquired from books alone. Actual first-hand observation of the properties of substances must be made by the student himself, but even these observations will be of small value unless he **appreciates** (意识到，懂得) the object of the experiment and the principle which it illustrates. Merely "doing" the experiments, that is, mechanically following directions, is as valueless as reading a book without comprehending its meaning. No experiment should be begun until the discussion is thoroughly understood.

Notes　Observations must be recorded immediately. To **facilitate** (使容易) clearness and **brevity** (简短) , they are to be entered on the blank pages facing the text, as answers to the numbered questions.

The data obtained in quantitative experiments should be carefully labeled and the result of each measurement should be placed on a separate line, as illustrated by the following example in the determination of the percentage of oxygen in potassium chlorate.

Wt.crucible empty...................................20.30gms
Wt.crucible and $KClO_3$.........................22.03
Wt.$KClO_3$..1.73
Wt.crucible and $KClO_3$ after first heating.........21.42
Wt.crucible and $KClO_3$ after second heating......21.35
Wt.crucible and $KClO_3$ after third heating.........21.35
Loss of weight = weight of oxygen..................0.68
Percentage of oxygen in $KClO_3$...................39.3

The data must be entered at once directly in the book, never on loose pieces of paper. All calculations should be made in the book.

Individual Work (一个人的工作)　All experiments should be done entirely independently. A student can get no more out of his laboratory exercises than he puts into them. If he allows another student to perform the experiments for him or copies his notes, he is **deriving** (得到) no benefit from his work; he is wasting his time and, moreover, he is dishonest.

Use of the Text Book　The numbers given in the references are those of the paragraphs in "A Text Book of Inorganic Chemistry" by James F. Norris. The text book should be brought to the laboratory at each exercise, the paragraphs **pertaining to** (与……有关系的) the exercise of the day should be read, and reference to the book should be made whenever the student is in doubt.

Repetition (重复) **of Experiments**　If the results obtained are not those which you have been led to expect, search for the cause, making sure particularly that you have followed the directions precisely. Not until a possible cause of error has been detected should the experiment be repeated.

Cleanliness (清洁)　Apparatus must be clean in order to obtain trustworthy results. Always have a clean towel at hand.

The desk top should be frequently **wiped** (擦净) with a wet **sponge** (海绵) and any material which is spilled upon it should be washed off immediately with plenty of water.

General apparatus must be cleaned and returned to its place at the close of each exercise.

Materials　All the materials required for an experiment and listed in the text should be obtained before beginning work. It is assumed that each desk is supplied with solutions of **dilute** (稀释的) hydrochloric, sulphuric, and nitric acids, of sodium and ammonium hydroxides, and with concentrated sulphuric acid. These substances therefore are not included in the lists. It is also assumed that each student is provided with the set of apparatus given in the appendix and only additional apparatus is included in the list of materials.

Only such amounts of substances as are actually required should he obtained. **Stock bottles** (储存瓶) should never be carried to the student's desk. Solids may be **measured out** (量出, 称出) on a watch glass or on a piece of filter paper and liquids may be carried in a test tube or beaker.

Always read the label on a stock bottle before taking material from it, and **verify** (核实) your reading by rereading after you have taken the substance.

Unused material must never be returned to the stock bottle but thrown away into the **crock** (瓦罐) or **sink** (污水池).

Waste Material　Solids, particularly matches, should be disposed of in the crock, not in the sink. Liquids should be poured into the sink, and in the ease of concentrated acids, the sink should be **flushed** (冲洗) immediately with a large amount of water.

Accidents　Any corrosive liquid on the skin should be washed off immediately with plenty of water.

Burns and Cuts (烧伤和割伤)　Even when slight, should be reported to an instructor for treatment.

Acid spilled upon the clothing should be neutralized with ammonia.

Fires should be extinguished by throwing a wet towel over the **blaze** (火焰).

Laboratory Processes

1. Heating

The Bunsen burner is lighted by first turning on the gas and then applying a burning match. This order of procedure allows time for the air in the tube and burner to be **expelled** (排出) while the match is being "struck". The burner should always be lighted before it is placed under the apparatus to be heated, never after it has been put in position.

The character of the flame is **regulated** (调节) by adjusting the quantity of air which enters through the holes at the bottom of the burner. Ordinarily a clear **bluish** (有点蓝的) flame should be used. Sometimes the flame "strikes back", that is, begins to burn at the base where the air enters. The occurrence is usually due to the admission of too much air. Turn off the gas, and after the burner has cooled, **diminish** (减少) the amount of air by adjusting the movable ring and relight the burner.

Thick vessels, like a bottle or a mortar, should never be heated because the uneven expansion of the material causes them to break.

Porcelain evaporating dishes and crucibles can be heated directly in the flame, but beakers should rest on a piece of wire gauze.

Test tubes should be held in the test tube holder. When a liquid is being heated, the test tube should be inclined and held in such a position that the flame strikes the glass opposite the upper part of the liquid. The tube should be slightly shaken constantly. If the tube is held still and the bottom heated, a large amount of steam may be formed suddenly and throw the contents out of the tube.

All vessels must be dry on the outside when heated; otherwise they may crack.

2. Filtering

Fold a disc of filter paper just in halves and then again in quarters. Open one of the segments, leaving three thicknesses of paper on one side and one oil the other. A paper **cone** (圆锥形) will be formed which should be placed in a glass funnel. If water solutions are to be filtered, wet the paper and press it firmly to the glass with your finger. When a filtration is to be made into a beaker or an evaporating dish, the funnel should be supported by a ring or by a filter arm and the apparatus adjusted so that the stem of the funnel touches the side of the vessel which is to catch the liquid which comes through; otherwise the liquid will **spatter** (溅洒) out as it falls.

3. Drying Test Tubes

Slip the washed test tube over the end of a glass tube about 1 foot in length and, inclining it mouth downward, wave it back and forth through the flame of a burner. When the test tube is hot blow through the glass tube for an instant. Repeat the procedure if necessary.

4. Preparing Glass Tubing

(1) Cutting: To cut a piece of glass tubing, first make a slight scratch on the tube with the edge of a **triangular file** (三角锉) . Then, holding the tube in both hands, place the thumbs together against the tube on the side opposite the scratch. Press the tube as though bending it away from the scratch. A clean break should result. The broken edge of the tube will be very sharp and must always be smoothed by holding the end of the tube in the flame until it becomes a **dull** (钝的) red color. This process, which is called "fire polishing", causes the sharp edges of the glass to melt and thereby become smooth.

(2) Bending: A good bend is a smooth curve rather than a sharp angle. Several inches of the tube, therefore, must be involved in the bend, and it becomes necessary to soften the glass by heating it in a wide flame. Put a "flame spreader" on the top of the Bunsen burner and close the holes at the bottom of the burner so as to give a **luminous** (明亮的) flame. Hold the tube in the yellow part of the flame, letting one end 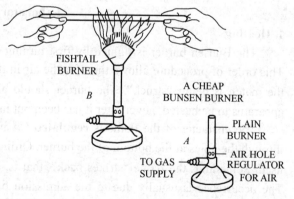 rest lightly between the fingers and thumb of the left hand and turning the tube slowly, but constantly, in one direction with the fingers and thumb of the right hand. When the tube feels **pliable** (易弯的) , remove it from the flame and make a bend of the desired angle.

The following bent tubes should be prepared as they will be used frequently in these experiments : two **right-angled** (直角的) bends each leg of which is 10 cm long; one right-angled bend one leg of which is 10cm, the other 20cm long; one 45℃ bend one leg of which is 10cm, the other 15cm long; and one 135℃ bend one leg of which is 10cm, the other 15cm long.

5. Preparing Corks

Before being used corks should be softened. This can be done by means of a press, which is made for this purpose, or the cork can be rolled on the desk while it is being pressed firmly by means of a block of wood. Sharp cork **borers** (钻孔器) should be used to make the holes of such a size that the tubes to pass through fit **snugly** (贴身地). In boring corks it is advisable to push the borer with a **rotary** (旋转的) motion half way through the cork, taking care that the hole is bored through the center of the cork; the borer is then removed and a hole made from the center of the other end of the cork to meet that first made. By proceeding in this way the edges of the holes; on the two sides of the cork will be clean cut, and thus make a tight joint with the tube to be passed through the hole; and the latter will run evenly through the axis of the cork.

6. Setting up Apparatus

Place the front of the ring stand parallel to the edge of the desk and so that the rod is away from you, not toward you. Arrange the clamps so the main weight of the apparatus is over the base of the stand. The movable jaw of the clamp should be on top. Be careful not to screw the clamp so tight as to crush thin apparatus. Keep vertical lines vertical and horizontal lines horizontal.

In putting a glass tube through a rubber stopper, wet the outside of the tube and then push it in with a twisting motion. Give three distinct pushes, pause long enough to count three and then give three more pushes. If you do not pause, you **involuntarily** (不知不觉地) keep pushing harder and harder, often with the result that the tube breaks and the **jagged** (锯齿状的) end cuts the hand. Be sure that all stoppers fit tightly and that there are no leaks around the holes of stoppers through which glass tubes pass.

7. Weighing

Objects and material to be weighed fall into two general classes, those in which only an approximate weight, to the nearest gram, is required and those in which an accurate weight, to the nearest centigram, is demanded. Cases in which a convenient amount of material for experimentation is to be weighed out belong to the first class ; the quantitative determination of the composition of substances belongs to the second class. **Platform scales** (台秤) are used in the first instance; **beam balances** (杠杆式天平) are necessary in the second.

(1) The platform scales must always be tested before using them because they are exposed to the **fumes** (烟气) of the laboratory, and easily become rusted. Be sure that the rider on the front of the scales is pushed to the extreme left, and then gently tap one of the platforms. The pointer should swing an equal number of divisions on each side of the middle of the pointer scale and should not come to rest until at least two swings on each side have been completed. If the scales are found to be out of adjustment, an assistant should be called to fix them.

In weighing out material for an experiment the empty container should be placed on the left-hand platform and a brass weight which is **presumably** (推测起来) too heavy to **counterpoise** (平衡) it should be placed on the right-hand platform. When this weight has been tried and found too great, it should be removed and the next smaller one substituted. If this is not heavy enough the next smaller one should be added to it. If this second weight is found to be too great, the next smaller should be substituted for it; or if it is found insufficient, the next smaller should be placed on the platform in addition. Thus all the weights in turn should be tried and rejected or kept as the case demands down to and including the 5-gram weight. The rider on the front of the scales should next be pushed to the right until the scales just balance. The weights on the platform should then be counted up, together with the weight indicated by the position to the rider and the number of grams immediately recorded in the notebook. The weights should then be returned to their proper holes in the weight box, beginning with the largest weight; and the number of grams again counted as they are removed from the platform. By this second counting errors are often discovered.

The material to be weighed is now placed in the container and the weight of it and the container together should be determined and recorded in the same way that the weight of the container alone has just been found.

Sometimes objects, such as a piece of wire, which do not require a container, can be weighed directly by placing a disc of filter paper on each platform, testing the scales, putting the material on the paper on the left and counterbalancing with weights on the right, as just described.

Never place any substance directly on the platform without a container or paper to protect it.

(2) The beam balances are delicate and should be handled with care. When not in use the beam is supported and held at rest by a mechanism that is controlled by a thumb screw in the front of the balance case. Free the beam by turning the screw, raise the door of the case, and set the balance in motion by waving your hand near one of the pans. The **disturbance** (扰动)of air should be enough to set the beam swinging. Draw down the door and observe the swing of the pointer. When the balance is perfectly adjusted, the pointer should swing through an equal number of divisions on each side of the middle of the scale. The balance can he used satisfactorily, however, even when not in perfect adjustment, by taking as the "zero" of the scale that **division mark** (分度) which is midway between the extremes of the swing. For example, if the pointer swings through two **divisions** (刻度) on the left of the middle of the scale and through four divisions on the right of the middle of the scale, the "zero"for this particular occasion is the first mark to the right of the middle. In other words, in making a weighing, the balance is in equilibrium if the pointer swings in the same way it did when nothing was on the pans. When the swing of the pointer on one side of the middle exceeds the swing on the other side by more than four divisions, it is best to call an assistant to make a readjustment.

Always test a balance before using it. After testing turn the **thumb screw** (指旋螺丝) so that the beam is again supported. Now the object to be weighed may be placed on the left-hand pan and a weight, presumably too heavy, on the other pan. Never place on the pans or remove from them objects or weights while the beam of the balance is free to swing. If the beam is not supported the sudden addition or subtraction of weight will make it **lurch** (突然倾斜) and may completely change the adjustment. The weights must always be handled with pincers, as the **perspiration** (出汗) on your fingers will **tarnish** (失去光泽) them.

After the object and a weight have been placed on the pans, as just directed, slowly turn the thumb screw and free the beam. Observe whether the weight chosen is too great or too small, again support the beam and make the proper substitution or addition of weights in the same way as directed in the use of the platform scales. Continue the systematic **trial** (尝试) of the weights down to the 1-centigram weight. If a certain number of centi-grams appears to be insufficient and one more to be too much, choose the one which makes the balance nearer to equilibrium as judged by the swing of the pointer. Count the weights as they lie in the pan and immediately make a record in your notebook. The number of centigrams must be recorded even when it is zero; otherwise it will appear that you did not weigh the objects to centigrams.

Never attempt to weigh a warm object. The upward current of warm air arising from it will make it appear lighter than it really is. Never place any material directly on the scale pan. Never add anything to a container or take anything out of it while it rests on the pan.

Lesson 37 An Experiment Case: Synthesis of Aspirin

Lecture and Lab Skills Emphasized
1. Synthesizing an organic substance.
2. Understanding and applying the concept of limiting reagents.
3. Determining percent yield.
4. Learning how to perform a vacuum filtration.
5. Understanding and performing **recrystallization** (重结晶).

In the Lab
1. Students will work in pairs.
2. Parts must be completed in order.
3. Record your procedure and original data in your lab notebook along with your calculations.
4. Report data collected and subsequent calculations.
5. All equipment should be returned to the correct location after use.

Waste
1. **Salicylic acid** (水杨酸) should be disposed in the solid waste container.
2. **Acetic anhydride** (乙酸酐) should be disposed in the organic waste container.
3. Filtrates (liquids) can be washed down the drain with excess water.
4. Solid product and filter paper can be disposed in the solid waste container.
5. Purity test solution should be disposed in the aqueous waste container.

Safety
1. Gloves and safety goggles are **mandatory** (强制性的) when anyone is performing an experiment in the lab.
2. Work in the hood when indicated in the procedure.
3. Wear long pants, closed-toed shoes, and shirts with sleeves. Clothing is expected to reduce the exposure of bare skin to potential chemical splashes.
4. Always wash your hands before leaving the laboratory.

 A local **pharmaceutical** (药物) company is looking at forming a partnership with your research company. As part of their research on your company, they have asked your supervisor to look at researcher performance and have asked that you synthesize a simple substance and demonstrate its purity.

Synthesis of Aspirin
Aspirin (阿司匹林) is the single most manufactured drug in the world. Aspirin's chemical name is **acetylsalicylic acid** (乙酰水杨酸), and it is synthesized from the reaction of acetic anhydride with salicylic acid in the presence of phosphoric acid as a catalyst. The by-product is acetic acid (Figure 37-1).

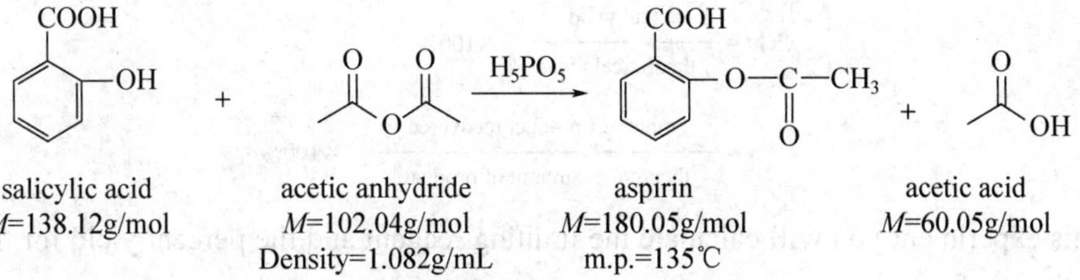

```
        COOH                                    COOH
         OH        O    O     H₃PO₅              O—C—CH₃        O
              +                         →                 +
                       O                          O              OH

     salicylic acid   acetic anhydride         aspirin       acetic acid
     M=138.12g/mol    M=102.04g/mol        M=180.05g/mol   M=60.05g/mol
                     Density=1.082g/mL       m.p.=135℃
```

Figure 37-1 Synthesis of aspirin

Salicylic acid has the same **analgesic** (止痛的) properties as aspirin and was used for many years as a headache medicine. However, salicylic acid is more acidic than aspirin, and is especially **irritating** (起刺激作用的) to the mouth and stomach. Some people find even aspirin too acidic and prefer to use aspirin substitutes like **acetaminophen** (对乙酰氨基酚) **and ibuprofen** (布洛芬).

Aspirin is produced from a reaction between acetic anhydride and salicylic acid. The structures of the reactants and products are shown in Figure 37-1. Examine them. What is changing about the salicylic acid? Being able to identify what is reacting helps organic chemists determine how a reaction occurs.

Limiting Reagents

In order to determine your efficiency [expressed as percent **yield** (产率) in chemistry], you will need to understand what is controlling your chemical reaction in terms of how much material you will produce.

When you perform a chemical synthesis in lab, you measure out quantities of all reagents. These reagents react with each other, using up material as they react in order to produce products. The reagent that is used up first and controls the amount of product formed is known as the limiting reagent. The limiting reagent is the one that is completely consumed in the reaction. The limiting reagent is determined by the relative amounts of starting materials, and must be calculated for every reaction. For every reagent, you must calculate the mole quantities present, and then using the **stoichiometry** (化学计量关系) of the equation, how many moles of product will form. The reagent that gives the least amount of product is the limiting reagent.

Theoretical (理论的) and Percent Yields

When calculating the yield of product, the calculation must always be based on the limiting reagent, not the reagents that are in excess. The theoretical yield is the maximum amount of product that you can recover from your experimental conditions. In order to obtain the theoretical yield, convert the amount of moles of product (from the limiting reagent) to grams, using the product's **formula weight** (分子量).

In the laboratory, however, you will find that you will **invariably** (总是) lose small quantities of your material along the way. For these reasons, the actual yield of a compound (the quantity you obtain from your experiment) may be less than what you calculated. The efficiency of a chemical reaction and the techniques used to obtain the compound can be calculated from the ratio of the actual yield to the theoretical yield, to give the percent yield:

$$\text{yield} = \frac{\text{actual yield}}{\text{theoretical yield}} \times 100\%$$

$$= \frac{\text{amount of product recovered}}{\text{theoretical amount of product}} \times 100\%$$

In this experiment you will calculate the limiting reagent and the percent yield for the reaction in the synthesis of aspirin.

Recrystallization

Before you can determine how much material you produced, you need to make sure you are working with a pure sample. How does your sample become impure? Generally, chemists are looking at separating out any unreacted starting materials. Other impurities, like **contamination** (污染) caused by unclean glassware, should have been fixed by using better laboratory technique (like always cleaning your glassware before you start a reaction).

Recrystallization is the most common method of purification of solids. Recrystallization is a separation process. The object is to obtain one component of a solid mixture free from contamination by any other components originally present in the mixture. Recrystallization involves three basic steps:

(1) Dissolution of the solid to be purified in some hot solvent.
(2) Formation of crystals as the solution cools.
(3) Recovery of the purified crystals, usually by filtration. The impurities are left in solution.

Purification by recrystallization does not involve formation of crystals from the hot solution by evaporation of the solvent. Such a process would simply **precipitate** (沉淀) the impurities. The success of recrystallization depends on the characteristics of the solvent toward both the desired component and the impurities. The solvent should dissolve little or none of the desired component in the mixture at low temperatures, while it should completely dissolve the component at high temperatures. Recrystallization also does not involve any chemical reaction that changes the chemical structure of your substance.

If the limiting reagent for your reaction was acetic anhydride, there is a good chance that you will have an excess of salicylic acid left at the end of the reaction, contaminating your aspirin product. We will purify our product by recrystallization of the aspirin from hot ethanol, leaving any impurities in the ethanol and recovering the aspirin by filtration.

Testing the Purity of Aspirin

While you may have added all of the ingredients (reagents) that when reacted together form aspirin, there is no **guarantee** (保证) that you successfully made what you intended. Chemists use a variety of tests to determine whether they made the intended product or not. Often, reactions do not completely react, so some starting material may remain.

In this lab, we will be using a chemical test to determine the purity of the aspirin you synthesized. This particular chemical test takes advantage of one of the differences between salicylic acid and aspirin, namely the functional group called phenol. A functional group is a group of atoms that have characteristic reactivity. The **phenol** (苯酚) functional group is defined as being

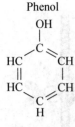

Figure 37-2

an —OH group attached to a benzene ring. Benzene, another functional group, consists of a 6-atom carbon ring, connected by alternated single and double bonds.

A phenol is a type of functional group that is present in a variety of organic compounds, see Figure 37-2. Compounds that have the same functional group often have similar properties and react in similar ways. A phenol is present in salicylic acid but absent in aspirin. The relative purity of the synthesized aspirin can be determined by reaction of the product with Fe^{3+}, iron(III) reacts with phenols to form a violetcolored complex:

$$6C_6H_5OH + Fe^{3+} \longrightarrow [Fe(OC_6H_5)_6]^{3-} + 6H^+$$
phenol   violet complex

If your product is pure, no violet color will be observed, since the aspirin cannot form the violet complex (not a phenol). However, if salicylic acid is present you will see a violet color develop. The amount of violet is directly related to the amount of salicylic acid present in your aspirin.

Materials and Procedures

100mL beaker
400mL beaker
10mL graduated cylinder
stirring rod
wash bottle
watch glass
Buchner funnel
filter flask
iron clamp
ring stand

filter paper
spatula
salicylic acid
acetic anhydride
H_3PO_4 (concentrated)
ethanol
0.1mol/L $Fe(NO_3)_3$
ice
semi-micro test tube

Figure 37-3 Hot water bath

Synthesis of Aspirin

(1) Obtain approximately 2g of salicylic acid and determine its mass accurately, record the exact value in your data table. Transfer to a 100mL beaker.

(2) Do this step in the hood. Measure 3mL of acetic anhydride with your small graduated cylinder, record the exact volume and add this to the 100mL beaker with salicylic acid. Add 5 drops of concentrated phosphoric acid.

(3) Do this step in the hood. Place the 100mL beaker in the 400mL beaker of warm water (70~80°C). Use a stirring rod to get the reactants into solution. Allow the mixture to react for about 5 minutes in the warm water.

(4) Remove the 100mL beaker, and while the mixture is still warm, carefully add about 1mL of cold water, drop by drop to the 100mL beaker. You will now hydrolyze any excess acetic anhydride. Use caution: the hydrolysis may cause spattering.

(5) Add 15mL of water and cool the 100mL beaker in a larger beaker with ice water. Stir with a stirring rod. Allow 5 minutes for the aspirin to crystallize from the solution. In the meantime, put a wash bottle filled with distilled water in ice water to cool. You will need cold water in the

filtration step to follow.

(6) Set up a Buchner filtration system as shown in Figure 37-4. Place a filter paper in the funnel, and wet the paper thoroughly with distilled water. Turn the water on to allow the paper to be sucked onto the funnel. In order to ensure a good seal, you may want to add a small amount of water into the funnel.

(7) Collect the crystals by pouring the solution onto the filter paper. Make sure that the water is fully open for maximum suction filtration. Do not disturb the crystals on the filter paper or you will break the filtration vacuum. Wash any product in the beaker into the funnel with cold distilled water, rinse your stirring rod as well. Remember, any product you lose here will affect your percent yield.

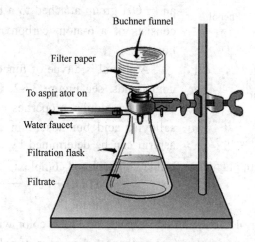

Figure 37-4   Separation technique: suction filtration

(8) Allow the filtration to continue until you see no more water droplets fall from the funnel. Turn the water off. Insert a spatula between the edge of the paper and the side of the funnel and lift the paper from the funnel. Transfer the product to a 100mL beaker. Scrape all the product from the filter paper with a spatula. Work carefully as you don't want to lose any product in this step. Note the color and texture of your crystals in your data table.

Recrystallization of Aspirin

(1) Add about 3mL of ethanol to the crystals. Warm the solution in a hot water bath. If all the crystals do not dissolve, add another 1~2mL of ethanol. You want to use as little ethanol as possible in the recrystallization step.

(2) Pour 15mL of warm water into the alcohol solution. Cover the beaker with a watch glass and set aside to cool. (The crystallization process can be sped up by putting the beaker in an ice bath.) Allow 5 minutes for the crystals to crystallize from the solution.

(3) Record the mass of your clean, dry filter paper and filter the crystals using suction filtration as before. While the filtration is in progress, determine the mass of a dry, clean watch glass.

(4) When the filtration is complete, remove the filter paper and place the paper and crystals on the watch glass. Label your watch glass before putting it in the oven and dry your aspirin in the oven at 110°C for 10~15 minutes.

(5) When your product is dry, remove your sample from the oven and allow it to cool. Weigh the watch glass with the aspirin product and record the value in your data table. Record the color and texture of the recrystallized aspirin.

Testing the Purity of Aspirin

Dissolve a small spatula point of your aspirin in about 50mL of warm water. Transfer 1mL of this solution to a test tube. Add 3 drops of the iron reagent [$Fe(NO_3)_3$]. Shake well for 2 minutes. Record all observations.

Data Analysis and Experimental Report

Make sure to show all of your calculations in your lab notebook as a record of how you completed your calculations. Don't forget to include your units and correct number of significant figures! Then report your results:

1. What was the mass of aspirin synthesized?
2. What is the maximum mass of aspirin possible if salicylic acid is the limiting reagent?
3. What is the maximum mass of aspirin possible if acetic anhydride is the limiting reagent?
4. What is the theoretical yield of aspirin?
5. What is the percent yield of aspirin?
6. What happened when the iron nitrate solution was added to the aspirin? Make sure to respond with both an observation and the interpretation of your result.

Sample Report for Reference Only

Synthesis of Salicylic Acid from Methyl Salicylate
C. Diver
Tues Lab, Box 007

Abstract

Salicylic acid was obtained from methyl salicylate in 78% yield. The product was identified by its melting point.

Results and Discussion

Salicylic acid was obtained in 78% yield by refluxing methyl salicylate with aqueous NaOH. Recrystallization of the acid gave white needles, m.p. 157~158.5℃ (lit. 159℃).

Experimental

Preparation of salicylic acid. Methyl salicylate (4mL, 31mmol) and 2mol/L NaOH aq. (10mL, 20mmol) were refluxed for 30min. The solution was acidified with 6mol/L H_2SO_4 aq. (~25mL) and filtered. The crude product was recrystallized from boiling ethanol and dried for three weeks to give green crystals of salicylic acid: 3.33g (78% from methyl salicylate), m.p. 157~158.5℃ (lit. 159℃).

References

1. Kaplan R., ed. "Bob's Big Book of Chemicals", 1st Ed., XYZ Press, New York, 1984.

E-factor

This experiment used the following materials: methyl salicylate (3g), NaOH (3g), sulfuric acid (4g), ethanol (40g); total consumables (50g). The product (3.33g) does not count as waste, total disposables (46.7g). E-factor = 14.

Exercise

Try to write an English chemical experiment report according to a laboratory text.

Part 5
Reference Reading for Graduate Admission Examination

Lesson 38　研究生复试英语考试参考样题

◆ **Written Exam (笔试部分): Choose passages and translate them into Chinese.**

Passage 1
Metal lattice has a very number of free, delocalized outer electrons in it. When a potential gradient is applied, these electrons can move towards the positive end of the gradient carrying charge.
　　A metallic bond is the electrostatic force of attraction that two neighbouring nuclei have for the delocalized electrons between them. Both ions attract the delocalized electrons between the leading to metallic bonding.

Passage 2
(1) Words: methyl, ethyl, molecule, ethane, trichloromethane.
(2) Reduction and oxidation reactions involve the transfer or less and gain of electrons.
(3) Redox reactions involve electron transfer. Because electrons do not appear in most written equations, it is not always obvious which particle has lost or gained electrons. Oxidation is defined as the loss of electrons; reduction is defined as the gain of electrons.

Passage 3
The chemical industry is a very high technology which takes full advantage of the latest advances in electronics and engineering. Computers are very widely used for all sorts of application, from automatic control of chemical plants, to molecular modeling of structures of new compounds, to the control of analytical instruments in the laboratory.

Passage 4
Covalent bonding is the bonding between non-metallic atoms. A covalent bond is the electrostatic force of attraction that two neigbouring nuclei have for a localized pair of electrons shared between them. Covalent bond form when the orbital of the two neighbouring atoms overlap so that nuclei attract the pairs of electrons between them. This can happen in two different ways making two different kinds of bond: σ-bonds and π-bonds.

Passage 5
(1) Words: stereochemistry, isomer, optical activity, synthesis, analyse.
(2) Isomers are different compounds with the same molecular formula. Isomers are molecules that contain the same number of atoms and also the same kind of atoms. However, they have different bonding arrangements. There are two major classes of isomers: constitutional and stereoisomers.

Passage 6

The homologs of benzene are those containing an alkyl group or alkyl groups in place of one more hydrogen atoms.

The methyl group on the benzene ring greatly facilitates the nitration of toluene.

◆ **Oral Exam(口试部分): self-introduction (Sample)**

Good afternoon, my dear professors. I am glad to be here for this interview. It is my great pleasure to introduce myself to all of you. My name is×××, 27 years old, born in Ganzhou, Jiangxi Province. I graduated from Jiangxi Medical College, my major is clinical medicine. Because of my intelligence in study, I gained the scholarships each year, and in 2000, I gloriously took part in the Chinese Communist Party.

After graduation, I went to work in Taizhou hospital. It lies in Luqiao, Zhejiang Province. It is a general hospital and the center of medicine for Taizhou people. In the first two years, as a house surgery, I accepted the normal and strict training. In 2003, I acquired the license for practitioner, in the same year, I am cheerful to be distributed to the department of cancer surgery. During the clinical practice, I have more and more interest in oncology, but, I knew deeply that I have no adequate knowledge to be competent to the clinical practice. Furthermore, I believe that a qualified cancer surgeon should have aboard, systematic and full-scale medical knowledge and must experience more normal and strict training. So, I determined to pursue the master's degree in ××× University. In order to prepare well to the tests for entrance for post graduation, I quit my job in July, 2005 and try my best to be ready for the tests. Now all my hard work has got a good result since, I have got a chance to be interviewed by all of you.

I always believe that a doctor will easily lag behind unless he keeps on learning. I am a delight, studious, and creative person. Of course, if I am given a chance to study in ××× University, I will study harder, and, with your guidance and help, I wish I can, someday, like you, devote myself to the great cause of conquering the tumor.

That's all. Thank you for your attention.

Lesson 39 Letters of Application to Graduate Schools

Sample 1

Dear Sir,

　　As a senior in Department of Chemistry, ×××University, I thought that now was the time to inquire about application materials for your Graduate School in the fall semester of 2018.

　　After graduation I would like very much to work towards my M.S. in your Department of Chemistry. During my undergraduate career I have maintained a very high scholastic average. I am in the top 5% of my class, and feel certain that I will be able to do the course work required for an M.S. degree.

　　I have a reasonable command of English. I am planning to take the TOEFL test soon and when the results are made known, I will inform the ETC at Princeton, New Jersey, to send them directly to you. I hope this will help you decide if I am acceptable as an M.S. candidate.

　　If you could send all application materials and any other pertinent information as soon as possible, I would appreciate it very much.

　　Thank you for all your kind assistance.

　　Sincerely yours,

　　H. Zhang

Sample 2

Dear Prof Anderson,

　　This letter is to request that you send me all the materials necessary for application to your graduate school. I am also interested in any information you have on financial aid available through your university.

　　I will graduate from the Graduate School of ×××University in July 2018 with an M.S. in Chemistry. Several professors here have taken a personal interest in my academic career and suggested that your university is among the best in this field. It is my wish to complete the PhD Program in chemistry.

　　I understand that I must take the TOEFL examination as a prerequisite for entrance into your university. I am planning to take the exam on March 3 and have instructed the ETC at Princeton, New Jersey to forward my scores directly to you.

　　I would appreciate it very much if you could send me a brochure listing research in program in the graduate school. I hope to be corresponding with you again in the very near future.

　　Yours sincerely,

　　F. Wang

Lesson 40 诺贝尔化学奖（2014）颁奖晚宴演讲
——斯特凡·赫尔

Your Majesties, Your Royal Highnesses, Ladies and Gentlemen,
What a week, what a day, and what a night…!

I cannot imagine anything more exhilarating than to stand here this evening—also on behalf of my colleagues W E Moerner and Eric Betzig—thanking the Swedish Academy and the Nobel Foundation for the honor that has been bestowed upon us. We are so grateful to all who have supported us on our path and—above all—we feel very, very humbled.

Like all laureates, each of us three has his own road to this magnificent hall. Our personal stories have been quite different.

Yet, we have much in common: passion for what we do, and fascination with things that cannot be done, or, let's say, things that cannot be done…supposedly.

Erwin Schrödinger, who spoke at this banquet eighty-one years ago tonight, wrote: "it is fair to state that we are not going to experiment with single particles any more than we will raise dinosaurs in the zoo."

Well, one of us, W E, discovered just the opposite, single molecules can indeed be seen and played with individually.

Now, ladies and gentlemen, what do we learn from this?

First, Erwin Schrödinger would never have gone on to write "Jurassic Park"…

Second, as a Nobel Laureate you should say "this or that is never going to happen", because you will increase your chances tremendously—of being remembered—decades later—in a Nobel banquet speech. And so, on to super resolution fluorescence imaging. According to the belief, molecules closer together than 200 nanometers could not be told apart with focused light. This is because, in a packed molecular crowd, the molecules shout out their fluorescence simultaneously, causing their signal, their voices, to be confused.

But, believe it or not, Eric found a way to discern the molecules by calling on each one of them individually, using a microscope so simple-that he built it with a friend—in his living room.

As for myself, I never had that kind of patience. Calling on each molecule one by one? No way. I just told all of them to be quiet, except for a selected few.

Just keep the molecules quiet, and let only a few speak up. …A simple solution to a supposedly unsolvable problem. It made the resolution limithistory.

Now have a guess, where did this idea occur to me?

Not very far from here, actually: in a student dorm in Finnish Åbo—in what you may kindly call– a living room.

So, what does it take, ladies and gentlemen, to end up standing here, telling you a story of important discoveries or improvements?

Well You definitely need a living room. At the very least, you need a place to sleep. And when you fall asleep you may forget that others consider you—too daring or too foolish.

But when morning comes, you would better find yourself saying: "I have so many choices of what to do or what to leave every morning, every day. I better judge for myself, go ahead and do it."

Because nothing is more powerful than an idea whose time has come even if it came in a living room, or to someone, with a humble living.

And if you feel we'll never raise dinosaurs. Who knows? One day someone may be actually standing here, giving a banquet speech.

So, let us embrace a culture that addresses problems deemed impossible to solve, and let us now honor those who will do so with a toast.

Appendix: Stefan W Hell Facts

Stefan W Hell
Born: 23 December 1962, Arad, Romania.
Affiliation at the time of the award: Max Planck Institute for Biophysical Chemistry, Göttingen, Germany, German Cancer Research Center, Heidelberg, Germany.
Prize motivation: "for the development of super-resolved fluorescence microscopy."

Life
Stefan Hell was born in Arad, Romania. His father was an engineer and his mother was a teacher. When Hell was 16, the family emigrated to Germany and after studies in physics at the University of Heidelberg, he received his doctorate in 1990. After a few years at the European Molecular Biology Laboratory in Heidelberg, the University of Turku in Finland and Oxford University in the UK, he moved to the Max-Planck-Institut für biophysikalische Chemie, in Göttingen, Germany, where he has worked since 1997, and at present he also works at the German Cancer Research Center in Heidelberg.

Stefan Hell at the Department of Medical Physics in 1993

Work
In normal microscopes the wavelength of light sets a limit to the level of detail possible. However this limitation can be circumvented by methods that make use of fluorescence, a phenomenon in which certain substances become luminous after having been exposed to light. In 1994, Stefan W Hell developed a method in which one light pulse causes fluorescent molecules to glow, while another causes all molecules except those in a very narrow area to become dark. An image is created by sweeping light along the sample. This makes it possible to track processes occurring inside living cells.

Part 6
Guide for Scientific Paper Writing

Lesson 41 Notes on the Structure of a Scientific Paper

A scientific paper is a written report describing original research results. The format of a scientific paper has been defined by centuries of developing tradition, editorial practice, scientific ethics and the interplay with printing and publishing services. A scientific paper should have, in proper order, a title, abstract, introduction, materials and methods, results, and discussion.

Title

A title should be the fewest possible words that accurately describe the content of the paper, omit all waste words such as "A study of…" "Investigations of …" "Observations on…", etc. Indexing and abstracting services depend on the accuracy of the title, extracting from it keywords useful in cross-referencing and computer searching. An improperly titled paper may never reach the audience for which it was intended, so be specific. If the study is of a particular species, name it in the title. If the inferences made in the paper are limited to a particular region, then name the region in the title.

Keyword List

The keyword list provides the opportunity to add keywords, used by the indexing and abstracting services, in addition to those already present in the title. Judicious use of keywords may increase the ease with which interested parties can locate your article.

Abstract

A well prepared abstract should enable the reader to identify the basic content of a document quickly and accurately, to determine its relevance to their interests, and thus to decide whether to read the document in its entirety. The abstract should concisely state the principal objectives and scope of the investigation where these are not obvious from the title. More importantly, it should concisely summarize the results and principal conclusions. Do not include details of the methods employed unless the study is methodological, i.e. primarily concerned with methods.

The abstract must be concise, not exceeding 250 words. If you can convey the essential details of the paper in 100 words, do not use 200. Do not repeat information contained in the title. The abstract, together with the title, must be self-contained as it is published separately from the paper in abstracting services such as Biological Abstracts or Current Contents. Omit all references to the literature and to tables or figures, and omit obscure abbreviations and acronyms even though they may be defined in main body of the paper.

Introduction

The introduction should begin by introducing the reader to the pertinent literature. A common mistake is to introduce authors and their areas of study in general terms without mention of their major findings. For example:"Parmenter (1976) and Chessman (1978) studied the diet of Chelodina longicollis at various latitudes and Legler (1978) and Chessman (1983) conducted a similar study on Chelodina expansa."

Compares poorly with:

"Within the confines of carnivory, Chelodina expansa is a selective and specialized predator feeding upon highly motile prey such as decapod crustaceans, aquatic bugs and small fish (Legler, 1978; Chessman, 1984), whereas C. longicollis is reported to have a diverse and opportunistic diet (Parmenter, 1976; Chessman, 1984)."

The latter is a far more informative lead-in to the literature, but more importantly it will enable the reader to clearly place the current work in the context of what is already known. An important function of the introduction is to establish the significance of the current work: Why was there a need to conduct the study?

Having introduced the pertinent literature and demonstrated the need for the current study, you should state clearly the scope and objectives. Avoid a series of point-wise statements—use prose. A brief description of the region in which the study was conducted, and of the taxa in question, can be included at this point. The introduction can finish with the statement of objectives or, as some people prefer, with a brief statement of the principal findings. Either way, the reader must have an idea of where the paper is heading in order to follow the development of the evidence.

Materials and Methods

The main purpose of the materials and methods section is to provide enough detail for a competent worker to repeat your study and reproduce the results. The scientific method requires that your results be reproducible, and you must provide a basis for repetition of the study by others. Often in field-based studies, there is a need to describe the study area in greater detail than is possible in the introduction. Usually authors will describe the study region in general terms in the introduction and then describe the study site and climate in detail in the materials and methods section. The sub-headings "Study Site", "General Methods" and "Analysis" may be useful, in that order.

Equipment and materials available off the shelf should be described exactly (Licor underwater quantum sensor, Model LI 192SB) and sources of materials should be given if there is variation in quality among supplies. Modifications to equipment or equipment constructed specifically for the study should be carefully described in detail. The method used to prepare reagents, fixatives, and stains should be stated exactly, though often reference to standard recipes in other works will suffice.

The usual order of presentation of methods is chronological, however related methods may need to be described together and strict chronological order cannot always be followed. If your methods are new (unpublished), you must provide all of the detail required to repeat the methods. However, if a method has been previously published in a standard journal, only the name of the method and a literature reference need be given.

Be precise in describing measurements and include errors of measurement. Ordinary statistical methods should be used without comment; advanced or unusual methods may require a literature citation.Show your materials and methods section to a colleague. Ask if they would have difficulty in repeating your study.

Results

In the results section you present your findings. Present the data, digested and condensed, with important trends extracted and described. Because the results comprise the new knowledge that you are contributing to the world, it is important that your findings be clearly and simply stated. The results should be short and sweet, without **verbiage**(冗词). Do not say "It is clearly evident from Fig. 1 that bird species richness increased with habitat complexity." Say instead "Bird species richness increased with habitat complexity (Fig. 1)."

However, do not be too concise. The readers cannot be expected to extract important trends from the data unaided. Few will bother. Combine the use of text, tables and figures to condense data and highlight trends. In doing so be sure to refer to the guidelines for preparing tables and figures below.

Discussion

In the discussion you should discuss the results. What biological principles have been established or reinforced? What generalizations can be drawn? How do your findings compare to the findings of others or to expectations based on previous work? Are there any theoretical/practical implications of your work? When you address these questions, it is crucial that your discussion rests firmly on the evidence presented in the results section. Continually refer to your results (but do not repeat them). Most importantly, do not extend your conclusions beyond those which are directly supported by your results. Speculation has its place, but should not form the bulk of the discussion. Be sure to address the objectives of the study in the discussion and to discuss the significance of the results. Don't leave the reader thinking "So what?". End the discussion with a short summary or conclusion regarding the significance of the work.

References

Whenever you draw upon information contained in another paper, you must acknowledge the source. All references to the literature must be followed immediately by an indication of the source of the information that is referenced. For assignments in the Faculty of Applied Science, we expect you to use the Harvard system, for example:

"A drop in dissolved oxygen under similar conditions has been demonstrated before (Norris, 1986)."

"Williams (1921) was the first to report this phenomenon."

"…as discussed in detail by Ramsay (1983)."

If two authors are involved, include both surnames, "The dune lakes of Jervis Bay are not perched in the generally accepted sense (Smith and Jones, 1964)." However if more than two authors are involved, you are encouraged to make use of the et al. convention. It is an abbreviation of Latin meaning "and others".

"The significance of changes in egg contents during development is poorly understood (Webb et al., 1986). "

"Williams et al. (1921) were the first to report this phenomenon."

Do not use the et al. abbreviation in the reference list at the end of the paper.

If two or more articles written by the same author in the same year are cited, then distinguish between them using the suffixes a, b, c, etc., in both the text and the reference list (e.g. Smith and Jones, 1982b).

If you include in your report, phrases, sentences or paragraphs lifted **verbatim**(一字不差的) from the literature, it is not sufficient to simply cite the source. You must include the material in quotes and you must give the number of the page from which the quote was lifted. For example:

"Day (1979) reports a result where 33.3% of the mice used in this experiment were cured by the test drug; 33.3% of the test population were unaffected by the drug and remained in a moribund condition; the third mouse got away."

A list of references ordered alphabetically on author's surname, must be provided at the end of your paper. The reference list should contain all references cited in the text but no more. Include with each reference details of the author, year of publication, title of article, name of journal or book, volume and page numbers. Formats vary from journal to journal, so when you are preparing a scientific paper for an assignment, choose a journal in your field of interest and follow its format for the reference list. Be consistent in the use of journal abbreviations.

Appendices

Appendices contain information in greater detail than can be presented in the main body of the paper, but which may be of interest to a few people working specifically in your field. Detailed ANOVA tables for example may be relegated to an appendix. Only appendices referred to in the text should be included.

Miscellaneous Formatting Conventions

The manuscript should be typed with double spacing throughout and a 3cm left margin and 2cm margins to the right top and bottom, to enable detailed comments by the examiner (or reviewers and editors). To assist the typesetter, **indent**（缩进）paragraphs and do not **hyphenate**（用连字号连接）words at the right margin. A ragged right margin with no superfluous spaces between words may also be preferred by typesetters.

When Constructing Tables

Do include a caption and column headings that contain enough information for the reader to understand the table without reference to the text. The caption should be at the head of the table.

Do organize the table so that like elements read down, not across.

Do present the data in a table or in the text, but never present the same data in both forms.

Do choose units of measurement so as to avoid the use of an excessive number of digits.

DON'T include tables that are not referred to in the text.

DON'T be tempted to "dress up" your report by presenting data in the form of tables or figures that could easily be replaced by a sentence or two of text. Whenever a table or columns within a table can be readily put into words, do it.

DON'T include columns of data that contain the same value throughout. If the value is important to the table include it in the caption or as a footnote to the table.

DON'T use vertical lines to separate columns unless absolutely necessary.

When Constructing Figures

DO include a legend describing the figure. It should be **succinct**（简洁的） yet provide sufficient information for the reader to interpret the figure without reference to the text. The legend should be below the figure.

DO provide each axis with a brief but informative title (including units of measurement).

DON'T include figures that are not referred to in the text, usually in the text of the results section.

DON'T be tempted to "dress up" your report by presenting data in the form of figures that could easily be replaced by a sentence or two of text.

DON'T fill the entire A4 page with the graph leaving little room for axis numeration, axis titles and the caption. The entire figure should lie within reasonable margins (say 3cm margin on the left side, 2cm margins on the top, bottom and right side of the page).

DON'T extend the axes very far beyond the range of the data. For example, if the data range between 0 and 78, the axis should extend no further than a value of 80.

DON'T use colour, unless absolutely necessary. It is very expensive, and the costs are usually passed on to the author. Colour in figures may look good in an assignment or thesis, but it means redrawing in preparation for publication.

Source

These guidelines were prepared with the aid of Robert Day's entertaining book "How to Write and Publish a Scientific Paper" (ISI Press, Philadelphia, 1979). It would be a valuable addition to your library.

Exercises

1. Look for some well-organized papers in a familiar journal in your own discipline, and study the divisions of each paper.
2. Reconstruct one of your previous written English papers into clearly delineated sections, or write an English paper with the suggested outline of the divisions as a guide.

Lesson 42 Style of Writing and Use of English in Essays and Scientific Papers

Scientific writing is the basis of going to institutions, libraries, and universities to gain information and present it in a concise manner abiding to the generally accepted template of most scientific papers. The scientific format used to acknowledge bibliography resources is CSE. The following advice may be of help to students writing an essay or a scientific paper.

"Professor, why do you think that ordinary people have such a difficult time understanding Scientific English?"

"I cognate that it is do to the overstratification of auxilary phrases combined with the introduction of functional inflections…"

Three aspects of style seem to cause problems.

(1) Division of the text into sentences and paragraphs. Sentences should have only one idea or concept. In general, sentences in scientific prose should be short, but full stops should not be added so liberally that the writing does not flow. The use of paragraphs helps the reader to appreciate the sense of the writing.

(2) Superfluous phrases and words should be avoided. Do not write phrases such as "It is also important to bear in mind the following considerations". Most woolly phrases can be omitted or replaced by a single word.

(3) Try to use familiar, precise words rather than far-fetched vague words. For example, "cheaper" may replace "more economically viable".

A good style is helped by logical planning. Decide what you want to write, then write it simply and in a sensible order. Put the first draft away for a few days and then rewrite it.

Tense and Mood

Write in past tense unless you are describing present or future situations. Use the active voice rather than the passive voice. For example, instead of writing "The food was eaten by the pig", write "The pig ate the food". The active voice is easier to read and reduces the sentence length; this is particularly important, since most scientific journals have a word-count limit. Indiscriminate changes in tense are confusing and can give meaning to a statement which does not accord with the facts. However it can be acceptable to write in more than one tense in the literature review, e.g. "Brown (1995) showed that the brain is more fully developed at birth than other organs". In this case the present tense can be used for the second half of the sentence because its gives knowledge that is universally accepted. Materials and methods should be written in the past tense. "The experiment was designed in the form of a 6×6 Latin square." Remarks about results should mainly be in the past tense. "When a high protein diet was fed to rabbits they grew rapidly." Any conclusions drawn should be in the past tense, e.g. Pigs in this experiment grew most rapidly when fish meal was added to their diets. When referring to the conclusions of a particular experiment, it is incorrect to state pigs grow most rapidly when fish meal is added to the diet.

However, for the mathematical sciences, it is customary to report in the present tense.

Sentence Construction

The purpose of any paper is to convey information and ideas. This cannot be done with long involved sentences.

Keep sentences short, not more than 30 words in length. A sentence should contain one idea or two related ideas. A paragraph should contain a series of related ideas.

Choice of Words

Words have precise meanings and to use them correctly adds clarity and precision to prose. Look at the following pairs of words that are often used in scientific texts. Learn how to use them correctly: Fewer, less; infer, imply; as, because; disinterested, uninterested ; alibi, excuse ; data, datum; later, latter; causal, casual; loose, lose; mute, moot; discrete, discreet. Example to show difference between less and fewer by using the two words in the same sentence. Less active blood cells Fewer active blood cells.

Use a standard dictionary and Roget's Thesaurus of English Words and Phrases to find the correct meaning of words.

Use of Pronouns

When you write "it", "this", "which" or "they" are you sure that the meaning is plain? A pronoun usually deputizes for the nearest previous noun of the same number (singular or plural) — The cows ate the food; they were white. The cows ate the food; it was white.

Correct Spelling, Including the Use of Plurals

Some words have alternative spelling e.g. tyre, tire, grey, gray; draft, draught; connexion, connection, plow, plough, often the difference is between the American and British spelling. In other cases an apparent misspelling is a misuse of a word e.g., principle and principal; practice, practise (The former is a noun, the latter is a verb) The plural of many words in the English language is achieved by adding an s (or es) to the single. For example car becomes cars and potato becomes potatoes. However some words have the same form in both the singular and plural. For example sheep—there is no such word as sheeps. Other words are already plural such as people and equipment, so don't use peoples (unless you are referring to different groups of people or different ethnic groups) and equipments. Adopted words sometimes take on the plural of the original language, for example datum becomes data and fungus become fungi.

Use of Abstract Words

Use the concrete and not the abstract to achieve clarity and precision. "Cessation of plant growth operated in some of the plots." Obviously a cessation cannot operate (Some plots of plants did not grow during the trial) The abstract noun basis is commonly overworked. "Measurement of storm intensity involves recording staff to be available both day and night on a 24 hour basis." "To measure storm intensity recording staff have to be on duty throughout the day and night."

Be Careful with the Use of the Present Participle (Gerund)

Having said that the cow stood up. After standing in boiling water for an hour, examine the flask.

The gerund always ends in 'ing'. If the sentence is left without a subject (a hanging participle) then the action of the verb is transferred to the object of the sentence (first sentence) or to the person taking the action (second sentence).

Misuse of Emotional Words (avoid)

One cannot develop a logical argument using emotional words: e.g. progressive, reckless, crank, sound, good, correct, terrorist, freedom fighter, insurgent, sexist, imperialist, improved, superior, deviationist, fascist.

Superlatives

Very, more, much, have a place when used economically. As superlatives they are out of place in scientific writing. Superlatives such as gigantic, earth shattering or fantastic should never be used.

Qualifying the Absolute

Some adjectives are absolute and cannot be modified such as: sterile or unique. Other adjectives, such as "pregnant", have to be qualified with care. A petri dish is either sterile or not sterile. It cannot be very sterile, quite sterile or fairly sterile; An object is unique, and although a woman can be recently pregnant, she can not be slightly pregnant.

Loose Expressions (avoid)

■Bulls with a high milk yield tend to have fat percentages below average. Do bulls produce milk?
■Ewes were fed on a pregnant maintenance ration. Clearly the ration was not pregnant.
■In each selected village 30 farmers were interviewed, namely 10 large, average and small farmers. Is the reference here to the size of the farmers or to the size of their farms?

Grandiloquence [Examples quoted by J Oliver (1968)]

Avoid the use of scientific jargon. The aim in scientific writing is to inform using simple language not to confuse by the use grandiose sounding words and phrases.

Grandiloquent Phrase: The ideal fungicide ⋯ must combine high fungitoxicity with low mammalian toxicity and phytotoxicity, and with an absence of tainting or other deleterious side effects when the fruit is processed.

Simple Replacement: The ideal fungicide ⋯ must kill fungus effectively, but must be harmless to animals and plants, and must cause no tainting or other harmful side effects when the fruit is processed.

Grandiloquent Phrase: The phenomenon can be macroscopically observed upon laparotomy.
Simple Replacement: Visible to the naked eye on opening bird's cavity.

Genteelism

"I" is not immodest in a research worker and therefore use it(although not to excess), NOT "The present writer" or "The author of this communication".

The Misuse of the Definite Article "The"

Avoid Overuse of the Word "the" Only use when it applies to a particular item that has been referred to before, e.g. 'the various patients' may have been mentioned before. All others could be omitted.

The Excessive Use of the Pronoun "It" in a Sentence

Avoid excessive use of the indefinite pronoun "it".
- "It would thus appear that" can be replaced by "apparently".
- "It is evident that" by "evidently".
- Other commonly used phrases such as: "It will be seen that"; "It is interesting to note that" and "It is thought that", can be left out without any meaning being lost.

Avoid Verbal Obscurantisms and Use Simple Words

Always say what you mean NOT for example: Some phrases show sloppy thinking. For example, the phrase 'It has long been known that' usually means that the writer has not bothered to look up the reference. Correct to an order of magnitude probably means that the answer was wrong. Almost reached significance at the 5 % level usually means a selective interpretation of results. Text is easier to understand if simple words and phrases can be used to replace more complex or foreign ones. For example ameliorate can be replaced by improve; analogous by similar ; anthropogenic by human; Ceteris paribus by other things being equal; component by part; ingenuous by innocent; ingenious by clever; inter alia by amongst other things; utilise by use; Prima facie by at first glance; remunerate by pay; terminate by end; pari passu by at the same rate, pace or time and peruse by read.

Punctuation

Colon (:) and Semi Colon (;)
A colon is used when a list or explanation follows, a semi colon is used to separate two or more related clauses provided each clause forms a full sentence. Note its use in the sentence below and in the section above on choice of words.

The Apostrophe

These are used either to indicate the absence of a letter e.g. isn't it (for is not it) doesn't (for does not). Note the difference between (it is) it's a boy and its, which is the possessive adjective of it (everything in its place) or to denote possession (the boy's bike). If a word ends in s, the apostrophe may be placed after the s and the final s omitted (the calves'eyes).

Commas

Examples of the most important uses of commas in scientific writing are given below. A comma is put in a sentence to denote a brief pause between groups of words.

I will show you the paper about which I was speaking, but it is not as useful as I first thought. Or to separate subclauses:

Professor Brown, who is in charge of recruiting for the university, said that the latest estimates were higher than those for this time last year.

Finally to separate all items in a list except for the last two.

The following items may be imported duty free into Azania: animals, cereals, plants, fruit, trees, legumes and nuts.

Observe the importance of the comma paced between fruit and trees in this particular list.

Other Points Concerning the Use of English

One mistake commonly made, particularly by students whose first language is not English, is to not match the verb with the noun. A singular verb must always be associated with a singular noun, and similarly a plural verb with a plural noun, although a number of exceptions exist where a singular noun is used in a plural sense (for example, 'number' in this sentence) or, less commonly, a plural noun is used in a singular sense (for example, 'headquarters'), and the verb then can, and usually should, agree with the sense of the noun's usage. Difficulties arise especially with nouns which do not end in 's' in the plural form. For example livestock and data are plural.

Numbers and Units

Quantities should be given only as many significant figures as can be justified. For example the metabolic rate of an animal should not be quoted as 326.18W if it can be measured to only within about 5%. It should be written as 330W.

The figures within a number should be grouped in threes (with a small space between each group) so that they are easier to read. Commas should be avoided. For example: 21306.1 not 21,306.1.

Some require units, and the Systeme International (SI) should be used where possible. Some common units and their abbreviations are given below. The full stop is not used in the SI system.

When incorporating statistical data into the text, the test used (e.g. chi squared) should be included, along with the degrees of freedom, the calculated value and the P value.

Lesson 43 The Abstract

General Functions of an Abstract

For conference papers, research papers, theses and **dissertations**(学位论文), you will almost always be asked to write an abstract. An abstract is the most important component of a professional paper, the very first part of a professional paper and a self contained entity. It has many important functions:

"It's a one-year timer. It gives an added sense of urgency to my research grant."

1. Miniaturizing(缩小) the Text

An abstract is a condensed statement of the contents of a paper, a short, concise and highly generalized text, a mini-version of the main body.

2. Deciding Yes or No

For busy readers, the abstract may be the only part of the paper they read, unless it succeeds in convincing them to take the time to read the whole paper. It may directly influence the paper's acceptance to a learned journal. A good abstract should attract audience and convince them to read the whole paper, and promote thorough discussion on the relevant subject.

3. Expanding the Circulation（发行量）

Because on-line search databases typically contain only abstracts, it is vital to write a complete but concise description of your work to entice potential readers into obtaining a copy of the full paper. Abstracting services may use the text of the title plus the abstract and keywords for their searchable databases. For readers in developing countries with limited access to the literature, the abstract may be the only information on your work that is available to them.

Linguistic（语言）Features

1. Limited Length

200 words is a sensible maximum for a relatively longer paper, never more than 500 words, 50~100 words may suffice for a short article. As a general rule, an abstract will be about 3%~5% of the length of the paper, seldom longer than 2/3 of a page.

2. Categories of Abstracts

There are three kinds: descriptive, informational and informational-indicative.

    A descriptive or indicative abstract usually states the general subject matter of the document that follows. It tells in a qualitative way what the paper or report contains. These tell readers what information the dissertation contains, and include the purpose, methods, and scope of the report, article, or paper. A descriptive abstract will not provide results, conclusions, or recommendations, and is usually shorter than an informative abstract—usually under 100 words. Its purpose is to merely introduce the subject to reader, who must then read the dissertation to find out your results, conclusions, or recommendations.

An informational or informative abstract highlights the findings and results, briefly and quantitatively. It is a condensed version of the research work, without discussion or interpretation. It should be specific and quantitative, giving only essential data. These communicate specific information from the dissertation, including the purpose, methods, and scope of the report, article, or paper. They provide the dissertation results, conclusions, and recommendations. They are short but not as short as a descriptive abstract—usually anything from a paragraph to a page or two, depending upon the length of the original work being abstracted. In any case, informative abstracts make up 10% or less of the length of the original piece. The informative abstract allows your reader to decide whether they want to read the dissertation.

The informational-indicative abstract is a combined form that bears specific information about principle findings and results and general information about the rest of the document. This type of abstract offers fewer details, instead, giving emphasis to the author's chief contribution. In doing so, the following abstract highlights the same technical information.

3. Complete Content

The abstract is self-contained, unified and coherent in content. An informational abstract should at least contain the following three elements: a statement of the problem; a statement of the approach to solving the problem; the principal result.

Abstract should include:
① Background: introduce the topic (optional)
② Purpose of the study: study aim
③ Methods of the study: overview of the methods
④ Finding/Results of the study: overview of result
⑤ Conclusion, implication of the study

(1) Motivation

Why do we care about the problem and the results? If the problem isn't obviously "interesting" it might be better to put motivation first, but if your work is **incremental**（增加的） progress on a problem that is widely recognized as important, then it is probably better to put the problem statement first to indicate which piece of the larger problem you are breaking off to work on. This section should include the importance of your work, the difficulty of the area, and the impact it might have if successful.

(2) Problem Statement

What problem are you trying to solve? What is the scope of your work (a generalized approach, or for a specific situation)? Be careful not to use too much jargon. In some cases it is appropriate to put the problem statement before the motivation, but usually this only works if most readers already understand why the problem is important.

(3) Approach

How did you go about solving or making progress on the problem? Did you use **simulation**（模拟）, analytic models, **prototype**（样本） construction, or analysis of field data for an actual product? What was the extent of your work (Did you look at one application program or a hundred programs in twenty different programming languages?) What important variables did you control, ignore, or measure?

(4) Results

What's the answer? Specifically, most good computer architecture papers conclude that something is so many percent faster, cheaper, smaller, or otherwise better than something else. Put the result there, in numbers. Avoid vague, hand-waving results such as "very", "small", or "significant". If you must be vague, you are only given license to do so when you can talk about orders-of-magnitude improvement. There is a tension here in that you should not provide numbers that can be easily misinterpreted, but on the other hand you don't have room for all the caveats.

(5) Conclusions

What are the implications of your answer? Is it going to change the world (unlikely), be a significant "win", be a nice hack, or simply serve as a road sign indicating that this path is a waste of time (all of the previous results are useful). Are your results general, potentially generalizable, or specific to a particular case?

4. Formalized Structure

Three major parts: topic sentence, supporting sentences and concluding sentences.

(1) Topic Sentence — introducing the purpose

The purpose of this paper is …
The primary goal of this research is …
The intention of this paper is to survey …
The work presented in this paper focuses on several aspects of the following …
The overall objective of this study is …
In this paper, we propose/aim at …
Our goal has been to provide …
The emphasis of this study lies in …
The chief aim of the present work is to investigate the features of …
The author's endeavor is to explain …

(2) Supporting Sentences—introducing the methods

The method used in our study is known as …
The technique we applied is referred to as …
A number of experiments were performed to check …
This research has recorded valuable data using the newly developed method of …
The experiment consisted of three steps, which are described as …
The approach adopted extensively is called …
Included in this experiment were …
Special attention is given here to …
Recent experiments in this area suggested that …

(3) Concluding Sentences

In conclusion, we state that …
In summing up it may be stated that …
It is concluded that …
From our experiment, the author came to realized that …
These findings of research have led the author to the conclusion that …
The result of the experiments indicates that …
Our work involving … proves to be encouraging.

Writing Requirements

1. Integrity

Include what the writer has done and what he has achieved within the scope of the topic, such as the research theories, research methods, investigations and experimental results and conclusions, and to stress his own contribution.

2. Concise

Use only essential information. Avoid displaying mathematical expressions, never number equations, omit tables.

3. Consistency

Never included what has not been mentioned in the paper and no modification in meaning is permitted.

4. Concentration

Do not use figures, tables or literature references in this part.

Some journals provide a list of questions or headings for authors to respond to in writing their abstracts, and others do not. All provide a maximum number of words that an abstract may contain. Based on analyses of many abstracts in science and technology fields, the following information elements can be proposed as constituting a full abstract.

Some background information	Background
The principal activity (or purpose) of the study and its scope	Purpose
Some information about the methods used in the study	Methods
The most important results of the study	Results
A statement of conclusion or recommendation	Conclusion

Other Considerations

An abstract must be a fully self-contained, capsule description of the paper. It can't assume (or attempt to provoke) the reader into flipping through looking for an explanation of what is meant by some vague statement. It must make sense all by itself. Some points to consider include:

(1) Meet the word count limitation. If your abstract runs too long, either it will be rejected or someone will take a chainsaw to it to get it down to size. Your purposes will be better served by doing the difficult task of cutting yourself, rather than leaving it to someone else who might be more interested in meeting size restrictions than in representing your efforts in the best possible manner. An abstract word limit of 150 to 200 words is common.

(2) Any major restrictions or limitations on the results should be stated, if only by using "weasel-words" such as "might", "could", "may", and "seem".

(3) Think of a half-dozen search phrases and keywords that people looking for your work might use. Be sure that those exact phrases appear in your abstract, so that they will turn up at the top of a search result listing.

(4) Usually the context of a paper is set by the publication it appears in (for example, IEEE Computer magazine's articles are generally about computer technology). But, if your paper appears in a somewhat un-traditional venue, be sure to include in the problem statement the domain or topic area that it is really applicable to.

(5) Some publications request "keywords". These have two purposes. They are used to facilitate keyword index searches, which are greatly reduced in importance now that on-line

abstract text searching is commonly used. However, they are also used to assign papers to review committees or editors, which can be extremely important to your fate. So make sure that the keywords you pick make assigning your paper to a review category obvious (for example, if there is a list of conference topics, use your chosen topic area as one of the keyword tuples).

How not to Write an Abstract

(1) Do not refer extensively to other works.

(2) Do not add information not contained in the original work.

(3) Do not define terms.

(4) Do not use "I" or "You", use third person (passive) in the abstract.

(5) Do not include abbreviations or acronyms in your abstract if you can help it, but if you must, don't use them without explaining them first. For example, the first time you use the abbreviation you must write out the full form and put the abbreviation in brackets. e.g. "Magnetic Resonance Imaging (MRI)". From then on you may use "MRI" for the duration of the abstract.

(6) Do not use headings for your abstract paragraphs. (e.g. Objectives, Methods, Results and Conclusions)

(7) Do not use both British English and American English, use either British English spelling conventions or American English spelling conventions throughout your abstract. Always check your grammar, spelling, and formatting.

Examples

Sample 1

Here is a screenshot from Journal of Catalysis. Please figure out the main information of the paper (Title of the article, *title of the journal, volume number,* page numbers, authors, author affiliation, abstract, key words).

Journal of Catalysis 364 (2018) 141–153

Contents lists available at ScienceDirect

Journal of Catalysis

journal homepage: www.elsevier.com/locate/jcat

Aerobic oxidation of alkanes on icosahedron gold nanoparticle Au55

Aleksandar Staykov [a,*], Tetsuya Miwa [b], Kazunari Yoshizawa [b,*]

[a] *International Institute for Carbon-Neutral Energy Research I2CNER, Kyushu University, Fukuoka 819-0395, Japan*
[b] *Institute for Materials Chemistry and Engineering IMCE and Integrated Research Consortium for Chemical Science IRCCS, Kyushu University, Fukuoka 819-0395, Japan*

ARTICLE INFO	ABSTRACT
Article history: Received 14 March 2018 Revised 16 May 2018 Accepted 16 May 2018 Available online 5 June 2018 *Keywords:* Gold nanoparticles Catalysis Aerobic oxidation Density functional theory C—H Activation	Aerobic oxidation of cyclohexane, propane, ethane, and methane to the corresponding alcohols was investigated over an Au55 gold nanoparticle with icosahedron symmetry using density functional theory. Reaction mechanisms were elucidated and activation barriers for catalytic C—H bond cleavage and corresponding alcohols' formation were estimated. Furthermore, on the basis of the reaction rate constants calculated for realistic reaction temperatures, the relative reaction rates for each alkane hydroxylation were discussed. The catalyst selectivity was investigated for the formation of primary and secondary alcohols. All reaction mechanisms for alkane hydroxylation are compared with the catalytic dissociation of dioxygen molecule over gold nanoparticle surface, which is an important precursor reaction for aerobic oxidation. We have further investigated overoxidation reaction mechanisms leading to formation of ketones. Our results are compared with experimental findings to provide important guidelines for the tuning of catalytic reactions towards the desired products and reaction conditions. © 2018 Elsevier Inc. All rights reserved.

Sample 2

Here is an abstract from a published paper. It is 150 words long. Read it through looking for the main purpose of each sentence (for example, presenting research problem, objective, methodology, main findings, or conclusion).

In this paper, a new learning rule and theoretical analysis of an extended bidirectional associative memory network (MLBAM) is presented, by using the maximum likelihood criterion based on two well recognized and essential criteria, i.e., the convergence of the learning rule, and the noise tolerance of this network. Traditional methods fail to distinguish highly approximate patterns. However, the method in our study can improve this by using the newly developed method of maximum likelihood criterion. In addition, the learning approach guarantees that correlated patterns could be associated as a stable state and the network possesses excellent anti-noise property by using likelihood function, namely, the learning approach specializes in the situation including stochastic disturbance. Additionally, the associative capability of the bidirectional associative memory is specifically discussed. Finally, two experiments are used to certify the validity and efficiency of our method, especially the method's excellent anti-noise property by using likelihood function.

Sample 3

Here is an abstract from a published paper. It is 187 words long. Read it through looking for the main purpose of each sentence (for example, presenting research problem, objective, methodology, main findings, or conclusion).

A nanocomposite film, chitosan(CS)-polyvinylpyrrolidone(PVP)-bentonite(BN) was fabricated to enhance wound healing processes as a new nanoplatform for wound dressing. Both physical properties and antibacterial activity of the proposed film were measured to validate its applicability and inhibitory effect for wound management. In vitro cytotoxicity was evaluated by using MTT assay on L929 and NIH3T3 cells to identify the toxicity level of the film. In vivo wound healing test assessed the wound healing performance in animal models. The results confirmed a strong interaction between surface functional groups among CS, PVP and BN with suitable surface morphology and high thermal stability. The CS-PVP-BN film improved various material features such as including mechanical property, tensile strength, pH and porosity, inhibitory activity on bacterial organisms, and collagen deposition. The animal study confirmed that the fabricated film yielded a rapid healing rate of 97%, less scarring, thick granulation at the 11th day, regeneration of epidermis at the 16th day, and abundant deposition of collagen and fibroblast, compared with control. The non-toxic nanocomposite film can be a promising antibacterial wound dressing with rapid wound healing effects in wound care management.

Exercises

1. Get a recent issue of one of the first class English journals in your field, read the "abstract" of all the papers in it. Study the organization of these abstracts and try to learn some useful sentence patterns used in them.
2. Write two abstracts for one of your papers—one descriptive, the other informative.

Appendix

Appendix 1 Sample Exams

Sample Paper A

General Chemistry Exam (I)

Part I Multiple-Choice or Short Response

Each multiple-choice question is worth 3 points. This part of the exam is worth 66% of the total points.

1. Manganese has the oxidation number of +5 in_____.
    (A) $[MnF_6]^{3-}$ (B) Mn_2O_7 (C) $[MnO_4]^{2-}$ (D) $[Mn(CN)_6]^-$

2. Which of the following statements is true about this process at 25°C and constant pressure?
    $$H_2O\,(g) \longrightarrow H_2O\,(l)$$
    (A) H is positive and the process is endothermic.
    (B) H is positive and the process is exothermic.
    (C) H is negative and the process is endothermic.
    (D) H is negative and the process is exothermic.

3. When 50mL of 0.1mol/L HCl is mixed with 50mL of 0.2mol/L NaOH, the resulting solution will be_____.
    (A) acidic   (B) basic
    (C) neutral   (D) not enough information to tell

4. A strong acid and a metal carbonate react to form_____.
    (A) a salt and hydrogen   (B) a salt and water
    (C) a salt, water and carbon dioxide   (D) a salt and a base
    (E) another acid and a base

5. If a 17.0g sample of impure nickel metal reacts under standard conditions with 1.12mol of CO gas to form 0.279mol of $Ni(CO)_4$ gas, what is the percentage of Ni in the metal sample? [Molar Masses: Ni, 58.7g/mol; $Ni(CO)_4$, 171g/mol]
    $$Ni\,(s) + 4\,CO\,(g) \longrightarrow Ni(CO)_4\,(g)$$
    (A) 24.1%   (B) 25.0%   (C) 96.4%   (D) 100%

6. Which change requires an oxidizing agent to produce the indicated product?
    (A) $2\,S_2O_3^{2-} \longrightarrow S_4O_6^{2-}$ (B) $Zn^{2+} \longrightarrow Zn$
    (C) $ClO^- \longrightarrow Cl^-$ (D) $SO_3 \longrightarrow SO_4^{2-}$

7. To determine experimentally whether the compound MCl_3 is ionic or covalent, one might _____.
    (A) test the solubility in water

(B) determine the percentage composition
(C) find the valence of M
(D) test the electrical conductivity in the fused (molten) state
(E) determine whether the compound has an electrical charge

8. Which pair represents compounds each of which dissolves in water to give solutions that are good conductors of electricity?
   (A) CH_3COOH and NaCl
   (B) NaCl and AgCl
   (C) HF and HCl
   (D) NaCl and H_2SO_4

9. The heat required to increase the temperature of 50.0g of H_2O (l) from 25.0°C to 45.0°C [specific heat, 4.18J/(g·°C) for water] is_____.
   (A) 116J
   (B) 209J
   (C) 4180J
   (D) 4180kJ

10. A solution that has pH of 8 is_____.
    (A) very acidic
    (B) very basic
    (C) slightly acidic
    (D) slightly basic
    (E) exactly neutral

11. Solution A was prepared by dissolving 0.60mol NaCl in 2.0L water. Solution B was prepared by dissolving 0.60mol Na_2SO_4 in 2.0L water. What is the concentration of Na^+ ions in solution C which is made by combining equal volumes of solutions A and B.
    (A) 0.30mol/L
    (B) 0.45mol/L
    (C) 0.60mol/L
    (D) 0.90mol/L

12. Which is an oxidation-reduction reaction?
    (A) Iron reacts with sulfur.
    (B) Sodium hydroxide reacts with hydrochloric acid.
    (C) Sodium chloride reacts with silver nitrate solution.
    (D) Calcium oxide reacts with hydrochloric acid.
    (E) Sodium carbonate reacts with hydrochloric acid.

13. On the basis of metal reactivities, which pair of substances may be expected to react? (All solutions are of the same molar concentration. Metal Reactivity: Mg > Al > Zn > Fe > H > Cu > Hg > Ag)
    (A) mercury (II) chloride in solution and aluminum metal
    (B) aluminum chloride in solution and silver metal
    (C) zinc sulfate in solution and hydrogen gas
    (D) copper nitrate in solution and silver metal
    (E) magnesium sulfate in solution and iron metal

14. A solution contains 0.400g NaOH in 20.0mL solution. What is its molarity?
    (A) 0.250mol/L
    (B) 0.400mol/L
    (C) 0.500mol/L
    (D) 1.00mol/L
    (E) 2.00mol/L

15. A precipitate will form from an aqueous solution of Fe^{3+} ion upon the addition of_____.
    (A) KOH
    (B) $NaNO_3$
    (C) HCl
    (D) HNO_3

16. Which statement is true for the following reaction?
    $$Fe (s) + Cu^{2+} (aq) \longrightarrow Cu (s) + Fe^{2+} (aq)$$
    (A) Cu^{2+} is oxidized
    (B) Cu^{2+} gains in oxidation state
    (C) Cu^{2+} is reduced
    (D) Fe (s) is reduced

17. Which equation best represents the net ionic reaction that occurs when sodium hydroxide and hydrochloric acid solutions are mixed?

(A) Na⁺ + HCl ⟶ NaCl + H⁺
(B) OH⁻ + HCl ⟶ H₂O + Cl⁻
(C) OH⁻ + H₃O⁺ ⟶ 2H₂O
(D) NaOH + H₃O⁺ ⟶ 2H₂O + Na⁺

18. 10.0g silver is heated to 100.0℃ and then added to 20.0g water at 23.0℃ in an insulated calorimeter. At thermal equilibrium, the temperature of the system was measured as 25.0℃. What is the specific heat, of silver? [specific heat for water, 4.18 J/(g·℃)]
    (A) 0.11J/(g·℃)   (B) 0.22J/(g·℃)   (C) 17J/(g·℃)   (D) 34J/(g·℃)

19. Complete and balance the equation for the reaction, where the reactants are in aqueous solution. Use no fractional coefficients.
    $$? \ Na_3PO_4 + ? \ Ba(NO_3)_2 \longrightarrow ? + ?$$
    The number of moles and the formula of the product containing Ba are
    (A) 3 NaNO₃   (B) BaPO₄   (C) Ba(PO₄)₂
    (D) Ba₂P₃   (E) Ba₃(PO₄)₂

20. What makes carbon such a unique element?
    (A) Elemental carbon comes in two forms, diamond and graphite.
    (B) Carbon forms four bonds, although the ground state configuration would predict the formation of fewer bonds.
    (C) Carbon forms covalent bonds rather than ionic bonds.
    (D) To a greater extent than any other element, carbon can bond to itself to form straight chains, branched chains and rings.
    (E) Carbon has two stable isotopes, carbon-12 and carbon-13.

21. A molecule with the formula C₃H₈ is a(n)_____.
    (A) hexane   (B) propane   (C) decane
    (D) butane   (E) ethane

22. Select the correct IUPAC name for:
    (A) 5-methyl-5-ethyloctane   (B) 5-methyl-5-propylheptane
    (C) 4-ethyl-4-methyloctane   (D) 3-methyl-3-propyloctane
    (E) 3-methyl-3-propylheptane

    $$CH_3CH_2CH_2CH_2\underset{\underset{CH_2CH_2CH_3}{|}}{\overset{\overset{CH_3}{|}}{C}}CH_2CH_3$$

Part II Problems

Points possible per question and per part are indicated in curly braces {···}.

1. {10 pts} What is the value of $H°$ for this reaction? Show your work.
    $$3 \ H_2O \ (l) \longrightarrow 3 \ H_2 \ (g) + O_3 \ (g)$$
    The following reaction enthalpies are given:
    $$3 \ O_2 \ (g) \longrightarrow 2 \ O_3 \ (g) \quad H° = +271 kJ$$
    $$H_2 \ (g) + 1/2 \ O_2 \ (g) \longrightarrow H_2O \ (l) \quad H° = -286 kJ$$

2. {24 pts, each part worth 6 pts} Be sure to provide answers with the correct significant digits.
    130.0mL of a 0.110mol/L AgNO₃ solution is mixed with 90.0mL of a 0.095mol/L CaCl₂ solution in an insulated Styrofoam cup (approximating a calorimeter). Both solutions begin at 25.00°C. The reaction that takes place is
    complete reaction: 2 AgNO₃ (aq) + CaCl₂ (aq) ⟶ 2 AgCl (s) + Ca(NO₃)₂ (aq)
    net ionic reaction: Ag⁺ (aq) + Cl⁻ (aq) ⟶ AgCl (s)
    After the reaction is complete and the white precipitate has settled, the temperature of the

entire 200.0mL of solution is 27.00°C.

Notes: - Parts (2) and (3) are independent of part (1).
- You must show your work to receive credit. Partially correct work will receive partial credit. A correct answer with no work shown will receive no credit.

(1) Determine the moles of AgCl (s) precipitate that form and identify the limiting reagent.
(2) How many kilojoules of heat energy were absorbed by the water during the process? (Assume that the solution is dilute enough that its density is the same as pure water: 1.000g/mL.)
(3) Is the net ionic reaction shown above endothermic or exothermic? How do you know?
(4) What is the molar enthalpy of reaction (in kJ/mol) for the net ionic reaction shown above? Be sure to indicate the correct sign (+ or −).

Sample Paper B

General Chemistry Exam (II)

Part I Multiple-Choice or Short Response

Each multiple-choice question is worth 3 points. This part of the exam is worth 60% of the total points.

1. Oxygen and ozone are _____.
   (A) the same substance and the same element
   (B) the same element, but two different substances
   (C) the same substance, but two different elements
   (D) two different substances and two different elements

2. What is the molar mass (in g/mol) of anhydrous iron (III) sulfate, to the nearest whole number?
   (Atomic molar masses for: Fe, 55.8g/mol; O, 16.0g/mol; S, 32.1g/mol)
   (A) 104   (B) 152   (C) 248
   (D) 336   (E) 400

3. The species designated as $^{56}_{24}X$ is _____.
   (A) Fe   (B) Ge   (C) Ba   (D) Cr

4. Which pair of particles has the same number of electrons?
   (A) F^-, Mg^{2+} (B) Ne, Ar   (C) Br^-, Se   (D) Al^{3+}, P^{3-}

5. The ions present in solid silver chromate, Ag_2CrO_4, are _____.
   (A) Ag^+ and CrO_4^{2-} (B) Ag^{2+} and CrO_4^{4-}
   (C) Ag^+, Cr^{6+} and O^{2-} (D) Ag^+, Cr^{3+} and O^{2-}

6. Which represents an isotope of element "E"? $^{27}_{13}E$ _____.
   (A) ^{26}Al (B) ^{27}Si (C) ^{27}Co (D) ^{25}Mg

7. What is the correctly reported mass of water based on these data?
   mass of beaker and water   29.62g
   mass of beaker only   28.3220g
   (A) 1.3g   (B) 1.30g   (C) 1.298g   (D) 1.2980g

8. Based on their positions in the periodic table, which is most likely to replace selenium, Se, in a biological system?
   (A) Te   (B) Br   (C) As   (D) I

9. Balance the equation
   $$?\ N_2H_4 + ?\ N_2O_4 \longrightarrow ?\ N_2 + ?\ H_2O$$
   How many moles of N_2 will be produced for every mole of N_2O_4 that reacts?
   (A) one   (B) two   (C) three   (D) four

10. Which procedure can be used to demonstrate experimentally the reaction below.
    $$2\ Mg + O_2 \longrightarrow 2\ MgO$$
    (A) Take 1.000g Mg ribbon, burn it in pure O_2, and compare the mass of the product with the original mass of the Mg.
    (B) Show that the sum of 2 atomic molar masses of Mg plus 1 molar mass of O_2 is equal to 2 molar masses of MgO.

(C) Determine the mass of a sealed flash-bulb containing magnesium and oxygen, ignite (light on fire) the mixture, cool, and compare the final mass of bulb plus contents with the original mass of the bulb plus contents.

(D) Burn 1.000g Mg ribbon in a tall beaker filled with air, scrape out all of the MgO formed, and compare with the original mass of the Mg.

Use the periodic table below for questions 11~12.

	\multicolumn{8}{c}{Main groups}
	1	2	3	4	5	6	7	8
First period								$_2$He
Second period								$_9$F
Third period	$_{19}$K	E		M		Q	T	$_{36}$Kr
Fourth period	X	Y						

11. Judging from its position in the periodic table, what type of element is element X?
    (A) a metal   (B) a nonmetal   (C) an amphoteric element
    (D) an inert gas   (E) unpredictable in character

12. How many electrons will an atom of element Q need to gain to form a stable ion?
    (A) 1   (B) 2   (C) 3
    (D) 4   (E) 7

13. If 1.0g samples of each compound were dehydrated, which sample would lose the greatest mass of water?
    (Molar masses for: LiCl · H$_2$O,   60g/mol;   MgSO$_4$ · H$_2$O,   138g/mol
                       FeSO$_4$ · H$_2$O, 170g/mol;   SrC$_2$O$_4$ · H$_2$O, 194g/mol
    (A) LiCl · H$_2$O   (B) MgSO$_4$ · H$_2$O   (C) FeSO$_4$ · H$_2$O   (D) SrC$_2$O$_4$ · H$_2$O

14. A compound containing only carbon and hydrogen has this composition: C = 80% and H = 20% by mass. What is the simplest formula of the compound? (Atomic molar masses for: C, 12.0g/mol; H, 1.0g/mol.)
    (A) CH$_4$ (B) CH$_3$ (C) C$_2$H$_6$ (D) C$_3$H$_8$ (E) C$_4$H

15. What is the percentage of nitrogen by mass in (NH$_4$)$_3$PO$_4$?
    (Atomic molar masses for: H, 1.0g/mol; N, 14.0g/mol; O, 16.0g/mol; P, 31.0g/mol.)
    (A) (14/62)×100%   (B) (21/80)×100%
    (C) (14/113)×100%   (D) (42/149)×100%

16. The element X occurs naturally to the extent of 20.0% ^{12}X and 80.0% ^{13}X. The atomic mass of X is nearest to_____.
    (A) 12.2   (B) 12.5   (C) 12.6   (D) 12.8   (E) 13.0

17. Why is the following equation incorrect?
$$Mg_3 + N_2 \longrightarrow Mg_3N_2$$
    (A) Some of the subscripts are incorrectly used.
    (B) The equation is not balanced.
    (C) The valence (charge) of the nitride ion is incorrect.
    (D) The valence (charge) of the magnesium ion is incorrect.
    (E) The coefficient of N$_2$ is incorrect.

18. Which of these atoms has the greatest number of neutrons in its nucleus?
    (A) $^{52}_{26}$Fe   (B) $^{56}_{25}$Mn   (C) $^{55}_{26}$Fe
    (D) $^{57}_{27}$Co   (E) $^{56}_{28}$Ni

19. Which of the following formulas represents an alkene?
    (A) $CH_3CH_2CH_3$ (B) CH_3CH_3 (C) $CH_3CH_2CHCH_2$
    (D) CH_3CH_2Cl (E) $CHCH$

20. Select the IUPAC name for $(CH_3)_2CHCH(OH)CH_2C(CH_3)_3$.
    (A) 2,5,5-trimethyl-3-hexanol
    (B) 1,1,4,4-pentamethylbutanol
    (C) 1,1-dimethylisopentanol
    (D) 2,5-dimethyl-4-hexanol
    (E) None of these

Part II Problems

Points possible per question and per part are indicated in curly braces {⋯}.

1. {20pts} Nomenclature: correct spelling and correct symbols matter.
   (1) Name the following compounds. {10pts}
           NaCl;   HCl; $HClO_2$; CH_4; $Ni(NO_2)_2$
   (2) Provide chemical formulas for the following compounds. {10pts}
     calcium fluoride;   dinitrogen tetroxide;   ammonium permanganate;
     hypobromous acid;   iron(II) sulfate

2. {16pts} Consider the following reaction:
   $$2\ NO_2\ (g) + Cl_2\ (g) \longrightarrow 2\ NO_2Cl\ (g)$$
   (Molar masses for: NO_2, 46.01g/mol; Cl_2, 70.90g/mol; NO_2Cl, 81.46g/mol.)
   Notes:
     - The parts of this problem are independent.
     - You must show your work to receive credit. Partially correct work will receive partial credit. A correct answer with no work shown will receive no credit.
   (1) How many molecules of Cl_2 are in 1.39g of Cl_2? {6pts}
   (2) If 1.39g Cl_2 reacts with sufficient NO_2 for the reaction to go to completion, how many grams of NO_2Cl will be produced? {10pts}

Extra credit on this problem: If 2.10g of NO_2 and 2.00g of Cl_2 were placed in a reaction vessel and this reaction occurred, which one would be the limiting reagent? {4pts}

Sample Paper C

Chemistry Speciality English Exam

Part I   Multiple-Choice or Short Response (20pts)

1. The number 10.00 has how many significant figures?
   (A) 1 (B) 2 (C) 3
   (D) 4 (E) 5
2. Which one of the samples contains the most atoms?
   (A) 1mol of $CO_2(g)$ (B) 1mol of $UF_6(g)$
   (C) 1mol of $CH_3COCH_3(l)$ (D) 1mol of $He(g)$
   (E) all contain the same number of atoms
3. The limiting reagent in a chemical reaction is one that_____.
   (A) has the largest molar mass (formula weight)
   (B) has the smallest molar mass (formula weight)
   (C) has the smallest coefficient
   (D) is consumed completely
   (E) is in excess
4. The precipitate formed when barium chloride is treated with sulfuric acid is _____.
   (A) BaS_2O_4 (B) $BaSO_3$ (C) $BaSO_2$
   (D) $BaSO_4$ (E) BaS
5. Which of the following electron configurations is correct for nickel?
   (A) $[Ar]4s^13d^8$ (B) $[Kr]4s^14d^8$ (C) $[Kr]4s^13d^8$
   (D) $[Kr]4s^23d^8$ (E) $[Ar]4s^23d^8$
6. All of the following properties of the alkaline earth metals increase going down the group except_____.
   (A) atomic radius (B) first ionization energy
   (C) ionic radius (D) atomic mass
   (E) atomic volume
7. Which is classified as nonpolar covalent?
   (A) the H—I bond in HI (B) the H—S bond in H_2S
   (C) the P—Cl bond in PCl_3 (D) the N—Cl bond in NCl_3
   (E) the N—H bond in NH_3
8. Choose the molecule that is incorrectly matched with the electronic geometry about the central atom.
   (A) CF_4-tetrahedral (B) $BeBr_2$-linear (C) H_2O-tetrahedral
   (D) NH_3-tetrahedral (E) PF_3-pyramidal
9. In the Bronsted-Lowry system, a base is defined as_____.
   (A) a proton donor (B) a hydroxide donor (C) an electron-pair acceptor
   (D) a water-former (E) a proton acceptor
10. What volume of 0.100mol/L HNO_3 is required to neutralize 50.0mL of a 0.150mol/L solution of $Ba(OH)_2$?
    (A) 50.0mL (B) 75.0mL (C) 100mL
    (D) 125mL (E) 150mL

11. A real gas most closely approaches the behavior of an ideal gas under conditions of_____.
    (A) high P and low T   (B) low P and high T   (C) low P and T
    (D) high P and T   (E) STP
12. What type of interparticle forces holds liquid N_2 together?
    (A) ionic bonding   (B) London forces   (C) hydrogen bonding
    (D) dipole-dipole interaction   (E) covalent bonding
13. Which one of the following thermodynamic quantities is not a state function?
    (A) Gibbs free energy   (B) enthalpy   (C) entropy
    (D) internal energy   (E) work
14. A catalyst_____.
    (A) actually participates in the reaction
    (B) changes the equilibrium concentration of the products
    (C) does not affect a reaction energy path
    (D) always decreases the rate for a reaction
    (E) always increases the activation energy for a reaction
15. The conventional equilibrium constant expression (K_c) for the system as described by the below equation is:
    $$2SO_3(g) \rightleftharpoons 2SO_2(g) + O_2(g)$$
    (A) $[SO_2]^2/[SO_3]$ (B) $[SO_2]^2[O_2]/[SO_3]^2$ (C) $[SO_3]^2/[SO_3]^2[O_2]$
    (D) $[SO_2][O_2]$ (E) None of these
16. Which one of the following is a strong electrolyte?
    (A) H_2O (B) KF   (C) HF
    (D) HNO_2 (E) $(CH_3)_3N$
17. Which of the following combinations cannot produce a buffer solution?
    (A) HNO_2 and $NaNO_2$ (B) HCN and NaCN   (C) $HClO_4$ and $NaClO_4$
    (D) NH_3 and $(NH_4)_2SO_4$ (E) NH_3 and NH_4Br
18. The solubility product expression for tin(II) hydroxide, $Sn(OH)_2$, is_____.
    (A) $[Sn^{2+}][OH^-]$ (B) $[Sn^{2+}]^2[OH^-]$ (C) $[Sn^{2+}][OH^-]^2$
    (D) $[Sn^{2+}]^3[OH^-]$ (E) $[Sn^{2+}][OH^-]^3$
19. Select the correct IUPAC name for $[Co(NH_3)_6]^{2+}$.
    (A) hexammoniacobaltate(II) ion
    (B) hexaamminecobaltate(II) ion
    (C) hexammoniacobalt(II) ion
    (D) hexaamminecobalt(II) ion
    (E) hexammoniacobalt ion
20. How old is a bottle of wine if the tritium (3H) content (called activity) is 25% that of a new wine? The half-life of tritium is 12.5 years.
    (A) 1/4 y   (B) 3.1 y   (C) 25 y
    (D) 37.5 y   (E) 50 y

Part II Common Skill (20pts)

1. Write the oral English expression of the following.
    (1) $\rightleftharpoons$ (2) Cu, Δ (3) $Cl^- + Ag^+ \longrightarrow AgCl$ (4) $-3\,°C$ (5) $Mg(OH)_2$
2. Write the numerical equivalences to the following prefixes.
    (1) tetra   (2) nona   (3) tri   (4) hexa   (5) penta

3. Write the corresponding Chinese of the chemistry abbreviation below.
   (1) approx.   (2) aq.   (3) fig.   (4) amt.   (5) etc.
4. Select only one from the following table that best matches the name of lab glassware figure.

| A. distillation flask | B. separating funnel | C. condenser |
| D. erlenmeyer flask | E. measuring cylinder | |

(1) ( )   (2) ( )   (3) ( )   (4) ( )   (5) ( )

Part III. Select only one from the following table that best matches the IUPAC name of following compound (1.5×8=12pts)

A. 2,5-dimethyl-2-heptene	B. cyclopentane
C. sodium hydroxide	D. calcium oxide
E. potassium hypochlorite	F. 3,4-dimethylheptane
G. boric acid	H. sodium nitrite

(1) $\text{H}_3\text{C}-\overset{\text{CH}_3}{\underset{\text{H}_3\text{C}\ \ \text{CH}_3}{|}}$ ( )   (2) $\overset{\text{H}_3\text{C}}{\underset{\text{H}_3\text{C}}{}}\text{C}=\overset{\text{CH}_3}{\underset{\text{CH}_3}{\text{C}}}$ ( )   (3) ⬠ ( )

(4) H_3BO_3 ( )   (5) $KClO$ ( )   (6) CaO ( )   (7) $NaOH$ ( )   (8) $NaNO_2$ ( )

Part IV Write the chemical symbol of the following chemical elements (10pts)
(1) helium   (2) sulfur   (3) fluorine   (4) boron   (5) oxygen
(6) magnesium   (7) hydrogen   (8) chlorine   (9) aluminum   (10) argon

Part V. 17.0g of aluminum reacts with 34.0g of chlorine gas to produce aluminum chloride. Its chemical equation is $2Al + 3Cl_2 \longrightarrow 2AlCl_3$. (The atomic weight of the element Al and Cl is 27 and 35.5 respectively. You may show all the chemical calculations in Chinese or in English) (8pts)
   (1) What type of chemical equation is that?
   (2) Which reactant is limiting, which is in excess, and how much product is produced?

Part VI Translate the sentences below into Chinese. (30pts)
1. When a negative ion is formed, one or more electrons are added to an atom, the effective nuclear charge is reduced and hence the electron cloud expands. Negative ions are bigger than the corresponding atom.

2. The homologs of benzene are those containing an alkyl group or alkyl groups in place of one or more hydrogen atoms.
3. Ketones are very closely ralated to both aldehydes and alcohols.
4. Evaporation differs from crystallization in that emphasis is placed on concentrating a solution rather than forming and building crystals.
5. The laws of thermodynamics are of prime importance in the study of heat.
6. Sulphuric acid is one of extremely reactive agents.
7. The table above shows it.
8. Organic compounds are not soluble in water because there is no tendency for water to separate their molecules into ions.
9. When a copper plate is put into the sulfuric acid electrolyte, very few of its atoms dissolve.
10. The melting point of alkanes is rather irregular at first, but tends to rise somewhat steadily as the molecules become larger.
11. Benzene can undergo the typical substitution reactions of halogenation, nitration, sulphonation and Friedel-Crafts reaction.
12. We call such a zinc atom an ion with a double positive charge, Zn^{2+}.
13. The higher degree of unsaturation is associated with somewhat greater chemical reactivity.
14. The presence of a substituent group in benzene exerts a profound control over both orientation and the ease of introduction of the entering substituent.
15. The functional group of a ketone consists of a carbon atom connected by a double bond to an oxygen atom.

Appendix 2 Prefixes and Suffixes in Chemistry Speciality English

-a	……化物	bi-	二，双，酸式
a-	无，不，非	bio-	生物的
ab-	脱离	bis-	双
-able	易于……的，可……的	bromo-	溴代
abs-	相反，不	but-	丁
acet(o)-	乙酰	carb(o)-	碳，羰
acetyl-	乙酰基	-carbaldehyde	甲醛
-acious	有……性质的，多……的	carbamoyl-	酰氨基
acyl-	酰基	-carbohydrazide	甲酰肼
-ad	（名词词尾）	-carbonyl, carbonyl-	羰基，酰基
-adiene	……二烯	carboxy-	羧基
-adiyne	……二炔	carboxylate-	……酸酯
aero-	空气	-carboxylic acid	……酸
-al, ald-, -aldehyde	醛	chemico-, chem(o)-	化学的
alk-	烃类	chloro-	氯化……，氯代……
allyloxy-	烯丙氧基	chromato-	色谱
-amide	酰胺	chrom(o)-	铬的，色
-amine	胺	-cide	除……剂，防……剂
amino-	氨基	circum-	周围，环绕，附加
ammine-	氨化，氨合	cis-	顺式
amphi-, ampho-	两个，两种	co-, col-, com-, cor-, con-	同，共，和
an-	无，不，非	counter-	反，逆
-ane	……烷	cyano-	氰合，氰基
-aneous	……性质的，属于……的	cyclo-	环
anhydrous(o)-	酐，无水，脱水	deca-	十，癸
ante-	前，先	deutero-	氘代
anti-	反抗	de-	消除，脱去，解，减，除
aquo(o)-	水合	dexter-, dextro-	右
-ar	（形容词词尾）	di-	双，二，偶
arsa-	砷杂	dia-	对穿，横穿，通过，完全
aryl-	芳基	diazo-	重氮，偶氮
-ate	含氧酸盐，……酯	dif-	逆，反，散
-atriene	……三烯	-dione	二酮
-atriyne	……三炔	dis-	否定，相反，分离
azo-	偶氮	dodeca-	十二
benz-	苯基	ef-	出，脱离
benzeno-	桥亚苯基	electr(o)-	电，电的
benzyloxy-	苄氧基	-en	变得……，放入……内部，进入……之中

续表

endo-	吸收，桥，内	iso-	同，等，异
-ene	……烯	-ite	亚酸盐，亚……酯
-enyl	……烯基	-ium	(金属元素词尾)
ep(i)-	环	-ize	使……化
epoxy-	环氧	-ketone	酮
equi-	相等，平等	kilo-	千
-er	……剂	kin(e)-	动，运动
-ether	醚	-lactone	内酯
ethoxy-	乙氧基	laevo-	左旋
ethyl-	乙基	-less	不，无
ex-, exo-	由……向外，放出	lipo-	酯的
extra-, extro-	外，外部，离	-lysis	分散，消除
ferri-	铁	mal-	恶，失，不
ferro-	亚铁	macro-, magni-	大，巨，宏观
fluoro-	氟代	mercapto-	氢硫基
-fold	倍	mercura-	汞杂
-form	仿	meso-	间位，片，杂
formyl-	甲酰基	meta-	间，偏，变
-free	无	-meter	……表，……仪，……针
gluco-, glyco -	甜味的	-metry	……测量法，……度量法
-gram	图，记录，图表，克	micro-	微，小，百万分之一
graphy-	……方法	milli-	千分之一
halo-	卤素	mis	坏，误，否定
-hedral	……面体的	mol(e)	摩尔，分子
-hedron	……面体	mono-	单，一，独
hemi-	半	multi-	多
hetero-	杂，异，多	nano-	纳
hex(a)-	六，己	nona-	九，壬
homo-	均，同，单	nor-	正，正常，降
holo	全	-oate	……酸盐，酸酯
hypo-	次，低	-ode	路，通道，极，管
-ic	酸的，高价金属离子	-oic acid	酸
-ic anhydride	酸酐	-oid	似……，像……
-ide	……化物，无氧酸盐，酐	-ol	醇
il-, im-,-in, ir-	无，非	-olate	……醇……
imino-	亚氨基	-ole	氮杂茂
infra-	在下，较低	olig (o)-	少，低，微
inter-	相互，中间，一起	-on	名词词尾,非金属元素词尾
intra- , intro-	内	-one	酮
-ish	略带……的，微……的	-or	剂，体，器

269

organo-	有机的	-stat	恒，不动，固定
ortho-	邻，正，原	stereo-	立体，空间
osm(o)-	渗透	sub-	低，亚，次
-ous	亚……的，亚	sulf-	硫
over-	过（度），超，在外	sulfo-	磺（酸）基
oxa-	氧杂	super-	上，超，过，高
-oxide	氧化物	sur-	在上，胜过
oxo-, oxy-	氧代，氧化	syn-	同，共，与，顺式
para-	对，顺，聚，旁，侧	tauto-	互变（异构）
pent(a)-	五，戊	tele-	远
per-	高，过，全，完全，十分	telluro-	碲基
peri-	周围，包围	tetra-	四
phenyl-	苯基	thermo-	热
-phile, -philic	亲……的，亲……体	thia-	硫杂，噻
phono-	声音	thio-	硫代
phospha-	磷杂	-tion	(名词词尾，表状态、行为、结果等)
photo-	光，摄像	trans-	反式
poly-	聚，多	tri-	三
pre-	预先，早，前	ultra-	极度，超，过度，外
pro-	早，前，先，代替	under-	在下，底下，不足，从属
pyro-	火，热，高温	-um	(金属元素词尾)
quadri-quadr(u)	四	un-	不，未，非，去
-quinone	醌	uni-	单，一
radio-	放射，辐射	-ure	(表示行为及结果)
re-	再，重新，反，回	vapori-	水蒸气，水雾
self-	自身，自发，自动	vinyl-	乙烯基
semi-	半，部分，不完全	volt(a)-	伏特
sens-	感觉，灵敏	-yl	……基
-side	苷	-ylene	亚基
sila-	硅杂	-yne	炔
spectr(o)-	光谱，视觉	-ynyl	炔基

Appendix 3 Chemical Abbreviations

abs. absolute 绝对的	etc. etcetera 等等
Ac acetyl 乙酰基	evap. evaporation 蒸发
addn. addition 添加	expt. experiment 实验
alc. alcohol 醇	extn. extraction 提取
alk. alkaline 碱性的	fig. figure 图
amt. amount 数量	hyd. hydrous 水的
AO atomic orbital 原子轨（道）函数	ibid. in the same place 在同一地方，同上
AP analytically pure 分析纯	IR infrared 红外线
app. apparatus 装置	K kelvin 开（尔文），绝对温度
approx. approximate(ly) 近似的，大概的	lab. laboratory 实验室
aq. aqueous 水的，含水的	liq. liquid 液体
asym. asymmetric(al)(ly) 不对称的	L.R. laboratory reagent 实验试剂
atm atmosphere 大气压=1.01325×10^5 帕……	manuf. manufacture 制造
av. average (except as a verb) 平均	max. maximum 最大的
BOD biochemical oxygen demand 生化需氧量	Me methyl 甲基
b.p. boiling point 沸点	min. minute 分钟
ca. circa(about) 大约	mixt. mixture 混合物
Cal calorie 千卡，大卡=4186.8 焦	MO molecular orbital 分子轨道函数
calc. calculate 计算	mol.wt. molecular weight 分子量
cp. compare 比较	m.p. melting point 熔点
chem. chemistry 化学	org. organic 有机的
clin. clinical(ly) 临床的	oxidn. oxidation 氧化
COD chemical oxygen demand 化学需氧量	Pa pascal 帕（斯卡）（压力单位）
conc. concentrate (as a verb) 提浓，浓缩	Ph phenyl 苯基
const. constant 常数	ppm parts per million 百万分之（几）
contg. containing 包含，含有	ppt. precipitated 沉淀，沉淀物
compd. compound 化合物	prep. prepare 制备
CP chemically pure 化学纯	resp. respectively 分别地
cryst. crystalline 结晶	r.t. room temperature 室温
decomp. decompose 分解	sec secondary 仲，第二的
degrdn. degradation 降级，降解，退化	soln. solution 溶液
derive. derivative 衍生物	solv. solvent 溶剂
detn. determination 测定	sp.gr. specific gravity 相对密度
dil. dilute 稀释	sq. square 平方
dissocn. dissociation 离解	sub. sublime 升华
distd. distilled 蒸馏的	susp. suspended 悬浮的
e.g. for example 例如	tech technical(ly) 技术的
elec. electric 电的	temp. temperature 温度
eq. equation 方程	tert tertiary 叔，第三的
equil. equilibrium 平衡	UV ultraviolet 紫外
equiv. equivalent 当量，克当量	vol. volume 体积
et al and others 等等（人、地方）	wt. weight 质量

Appendix 4 English Name and Pronunciation of Greek Alphabet

Greek Alphabet

Symbol		Name	Phonetic symbol
Lower case	Upper case		
α	A	alpha	[ˈælfə]
β	B	beta	[ˈbetə,ˈbi-]
γ	Γ	gamma	[ˈgæmə]
δ	Δ	delta	[ˈdɛltə]
ε	E	epsilon	[ˈɛpsəˌlan,-lən]
ζ	Z	zeta	[ˈzetə,ˈzi-]
η	H	eta	[ˈetə,ˈitə]
θ	Θ	theta	[ˈθetə,ˈθi-]
ι	I	iota	[aiˈotə]
κ	K	kappa	[ˈkæpə]
λ	Λ	lambda	[ˈlæmdə]
μ	M	mu	[muː]
ν	N	nu	[nu,nju]
ξ	Ξ	xi	[zai,sai,ksi]
o	O	omicron	[ˈamiˌkran,ˈomi-]
π	Π	pi	[pai]
ρ	P	rho	[ro]
σ	Σ	sigma	[ˈsigmə]
τ	T	tau	[ˈtau,tɔ]
υ	Υ	upsilon	[ˈʌpsəˌlan,ˈjup-]
φ	Φ	phi	[fai]
χ	X	chi	[tʃi]
ψ	Ψ	psi	[sai]
ω	Ω	omega	[oˈmɛgə,oˈmigə,oˈme-]

Appendix 5 English Speaking of Chemical Formula, Equations and Mathematical Expressions

化学式/方程式/数学式	常见的英语读法
Na_2SO_4	N – a – two – S – O – four ; sodium sulphate
$H_2C=CH_2$	H – two – C – double – bond – C – H – two; ethene
$Mg(OH)_2$	M – g – pause –O – H – pause – twice ； magnesium hydroxide
$Ca_2(PO_4)_3 \cdot 2H_2O$	C – a – two – pause – P – O – four pause three times – dot – two – H – two – O calcium phosphate two hydrate
$[Zn(NH_3)_4]^{2+}$	tetraaminezinc cation
$\longrightarrow$	give; yield; produce; form; become
$\rightleftharpoons$	reacts reversibly
$\uparrow$	evolved as a gas; give off a gas
$\downarrow$	is precipitated; gives × precipitate
$Cu, \triangle$	in the presence of a copper as a catalyst on heating
$CO_3^{2-} + Ca^{2+} = CaCO_3$	a carbonate anion with a valence of two plus a calcium cation with a valence of two produces carbonate precipitate
$C_6H_{12}O_6 \longrightarrow 2C_2H_5OH+2CO_2$	C six H twelve O six yields two C two H five O H plus two C O two
$NaOH \longrightarrow Na^+ + OH^-$	N sub a O H yields N sub a positive plus O H negative
$Cl^- + Ag^+ \longrightarrow AgCl$	C–l–negative plus A–g–positive yields A–g–C–l
$aA+tT \longrightarrow P$	a moles of A plus t moles of T yields P
$CO_3^{2-}+2H^+ \longrightarrow H_2O+CO_2$	C O three two negative plus two H positive yields H two O plus C O two
R'	R prime
R''	R double prime, R second prime
R_1	R sub one
$100^\circ C$	one hundred degrees centigrade

"0" 的几种读法

(1) 在小数中，"0" 可读成字母 O 的音，或 zero，或 naught / nought; 但在整数中，一般只读基数词的名称，不读出零。

607.08	six O seven point O eight
0.2	zero point two
10	ten
101	one hundred and one
1002	one thousand and two

(2) "0" 表示温度，零上零下一般都读成 zero。

$0^\circ C$	zero centigrade
$-3^\circ C$	three degrees below zero centigrade

(3) "0" 在号码及年份中，一般读成字母 O 音。

room 205	room two O five
in 1807	in eighteen O seven

(4) "0" 在运动会比分中，一般读成 nil 或 nothing。

4:0	four to nil, four to nothing

小数的读法

(1) 小数点读 point；(2) 小数点前面的整数，若两位数以内，则合读一个基数词；若三位数以上，一般将每个数分开读；
(3) 小数点后的数字均分开读

0.1	point one / O point one / zero point one / naught point one
0.01	O point O one / zero point zero one / nought point nought one
2.50	two point five O
58.44	fifty eight point four four
40.72	forty point seven two
105.99	one zero five point nine nine
+	plus; positive
−	minus; negative
×	multiplied by; times
÷	divided by
±	plus or minus
=	is equal to; equals
≡	is identically equal to
≈	is approximately equal to
( )	round brackets; parentheses
[ ]	square brackets
# / { }	pound / braces
$a \gg b$	a is much greater than b
$a \geqslant b$	a is greater than or equal to b
$a \propto b$	a varies directly as b
$\log_n X$	$\log X$ to the base n
$\sqrt[3]{x}$	the 3rd root of X
$\sqrt[n]{x}$	the nth root of X
X^2	X square, X squared, the square of X
X^n	the nth power of X; X to the power n
X^{-8}	X to the minus eighth power
$\|X\|$	the absolute value of X
$\overline{X}$	the mean value of X
Σ	the sum of the terms indicated; summation of
ΔX or δX	the increment of X
dx	differential X
dy/dx	the first derivative of y with respect to X
$\int$	integral
∞	infinity
1/2	a half; one half
2/3	two thirds
5/123	five over a hundred and twenty three
$8\frac{3}{4}$	eight and three over fourths; eight and three quarters
6%	6 percent
3‰	3 per mille
2:3	the ratio of two to three
$r = xd$	r equals x multiplied by d
$5 \times 2 = 10$	five times two equals ten
$X^3/8 = y^2$	X raised to third power divided by eight equals y squared
$(a+b-c \times d)/e = f$	a plus b minus c multiplied by d, all divided by e equals f
$Y = (Wt-W)/X$	Y equals W sub t minus W over x

Appendix 6　Chinese-English Comparison of Common Terms in College Teaching

学院院长 dean
系主任 department chair
副主任 associate chair
辅导员 assistant for political and ideological work
教研室主任 head of teaching and research office
班主任 class discipline adviser
教职工 teaching and administrative staff
教学人员 the faculty
专职教师 full-time teacher
兼职教师 part-time teacher
教授 professor
客座教授 visiting professor
副教授 associate professor
讲师 lecturer/ instructor
研究生导师 graduate advisor
研究员 research fellow
学生会 students' union
教学助理 teaching assistant/TA
研究助理 research assistant/RA
研究生 graduate student
本科生 undergraduate student
留学生 international student
住宿生 boarder
走读生 day student
旁听生 auditor
在职博士生 on-job doctorate
应届毕业生 this year's graduate
领取助学金的学生 a grant-aided student
公费生 a government-supported student
校友 alumnus
家庭教师 tutor
教员休息室 staffroom
教务处 academic affairs office
学工处 student affairs office
研究生招生办公室 graduate admissions office
在职进修班 in service training course
进修班 class for advanced studies
自学考试 self taught examination
短训班 short-term training course
包分配 guarantee job assignment
课程表 school timetable

教学大纲 syllabus
教学内容 content of a course
学习年限 period of schooling
学历 educational background
学期 semester
学费 tuition
奖学金 scholarship
助学金 stipend
学分 credit
学年 academic year
分数 mark
百分制 100-mark system
学习成绩 academic performance
成绩加权平均值 GPA/grade point average
成绩单 transcript
课程 curriculum
主修 major
副修 minor
考勤 checking attendance
公开课 open class
听课 visit a class
缺课 miss a class
必修课 compulsory course
选修课 elective course
基础课 basic course
专业课 specialized course
补考 make-up examination
退选一门课 drop
课外辅导 instruction after class
课外活动 extracurricular activity
课堂讨论 class discussion
学术讲座 academic seminar
毕业论文 thesis
毕业实习 graduation field work
毕业设计 graduation project
毕业典礼 graduation ceremony
毕业证书 graduation certificate
授予某人学位 to confer a degree on sb
博士学位 doctoral degree
硕士学位 master's degree
学士学位 bachelor's degree

Appendix 7 Chemical Reference Books and Online Resources in English

1. 化学英文工具书

   (1) D R Lide. Handbook of Chemistry and Physics. CRC Press.
   (2) J Buckingham. Dictionary of Organic Compounds: 6th ed. Chapman & Hall.
   (3) Maryadele J O'Neil. The Merck Index: an Encyclopedia of Chemicals, Drugs, and Biologicals: 13th ed. Merck.
   (4) B Prager, P Jacbson, P Schmidt. Beilstein's Handbuch der Organischen Chemie. Springer-Verlag.
   (5) J G Graseselli. Atlas of Spectral Data and Physical Constants for Organic Compounds: 2nd ed. CRC Press.
   (6) Aldrich Handbook of Fine Chemicals.
   (7) J G Speight. Lange's Handbook of Chemistry. McGraw Hill.
   (8) C J Pouchert. The Aldrich Library of NMR Spectra: 2nd ed. Aldrich.
   (9) C J Pouchert, J Behnke. The Aldrich Library of ^{13}C and ^{1}H FT-NMR Spectra. Aldrich.
   (10) The Sadtler Standard Spectra.
   (11) C J Pouchert. The Aldrich Library of Infrared Spectra: 3rd ed. Aldrich.
   (12) The Aldrich Library of FT-IR Spectra: 2nd ed. Aldrich.

2. 网上资源

   (1) 美国化学学会 ACS   http://pubs.acs.org
   (2) Elsevier Sciencedirect   http://www.sciencedirect.com
   (3) Wiley Interscience   http://www3.interscience.wiley.com
   (4) 英国皇家化学学会 RSC   http://www.rsc.org/
   (5) Nature 周刊 www.nature.com/nature
   (6) Science Online http://www.sciencemag.org
   (7) Kluwer Online Journals 电子期刊 http://www.kluweronline.com/
   (8) Blackwell 期刊 www.blackwell-synergy.com
   (9) knovel 网络版电子工具书   http://www.knovel.com/
   (10) Net Library 电子图书   http://_www.netlibrary.com
   (11) Springerlink   http://_www.springerlink.com
   (12) CSA《剑桥科学文摘》(Cambridge Scientific Abstracts，CSA)
   (13) Derwent Innovations Index，基于 Web 的专利信息数据库
   (14) CCC   http://ccc.calis.edu.cn/ （学术期刊、专业书籍、会议录）
   (15) http://apps.webofknowledge.com/ （检索工具）
   (16) https://scifinder.cas.org （CA 网络版）

Appendix 8   Key to Exercises and Sample Exams

Part 1
Lesson 1
(1) 动力装置是船舶的心脏。   (2) 我们知道，所有的物体都有重量并占据空间。   (3) 去除水中矿物质的过程称为软化。   (4) 爱因斯坦的相对论是能解释这种现象的唯一理论。   (5) 影响植物生长的因素中最重要的是水的供应。   (6) 医生对血样做了分析，看是不是贫血症。   (7) 人们发现，自然界里许多元素，都是各种不同的同位素的混合物。   (8) 同性电荷相斥，异性电荷相吸是电学的一个基本规律。   (9) 电学上最常用的两个单位是伏特和安培：前者是电压的单位，后者是电流的单位。

Lesson 2
1. (1) yield   (2) carbon steel   (3) makeup   (4) atomic weight   (5) periodic table   (6) pick-me-up   (7) out-of-door   (8) smog   (9) X-ray   (10) I-steel

2. (1) 科技英语   (2) 大学英语水平考试   (3) 转/分   (4) 呼救信号   (5) 计算机辅助设计   (6) 小时   (7) 中国   (8) 年   (9) U 形管   (10) 马蹄形磁铁

3. (1) a-   (2) dis-   (3) im-   (4) in-   (5) anti-   (6) counter-   (7) semi-   (8) kilo-   (9) hydro-   (10) photo-   (11) -or   (12) -ist   (13) -tion   (14) -age   (15) -ance   (16) -wise

Lesson 3
(1) 由各种各样的硬合金钢制成。   (2) 营养丰富的食品。   (3) 多方面的成功。   (4) 它们能与其他有机化合物自由地混合并能溶于多种有机溶剂中。   (5) 非金属的性质差异很大。   (6) 昆虫到冬天就躲藏起来。   (7) 他挂了一幅油画肖像。   (8) 当孩子生病时，医生通常测量他的体温。   (9) 这些产品是成百地计数的。   (10) 光又返回到原来其传播的媒介中，就是反射。   (11) 比硝酸更强的酸的存在可以加速异种分解生成 NO^{2+} 和 OH^-。   (12) 火箭在月球上着陆了。   (13) 由上述可知，太阳的热量可以穿过太阳与大气层之间的真空。   (14) 这些烃都比水轻，并且几乎不溶于水，所以就浮在水面上。

Lesson 4
(1) 电视图像大都受高层建筑干扰。   (2) 苯环上的甲基使甲苯非常易于硝化。   (3) 他们将尽最大努力为盲人和聋哑人建一所学校。   (4) 他非常熟悉这台发报机的性能。   (5) 由于漏电损耗了大量的输出功率。   (6) 在 19 世纪以前，大部分炼铁方法是非常原始的。   (7) 图能直观地表示这个关系。   (8) 他们对实验中获得的数据非常满意。

Lesson 5
(1) 不同的金属有不同的熔点。   (2) 拉紧的弹簧随时可以做功。   (3) 制造 TNT 简单而且比较安全。   (4) 可以证明该反应物含有杂质。   (5) 本说明所用的一些术语定义如下。   (6) 现代工业要求开发越来越多的天然气。   (7) 热血动物的体温是恒定的。   (8) 草图的作用就是把人们的想法用图表示出来。   (9) 这种石头的相对密度是 2.7。   (10) 物理变化不生成新物质，也不改变物质的成分。   (11) 影响设备可靠性的因素很多。   (12) 水星的大小和外观很像月亮。   (13) 钠的化学性质很活跃。   (14) 海平面上的大气压为每平方英寸 15 磅。

Lesson 6
(1) 他英语学得很好。   (2) 我们称这种带有两个正电荷的锌原子为锌离子 Zn^{2+}。   (3) 我们

277

都熟知这样一个事实,即自然界中没有一件东西会自行开始或停止运动。　(4) 十进制计算法有着悠久的历史。　(5) 无数的有机物,在无氧情况下,受热、压力和时间的影响而转化为沉积的矿物燃料——煤、石油和天然气。　(6) 测试片的长度要适合被测的机器。　(7) 热量是一种可以转换成其他所有形式的能量。　(8) 分子运动作用力往往使它们分离。　(9) 电流与电动势成正比,与电阻成反比。　(10) 用电解方法制造氢气不如用热化学方法有效。　(11) 生物化学研究生命的分子基础。　(12) 在一个体积一定的容器里,温度的增加将导致气体压强的增加。　(13) 在有机研究中,结构分析很普遍。

Lesson 7
(1) 热力学定律将在下文予以讨论。　(2) 因为甲烷是该系列的第一个成员,因此该系列也被称为甲烷系。　(3) 无线电波也被认为是辐射能。　(4) 使用电子显微镜,能获得更大的放大倍数。　(5) 大家都知道银是最好的导体。　(6) 光纤通信利用大量激光器来产生光发射。　(7) 当罐内溶液达到所要求的温度时,就卸料。　(8) 有人看见过这些工人在修理发电机。　(9) 因此,可以将吸收过程简单地分为两类:物理过程和化学过程。　(10) 醛是由醇脱氢制得的,也因此而得名。　(11) 反应中,酸与碱彼此中和而形成水和盐。　(12) 众所周知,电子带微量负电荷。　(13) 使温度快速从室温升到125℃,并至少保持15分钟。

Lesson 8
(1) 苯中所有的碳-碳键都一样,具有介乎单键和双键之间的性质。　(2) 你有什么话要说?
(3) 污泥中含有各种碳基分子的有机物。　(4) 发动机的动力来自燃料燃烧时产生的热。
(5) 本发明解释了环形结构所显示的不饱和性降低的原因。　(6) 我们使用的各种能量中,电能最方便。　(7) 这些元素本身也在变化,这是除了古代炼金术士以外,人们一直认为不可能的事。　(8) 某些金属不但能导电,而且能被磁化。　(9) 物质是由分子组成的,而分子又是由原子组成的。　(10) 这些研究的重大成果给有机化学整个领域带来了理论上的统一,从而使有机化学原理的教和学变得较为容易。　(11) 正如上面所指出的,固体的膨胀没有气体和液体那么大。　(12) 你知道声音传播有多快吗?

Lesson 9
(1) 能量和物质这两样东西具有什么样的性质,如何相互起作用,如何控制它们为人类服务,这些问题构成了物理和化学两门基础自然科学的主要内容。　(2) 这一简单事实表明,摩擦力越小,球会滚得越远。由此我们可以推论,如果把对引力起着阻碍作用的一切阻力排除掉,那么曾经处于运动状态的球就没有理由停下来。　(3) 由于铝总是和其他元素结合,最常见的是和有很强亲和力的氧结合,因而在自然界任何地方都找不到处于游离状态的铝,所以,直到十九世纪,人们才知道有铝。　(4) 燃烧是一种化学反应,在这个反应中,气体氧的原子与某些其他元素的原子,如氢或碳,相化合。　(5) 制造过程可分为单件生产和大量生产。单件生产就是生产少量的零件,大量生产就是生产大批相同的零件。　(6) 化学工程的多功能性源于把复杂的过程分解为独立的物理步骤(单元操作)以及化学反应。　(7) 氧可以和一个碳原子共用两对电子形成碳氧双键,也可以分别和两个碳原子或者一个碳原子与一个氢原子形成两个单键。　(8) 设计本身应该是安全的,但无须追求完美的安全系统。应尽量避免使用或减少危险的原料的用量,或在较低温度和压力条件下使用,或用惰性材料稀释。

Part 2
Unit 1
Lesson 10
1. (1) toxic chemicals　(2) chemical pollution　(3) the properties of substance　(4) natural changes
　(5) scientific fields　(6) isolate　(7) determine　(8) synthesize　(9) fundamental principles

2. (1) F (2) M (3) B (4) J (5) C (6) D (7) E (8) G (9) K (10) L (11) H (12) I (13) A
3. (1) An experiment is the observation of some natural phenomenon under controlled circumstances. (2) A law is a simple generalization from experiment. (3) A hypothesis is a tentative explanation of a law, or regularity of nature. (4) A theory is a tested explanation of some body of natural phenomena.
4. (e)
5. 略
6. 略

Lesson 11

1. (1) Al_2O_3 (2) $Sr_3(PO_4)_2$ (3) $Al_2(CO_3)_3$ (4) Li_3N
2. (1) calcium phosphate (2) tin(II) fluoride, stannous fluoride (3) vanadium(V) oxide (4) copper(I) oxide, cuprous oxide
3. (1) TiO_2 (2) $SiCl_4$ (3) CS_2 (4) SF_4 (5) Li_2S (6) SbF_5 (7) N_2O_5 (8) IF_7
4. (1) sulfur hexafluoride (2) dinitrogen pentoxide (3) nitrogen triodide (4) xenon tetrafluoride (5) arsenic tribromide (6) chlorine dioxide
5. (1) hydrochloric acid (2) sulfuric acid (3) nitric acid (4) acetic acid (5) sulfurous acid (6) phosphoric acid
6. (1) Na_2O (2) K_2SO_4 (3) AgF (4) $Zn(NO_3)_2$ (5) Al_2S_3
7. (1) sodium sulfite (2) iron(III) oxide or ferric oxide (3) iron(II) oxide or ferrous oxide (4) magnesium hydroxide (5) nickel(II) sulfate hexahydrate (6) phosphorus pentachloride or phosphorus(V) chloride (7) chromium(III) dihydrogen phosphate (8) diarsenic trioxide (9) ruthenium(II) chloride
8. (1) hexacyanoferrate(II) ion, +2 (2) hexaaminecobalt(III) ion, +3 (3) aquapentacyanocobaltate(III) ion, +3 (4) pentaaminesulfatocobalt(III) ion, +3
9. (1) $K_3[Cr(CN)_6]$ (2) $[Co(NH_3)_5(SO_4)]Cl$ (3) $[Co(NH_3)_4(H_2O)_2]Br_3$ (4) $Na[Fe(H_2O)_2(C_2O_4)_2]$
10. (a) 3, (b)1or2, (d) 2

Lesson 12

1. hydrogen
2. chlorine
3. (1) Mo and Te (2) B
4. (1) three (Sc, Sg and Ag) (2) three (Ta, Tc and Ti)
5. Manganese is a typical metal, it is far from the metal-nonmetal dividing line on the periodic table.
6. As_2Os_5 (P and As are in the same periodic group)
7. (1) In, indium (2) Si, silicon (3) Mn, manganese (4) Ti, titanium (5) As, arsenic
8. bismuth (Bi)
9. atomic number = 15, atomic weight= 30.97 amu, group VA, period 3, nonmetal
10. (1) Ru (2) Zr (3) Nd (4) Ti (5) Cm (6) Fe (7) I (8) Li (9) Au
11. (c)
12. (d)

Lesson 13

1. ionic and covalent bond
2. (1) An ionic bond is the strong attractive force that exists between a positive ion and a negative in an ionic compound. (2) A covalent bond is a chemical bond formed by the sharing of a pair

of electrons between the two atoms.   (3) A bonding pair is an electron pair that is shared between atoms.   (4) A lone pair is an electron pair that is on one atom, it is not involved in bonding.   (5) Electronegativity is a measure of the ability of an atom in a covalent bond to draw bonding electrons to itself.

3. (1) Na—O   (2) H—O   (3)F—H
4. (1) ionic   (2) covalent   (3) ionic   (4) covalent
5. (1) HF   (2) NF_3 (3) H_2O
6. (a)
7. (d)
8. (e)
9. (e)
10. (b)
11. (c)
12. (e)
13. (d)
14. (d)
15. (a)
16. (e)
17. (b)
18. (b)

Lesson 14

1. $[Co(NH_3)_4(H_2O)Br]Cl_2$
2. $K_2[PtCl_6]$
3. (1) coordination   (2) ionization
4. (1) 4   (2) 6   (3) 4   (4) 6
5. (1) +2   (2) +3   (3) +3   (4) +3
6. (1) +3   (2) Name: amine, chloro, oxalate; Formula: NH_3, Cl^-, $C_2O_4^-$ (3) 6   (4) −3
7. (1) potassium hexafluoroferrate(Ⅲ)   (2) diaminediaquacopper(Ⅱ)ion   (3) ammonium aquapentafluoroferrate(Ⅲ)   (4)dicyanoargentate(Ⅰ) ion
8. (1) pentacarbonyliron(0)   (2) dicyanobis(ethylenediamine)rhodium(Ⅲ) ion
   (3) tetraaminesulfatochromium(Ⅲ) chloride   (4) tetraoxomanganate(Ⅶ) ion
9. (1) $K_3[Mn(CN)_6]$ (2) $Na_2[Zn(CN)_4]$ (3) $[Co(NH_3)_4Cl_2]NO_3$ (4) $[Cr(NH_3)_6]_2[CuCl_4]_3$

Unit 2

Lesson 15

1. (1) Organic chemistry discusses about the structure, properties, and reactions of organic compounds.   (2) Compounds having only elements carbon and hydrogen are called hydrocarbons.   (3) C_nH_{2n+2}, C_nH_{2n}, C_nH_{2n-2} (4) In alkanes, there are only single bonds and alkanes contain the maximum number of hydrogen atoms possible, being saturated.   (5) alkyne.   (6) Alcohol, ether, ketone, aldehyde, acid, ester, carbohydrate, etc.   (7)Ketones have a C=O group attached to two hydrocarbon chains (or ring). Aldehydes have a —CHO group attached to a hydrocarbon chain (or ring).   (8)carbon, hydrogen, and oxygen.   (9) $C_6H_2CH_3(NO_2)_3$; TNT is explosive.   (10) Sulphur, metal elements, such as iron, magnesium, etc.
2. (1) 分子式   (2) 无机分子   (3) 化合价   (4) 烃   (5) 环己烷

(6) 萘 (7) 碳水化合物 (8) 三硝基甲苯 (9) 磷 (10) 单键

3. (1) organic molecules (2) carbon ring (3) fuel (4) flammable (5) double bond
(6) hydrogen (7) compound (8) ethyne (9) ether (10) formic acid

Lesson 16

1. (1) cyclopentyl-C(CH₃)₂-CH₃ structure (2) CH₃CHCH₂CHCH₂CH₃ with CH₃ and OH (3) ketone structure

(4) branched carboxylic acid with COOH (5) CH₃CHCHO with Cl (6) CH₃CHCH₂CH₂CH₂NH₂ with CH₃

(7) N,N-dimethyl amide (8) CH₃CH₂CH₂CH(CH₃)C(=O)OCH(CH₃)₂ (9) alkene structure

(10) H₃C-C(CH₃)₂-CH₂CH₂CH₃

2. (1) 2,3,5-trimethyl-4-propylheptane (2) (1R,2S)-1-chloro-2-methylcyclohexane (3) 4-ethoxyoctane
(4) (E)-4-isopropylhept-4-en-3-ol (5) 6,6-dimethyl-3-octyne (6) 2,2-dimethyl-1,3-cyclohexanedione
(7) N-methyl-N-propylcyclohexanamine (8) 2-methylcyclopentanecarboxamide (9) tert-butyl cyclohexanecarboxylate (10) (S)-2-bromobutanoic acid

Lesson 17

1. (1)(a) (2)(c) (3)(a) (4)(d)

2. In a S_N1 reaction, only one molecule is involved which dissociates spontaneously to generate a carbocation intermediate. Then the carbocation reacts with the added nucleophile to yield the product. In a S_N2 reaction the nucleophile attacks the substrate from a position 180° away from the leaving group and that the reaction takes place in a single step without intermediates.

3. Addition reactions are classified as free radical, electrophilic, or nucleophilic addition reactions.

4. (1) CH₃CH₂CHOH (2) H₃C-C(=O)-O-CH₃ (3) (H₃C)₂C=CH₂ (4) 1-methylcyclohexene

(5) (CH₃)₂C(Cl)-CH₃ (6) Ph-HC(N-piperidine)-CH₂C(=O)CH₃ (7) 2-allylphenol (8) 1,2-dimethylcyclohexene

Unit 3

Lesson 18

1. (1) quantum chemistry (2) thermodynamic system (3) closed system (4) material equilibrium

2. (1) S (2) S (3) NS (4) NS (5) S (6) S

3. (1) (c) (2) (b) (3) (b) (4) (a) (5) (c)

4. (1) The key difference between microscopic viewpoint and macroscopic viewpoint is that the former makes explicit use of the concepts of molecules while the latter focuses on large-scale properties of matter, such as volume, pressure, composition, without explicit use of the molecule concepts.　(2) The three types of equilibrium refer to mechanical equilibrium, material equilibrium, and thermal equilibrium. For thermodynamic equilibrium, all three types of equilibrium must be present.　(3)The three types of system are open system, closed system, and isolated system. The difference between them is whether there is transfer of energy and matter.

Lesson 19

1. (1) 化学平衡　(2) 平衡常数　(3) 正逆反应速率　(4) 化学动力学
2. (1) chemical kinetics　(2) concentration　(3) reaction chamber
3. (1) with　(2) of　(3) to, into　(4) with, in
4. (1) $K = \dfrac{[SO_3]^2}{[SO_2]^2[O_2]}$　(2) $K = \dfrac{[H_2O]^2}{[H_2]^2[O_2]}$　(3) $K = \dfrac{[Fe^{2+}]^2}{[Fe^{3+}]^2[I^-]^2}$
5. (1) reactants.　(2) The forward. The chemical equilibrium constant K can quantitatively describe the limits of reversible reaction. The larger the K value, the greater the proportion of the generated products in the equilibrium system.
6. (1) reverse　(2) forward　(3) no influence　(4) forward　(5) no influence　(6) no influence
7. (1) forward　(2) reverse　(3) no influence　(4) forward　(5) reverse　(6) no influence

Lesson 20

1. (1) 大气污染　(2) 非降解污染物　(3) 无害污染物　(4) 相互作用的化学物　(5) 化学反应　(6) 声压级　(7) 阈值水平　(8) 主客观性质　(9) 讲话干扰　(10) 传输路径
2. (1) (b)　(2) (c)　(3) (a)(b)(c)
3. (1) Pollution can be defined as an undesirable change in the physical, chemical, or biological characteristics of the air, water, or land that can harmfully affect health, survival, or activities of humans or other living organisms. There are four major forms of pollution—waste on land, water pollution (both the sea and inland waters), and pollution by noise.　(2) Water pollution contains many forms, it is contaminated by city sewage and factory wastes, the run off of fertilizer and manure from farms and feed lots, etc.　(3) Degradable pollutants can decompose slowly but eventually are either broken down completely or reduced to harmless levels, such as DDT and radioactive materials.　(4) Air pollution is normally defined as air that contains one or more chemicals in high enough concentrations to harm humans, other animals, vegetation, or materials.　(5) Noise is the byproduct of machine which spoils leisure and even working hours of people.　(6) Nondegradable pollutants are not broken down by natural processes, such as mercury, lead and some of their compounds and some plastics.
4. (1) There are a large number of interacting chemicals in water, many of them only in trace amounts. About 30000 chemicals are now in commercial production, and each year about 1000 new chemicals are added.　(2) There are two major types of air pollutants. A primary air pollutant is a chemical added directly to the air that occurs in a harmful concentration. It can be a natural air component, such as carbon dioxide, that rises above its normal concentration, or something not usually found in the air, such as a lead compound. A secondary air pollutant is a harmful chemical formed in the atmosphere through a chemical reaction among air components.　(3) No, because it is diffcult to jude the noise. For example, vehicles which are judged to be equally noisy may show

considerable difference on the meter.   (4) Waste on land, water pollution (both the sea and inland waters), and pollution by noise.   (5) The effective way to solve the noise pollution is to reduce the sound pressure level, either at source or on the transmission path.

5. 略

Unit 4

Lesson 21

1. (1) qualitative analysis   (2) analyte   (3) accuracy   (4) reaction-rate   (5) desorption   (6) absorption   (7) quantitative analysis
2. Analytical chemistry is the science of determining what a compound is, separating it out, and measuring how much of it there is.
3. the species being measured in a chemical analysis
4. qualitative analysis and quantitative analysis
5. One is the weight or volume of sample to be analyzed. The second is the measurement of some quantity that is proportional to the amount of analyte in that sample. This second step normally completes the analysis.
6. (1) (c)   (2) (c)   (3) (b)

Lesson 22

1. (1) titrate   (2) indicator   (3) molarity   (4) burette   (5) phenolphthalein   (6) endpoint   (7) buffer solution   (8) alkali burette   (9) extract   (10) quantitative analysis   (11) dropping funnel   (12) volumetric analysis
2. Molarity is the number of moles (gram-molecular weights) of substance per liter of solution. Normality is the number of equivalent weights of substance per liter of solution.
3. Volumetric analysis is a chemical analyses made by determining how much of a solution of known concentration is needed to react fully with an unknown test sample and generally consists of titrating the unknown solution with the one of known concentration (a standard solution).
4. One reason: to assign unambiguously a structure to a new molecule which has not been described previously in the chemical literature; the other reason: to characterize the properties of the compound for comparison purposes.
5. (1)选出你需要的分液漏斗的大小。通常你会使用125毫升或者250毫升的分液漏斗，对于大规模的反应(1~10克)，你需要使用500毫升或1升的分液漏斗。谨记你的分液漏斗中含有溶剂和洗涤液，必须予以彻底的混合。   (2)用你所选用的溶剂稀释反应粗产物的混合物，并且转移到你选择的分液漏斗中。原料越多需要的溶剂量越大。普通的反应(50~500毫克的产品)用25~100毫升溶剂即可稀释。   (3)蒸馏是一个非常有用的技术，用于分离纯化试剂和粗产品的混合物。一般有两种蒸馏方法：常压蒸馏和减压蒸馏。

Lesson 23

1. (1) UV/Vis spectroscopy   (2) electronic transition   (3) spectroscopy   (4) bathochromic shift   (5) hypsochromic shift   (6) infra-red spectrum   (7) standard curve   (8) chromophore   (9) baci-extraction   (10) auxo chrome
2. This type of spectroscopic instrumentation that often operates with wavelengths between 800 to 200 nm are called UV/Visible spectrometers.
3. An auxochrome is a substituent that contains unshared (nonbonding) electron pairs, such as OH, NH, and halogens. An auxochrome attached to a chromophore with π electrons shifts the absorption maximum to longer wavelengths.

4. (1)(a) (2)(c) (3)(c) (4)(a)
5. (1) $n \sim 10$ (2) 4.0×10^{-4}

Lesson 24
1. (1) probe (2) optimizing (3) console (4) nuclear magnetic resonance (5) mass spectroscopy (6) signal (7) sensitivity (8) resolution (9) detector (10) component peak (11) chemical shifts (12) gyromagnetic ratio
2. Early NMR machines relied on electromagnets had low sensitivity and poor stability, but a superconducting magnets machine provides high sensitivity and stability.
3. sensitivity and resolution
4. Mass spectrometry is combined with gas chromatography for routine analysis of mixtures of compounds, such as reaction mixtures or environmental samples.
5. (1) (a) (2) (b) (3) (d) (4) (a)

Unit 5
Lesson 25
1. (1) molecular genetics (2) molecular biology (3) nutritional deficiency (4) dehydration synthesis (5) amino acid (6) three-dimensional conformation (7) peptide bond (8) nucleic acid
2. (1) (c) (2) (d) (3) (d) (4) (d) (5) (b) (6) (b) (7) (a) (8) (d) (9) (d) (10) (d) (11) (a) (12) (a) (13) (b) (14) (a) (15) (d) (16) (b) (17) (a) (18) (d) (19) (a) (20) (d) (21) (a) (22) (a)
3. Biochemistry, sometimes called biological chemistry, is the study of chemical processes within and relating to living organisms. Much of biochemistry deals with the structures, functions and interactions of biological macromolecules, such as proteins, nucleic acids, carbohydrates and lipids, which provide the structure of cells and perform many of the functions associated with life.

Unit 6
Lesson 26
1. (1) A polymer is a material whose molecules contain a very large number of atoms liked by covalent bonds, usually consisting of identical or similar units joined together. (2) Proteins, cellulose, starch, and natural rubber. (3) Because graphite has a two-dimensional structure, which has a planar network. (4) Thermoplastic polymers can be melted into a "liquid" state when be heated, while thermosetting polymers cannot be soften, melted or molded non-destructively. (5) A rubber material is one which can be stretched to at least twice its original length and rapidly contract to its original length. A rubber has elasticity is due to the coiling and uncoiling of its very long chains. (6) Because before 1839 the curing reaction with sulfur had not been discovered. The natural rubber becomes "soft" and "sticky" on hot days. (7) A fiber is often defined as having an aspect ratio (length/diameter) of at least 100. (8) To prepare a giant polymer molecule, a step growth polymerization is used to yield branched as well as networked polymers. And a multifunctional monomer is usually used so the branches will reach from chain to chain while the length and frequency of braches on the polymer chain increases. When all the chains are connected together, the entire polymer mass becomes one giant molecule. (9) The glass transition temperature T_g is a temperature, or range of temperatures, below which an amorphous polymer is in a glassy state and above which it is rubbery. (10) The glass transition temperature depends on five factors: free volume of the polymer, the attractive forces between the molecules, the internal mobility of the chains, the stiffness of the chains, and the chain length.

2. (1) 热塑性高分子 (2) 交联 (3) 机械性能 (4) 合成高分子 (5) 大分子 (6) 结晶度(结晶性) (7) 硬度 (8) 纺织品 (9) 形貌 (10) 分子量 (11) 阳离子聚合反应

3. (1) protein (2) monomer (3) amorphous (4) soften (5) epoxy plastic (6) rubber (7) polymer chain (8) radical polymerization (9) linear polymer (10) melt spinning (11) liquid crystal

Lesson 27

1. (1) Because oxygen in the air had leaked into the apparatus and acted as a radical initiator to catalyze the polymerization by a chain-reaction mechanism. (2) Nylon was taken off the domestic market and used in military parachutes. (3) Because cold-drawing process leads a special orientation of the polymer molecules which produces extensive intermolecular hydrogen-bonding binds the individual polymer molecules together in much the same way that separate strands in a rope. The association of linear polymer molecules through hydrogen bonding causes the increase of the strength of nylon fibers.

(4) [structural formulas of polystyrene $(-CHCH_2-)_n$ with phenyl group, and a nylon-type polyamide structure]

2. (1) 聚乙烯 (2) 苯甲醛 (3) 自由基引发剂 (4) 人造丝 (5) 拉伸强度 (6) 聚酯 (7) 冷拉处理 (8) 细丝 (9) 聚酰胺 (10) 氢键

Unit 7
Lesson 28

1. (1) carbohydrate (2) protein (3) nutrition (4) polysaccharide (5) food industry (6) synthetic food (7) health (8) modification (9) food chemistry (10) natural food

2.《食品化学》出版原创性研究论文,涉及食品化学和生物化学研究进展以及食品分析方法。所有论文都聚焦研究的新颖性。主要议题/收录范围包括:(1) 食品中主要和次要成分的化学研究及其营养、生理、感官、风味和微生物相关方面的研究。(2) 食品的生物活性成分研究,包括抗氧化剂,植物化学成分和植物药,研究数据必须充分论证并证明其与食品和/或食品化学相关。(3) 在加工、运输和室内贮藏条件下,在分子水平上研究食品的化学和生物化学成分与结构的改变。(4) 加工对食品组分、食品质量与安全、其他生物基材料、副产品及加工废弃物的影响。(5) 食品中添加剂、污染物和其他农用化学品及其代谢、毒理和食品腐败的化学研究。

3. 略

Unit 8
Lesson 29

1. (1) (c) (2) (d) (3) (d) (4) (a)

2. (1) pesticide (2) feedstuff (3) growth regulator (4) desiccant (5) fruit thinning agents (6) deterioration (7) domestic animals (8) broad-spectrum activity (9) fungi (10) malaria (11) monoculture system (12) internally feeding insects (13) acute (14) chronic

Unit 9
Lesson 30

1. (1) thermodynamics (2) insulator (3) evaporator (4) filter (5) precipitation (6) petrochemical product (7) process simulation (8) unit operation

2. This means that the entire production chain must be planned and controlled for costs.

3. The individual processes used by chemical engineers (e.g. distillation or filtration) are called unit operations and consist of chemical reactions, mass-, heat- and momentum- transfer operations. A "unit process" is one or more grouped operations in a manufacturing system that can be defined and separated from others.

4. Three primary physical laws underlying chemical engineering design are conservation of mass, conservation of momentum and conservation of energy.

5. Chemical engineering is applied in the manufacture of a wide variety of products. The chemical industry scope manufactures inorganic and organic industrial chemicals, ceramics, fuels and petrochemicals, agrochemicals (fertilizers, insecticides, herbicides), plastics and elastomers, oleochemicals, explosives, detergents and detergent products (soap, shampoo, cleaning fluids), fragrances and flavors, additives, dietary supplements and pharmaceuticals.

Lesson 31

1. (1) convection (2) eddy currents (3) turbulence (4) temperature gradient (5) radial flow (6) unit medium

2. Heat energy is also accurately called thermal energy or simply heat. It is a form of energy transfer among particles in a substance (or system) by means of kinetic energy. In other words, heat is transferred from one location to another by particles bouncing into each other.

3. Heat transfer is a process by which internal energy from one substance transfers to another substance.

4. Heat transfer can take place in one or more ways: conduction, convection and radiation.

5. (1) (b) (2) (b) (3) (b) (4) (c) (5) (a) (6) (c) (7) (b) (8) (b) (9) (a) (10) (c) (11) (c)

6. 略

Unit 10

Lesson 32

1. (1) 多相反应 (2) 竞争反应 (3) 连续与批量过程 (4) 连续或半连续过程 (5) 卫生(保健)品及化妆品 (6) 高附加值产品 (7) 散装(大宗)化学品 (8) 专用化学品 (9) 精细化学品 (10) 固有特性 (11) 半通用商品 (12) 小批量化学品 (13) 活性成分 (14) 原料药 (15) 功能性化学品 (16) 间歇搅拌式反应器 (17) 热的释放 (18) 规章规范 (19) 现有设备 (20) 放大

2. (1) raw material (2) side product (3) desired product (4) protective measure (5) intermediates (6) physico-chemical properties (7) engineering solution (8) hazardous chemical (9) effluent disposal (10) bulk production (11) small capacity (12) residence time (13) molecular modelling (14) stereoselective catalyst (15) supercritical condition

3. (1) 化学品以批量和特征分类。大宗(或商用)化学品以大批量生产，并按工业级别出售。例如丙酮、环氧乙烷、苯酚。半通用化学品也以大批量生产，但按功能出售。很多情况下，产品有配方，并且供应商不同，性质也不同，例如大批量聚合物，表面活性剂，涂料等。

(2) 精细化学品是高纯产品，相对小批量的生产，售价相对较高。精细化学品可以分为两类。一类用作其他产品中间体，另一类根据性质有特殊的活性，基于其功能特征进行使用。功能化学品被用作活性成分或配方中的添加剂，或处理过程中的助剂。(3) 精细化学品通常以多步合成的方式生产，导致大量原料消耗，并有大量副产物和废物。在精细化学品中的高投入产出比值证明广泛寻找选择性催化剂有道理。有效催化剂的使用可以减少反应物的消耗和废物的产生，同时减少合成步骤。

4. (1) Very pure compounds must often be produced with impurities at ppm or ppb level. (2) Significant fluctuations in the demand exist for a variety of chemicals. (3) Other evaporation

processes will be discussed briefly.
5. 略

Part 3
Lesson 33
1. (1) (c), $Fe_3O_4(s) + 4H_2(g) \longrightarrow 3Fe(s) + 4H_2O(g)$
   (2) (b), $2KClO_3(s) \longrightarrow 2KCl(s) + 3O_2(g)$
   (3) (c), $H_2O(g) + C(s) \longrightarrow H_2(g) + CO(g)$
   (4) (a), $Cl_2O_7(g) + H_2O(l) \longrightarrow 2HClO_4(aq)$
   (5) (c), $Br_2(l) + H_2O(l) \longrightarrow HBr(aq) + HBrO(aq)$
   (6) (c), $Ca_3(PO_4)_2(s) + 3H_2SO_4(aq) \longrightarrow 3CaSO_4(s) + 2H_3PO_4(aq)$
   (7) (c), $2K + 2H_2O \longrightarrow 2KOH + H_2$
   (8) (b), $MgCO_3 \longrightarrow MgO + CO_2$

2. (1) 当两元素化合时生成二元化合物(只由两种原子组成)。如果一种金属与一种非金属反应，产物是离子型的，由元素原子形成离子的电荷确定化学式。如果两种非金属发生化合反应，产品是一种具有极性共价键的分子，由涉及的元素原子的正价确定化学式。 (2) 这些都是化合反应，通过加热产物这些反应可逆。金属氢氧化物加热时分解生成金属氧化物和水。含氧酸加热分解生成水和具有合适氧化态的非金属氧化物。 (3) 许多化合物在水溶液中具有有限的溶解度。当溶液中离子浓度高于溶解度限制时，离子结合形成沉淀颗粒，从溶液中析出。剩余离子在溶液中浓度由平衡常数 K_{sp} 控制，K_{sp} 叫做溶度积。 (4) 金属离子在溶液中与配体结合形成可溶性化合物。这些反应通常由配合物形成的方向的平衡表达式来描述。平衡常数 K_f 叫做形成常数。

Lesson 34
1. (1) $2\ C_4H_{10} + 13\ O_2 \longrightarrow 8\ CO_2 + 10\ H_2O$
       58   32   44   18
   (a) $(400g/58) \times (13/2) \times 32 = 1434.48g$
   (b) $(400g/58) \times (10/2) \times 18 = 620.69g$
   (2) $3\ HCl + Al(OH)_3 \longrightarrow 3\ H_2O + AlCl_3$
       36.5   78
   $(5.3/3) \times 78g = 137.8g$
   (3) $Ca(ClO_3)_2 \longrightarrow CaCl_2 + 3\ O_2$
       136   32
   $(1000/136) \times 3 \times 32 = 705.88g$
   (4) $Ca + 2H_2O \longrightarrow Ca(OH)_2 + H_2$
       40   18
   $(2.35/40) \times 2 \times 18 = 2.115g$
   (5) $2K + I_2 \longrightarrow 2KI$
       39   254
   Iodine is the limiting reactant, and 19.6g of potassium iodide are produced.
   K excess $15.0g - 4.62g = 10.38g$
   (6) Balanced chemical equation:
   $C_2H_6O + 3O_2 \longrightarrow 2CO_2 + 3H_2O$
   Calculations:

287

17.56g C_2H_6O	1mol C_2H_6O	2mol CO_2	44.01g CO_2
	46.07g C_2H_6O	1mol C_2H_6O	1mol CO_2

=33.54g CO_2

102.5g O_2	1mol O_2	2mol CO_2	44.01g CO_2
	32g O_2	3mol O_2	1mol CO_2

=93.98g CO_2

(30.00g CO_2/33.54g CO_2) ×100%=89.45%

The limiting reactant is the reactant that gives the smallest amount of product. In this case, that is ethanol. The theoretical yield is that amount of product, 33.54g CO_2. The percent yield is 89.45%.

(7) (a) 0.5mol NH_3 ×6mol H_2O/4mol NH_3 =0.75mol H_2O (b) 1.5mol NO×(4mol NH_3/4mol NO) ×(17.04g NH_3/1mol NH_3) = 25.6g NH_3 (c) 120g NH_3 × (1mol NH_3 /17.04g NH_3) × (4mol NO /4mol NH_3) ×(30.01g NO/ 1mol NO) =211g NO

2. (1) 当化学微粒(原子、分子、离子)参与作用，就发生化学反应-化学改变的过程。化学改变的量的关系的计算叫做化学计量计算。化学计量计算可以通过物质转化的量，包括将质量(g，kg)，体积(L)转换成摩尔。摩尔数取决于化学式或化学反应方程的书写。化学反应方程是化学计量计算的基础，但是当反应物不是化学计量混合物时，一些反应物就过量，而另一些反应物则适量。   (2) 为了得到以磅或吨等计而不是以克计的质量，可以以磅-摩尔或者吨-摩尔代替摩尔简化计算。对一些包括连续反应的过程，化学计量计算可以基于总方程计算，如果中间体能从方程式中取消，就去掉中间体。   (3) 大多数反应复杂，通过许多步发生。在这种情况下，可以写出描述体系的计量关系的总反应吗？考虑硫在氧气中燃烧同时生成二氧化硫(多数) 和一些三氧化硫的例子。

Part 4
Lesson 35
1. (1) beaker   (2) measuring cylinder   (3) burette   (4) volumetric pipettes   (5) graduated pipettes   (6) Buchner funnel   (7) dropping funnel   (8) round bottom flask   (9) drying tower   (10) stopper   (11) crucible tongs   (12) ring stand   (13) table balance   (14) Bunsen burners   (15) fume hood
2. (1) volumetric pipettes   (2) beaker   (3) crucible   (4) tongs   (5) volumetric flask   (6) brush   (7) clamp holder   (8) mortar and pestle   (9) Erlenmeyer flask   (10) burette   (11) test tube clamp   (12) evaporating dish   (13) desiccator   (14) transfer pipettes

Lesson 37   略

Lesson 41   略

Lesson 43   略

Appendix 1
Sample Paper A
Part Ⅰ   1. (D) 2. (D) 3. (B) 4. (C) 5. (C) 6. (A) 7. (D) 8. (D) 9. (C) 10. (D) 11. (B) 12. (A) 13. (A) 14. (C) 15. (A) 16. (C) 17. (C) 18. (B) 19. (E) 20. (D) 21. (B) 22. (C)

Part Ⅱ   1. H^o = +994kJ   2. (1) 0.0143mol AgCl   (2) 1.84kJ   (3) Reaction shown is exothermic because water temperature increased.　　(4) –129kJ/mol

Sample Paper B

Part Ⅰ　1. (B)　2. (E)　3. (D)　4. (A)　5. (A)　6. (A)　7. (B)　8. (A)　9. (C)　10. (C)　11. (A)　12. (B)　13. (A)　14. (B)　15. (D)　16. (D)　17. (A)　18. (B)　19. (C)　20. (A)

Part Ⅱ　1. (1) sodium chloride; hydrochloric acid; chlorous acid; methane; nickel (Ⅱ) nitrite
　　(2) CaF_2; N_2O_4; NH_4MnO_4; HOBr or HBrO; $FeSO_4$
　　2. (1) 0.0196mol　(2) 3.19g
　　Extra credit: NO_2 is the limiting reagent.

Sample Paper C

Part Ⅰ　1. (D)　2. (C)　3. (D)　4. (D)　5. (E)　6. (B)　7. (D)　8. (E)　9. (E)　10. (B)　11. (B)　12. (B)　13. (E)　14. (A)　15. (B)　16. (B)　17. (C)　18. (C)　19. (D)　20. (C)

Part Ⅱ　1. (1) reacts reversibly　(2) in the presence of a copper as a catalyst on heating　(3) C-l negnative plus A-g positive yields A-g-C-l　(4) three degrees below zero centigrade　(5) M-g-pause-O-H-pause-twice
　　2. (1) 4　(2) 9　(3) 3　(4) 6　(5) 5
　　3. (1) 近似的/大概的　(2) 水的/含水的　(3) 图　(4) 数量　(5) 等等
　　4. (1) D　(2) C　(3) E　(4) B　(5) A

Part Ⅲ　(1) F (2) A　(3) B　(4) G　(5) E　(6) D　(7) C　(8) H

Part Ⅳ　(1) He　(2) S　(3) F　(4) B　(5) O　(6) Mg　(7) H　(8) Cl　(9) Al　(10) Ar

Part Ⅴ　(1) combination　(2) Al is in excess; Cl_2 is limiting reactant; 42.72g

Part Ⅵ　1. 原子得到一个或多个电子形成负离子, 这样有效核电荷数就降低了, 电子云扩大, 因而负离子的半径大于相应的原子。　2. 苯的同系物是那些苯环上含有单烷基(取代一个氢) 或多烷基(取代多个氢) 的物质。　3. 酮与醛和醇的关系都很密切。　4.蒸发和结晶不同, 因为蒸发着重于将溶液浓缩而不是生成和析出晶体。　5.热力学定律在研究热方面是很重要的。　6.硫酸是强反应试剂之一。　7.上面表格可说明这一点。　8.有机化合物不溶于水, 因为水没有将它们的分子离解为离子的倾向。　9. 当将铜片置于硫酸电解液中时, 几乎没有铜原子溶解。　10. 烷烃的熔点起初很不规则, 但随着分子的增大, 则有些稳步上升的趋势。　11. 苯可以进行典型的取代反应, 如卤化、硝化、磺化和傅氏反应。　12. 我们称这种带有两个正电荷的锌原子为锌离子: Zn^{2+}。　13. 高不饱和度在某种程度与化学反应活性相关。　14. 苯中取代基的存在对进入基团的位置及难易程度起决定作用。　15. 酮的官能团是碳氧双键。

Appendix 9   The Periodic Table of the Chemical Elements

1	2											13	14	15	16	17	18
1 **H** hydrogen [1.007, 1.009]																	2 **He** helium 4.003
3 **Li** lithium [6.938, 6.997]	4 **Be** beryllium 9.012											5 **B** boron [10.80, 10.83]	6 **C** carbon [12.00, 12.02]	7 **N** nitrogen [14.00, 14.01]	8 **O** oxygen [15.99, 16.00]	9 **F** fluorine 19.00	10 **Ne** neon 20.18
11 **Na** sodium 22.99	12 **Mg** magnesium [24.30, 24.31]	3	4	5	6	7	8	9	10	11	12	13 **Al** aluminium 26.98	14 **Si** silicon [28.08, 28.09]	15 **P** phosphorus 30.97	16 **S** sulfur [32.05, 32.08]	17 **Cl** chlorine [35.44, 35.46]	18 **Ar** argon 39.95
19 **K** potassium 39.10	20 **Ca** calcium 40.08	21 **Sc** scandium 44.96	22 **Ti** titanium 47.87	23 **V** vanadium 50.94	24 **Cr** chromium 52.00	25 **Mn** manganese 54.94	26 **Fe** iron 55.85	27 **Co** cobalt 58.93	28 **Ni** nickel 58.69	29 **Cu** copper 63.55	30 **Zn** zinc 65.38(2)	31 **Ga** gallium 69.72	32 **Ge** germanium 72.63	33 **As** arsenic 74.92	34 **Se** selenium 78.97	35 **Br** bromine [79.90, 79.91]	36 **Kr** krypton 83.80
37 **Rb** rubidium 85.47	38 **Sr** strontium 87.62	39 **Y** yttrium 88.91	40 **Zr** zirconium 91.22	41 **Nb** niobium 92.91	42 **Mo** molybdenum 95.95	43 **Tc** technetium	44 **Ru** ruthenium 101.1	45 **Rh** rhodium 102.9	46 **Pd** palladium 106.4	47 **Ag** silver 107.9	48 **Cd** cadmium 112.4	49 **In** indium 114.8	50 **Sn** tin 118.7	51 **Sb** antimony 121.8	52 **Te** tellurium 127.6	53 **I** iodine 126.9	54 **Xe** xenon 131.3
55 **Cs** caesium 132.9	56 **Ba** barium 137.3	57-71 lanthanoids	72 **Hf** hafnium 178.5	73 **Ta** tantalum 180.9	74 **W** tungsten 183.8	75 **Re** rhenium 186.2	76 **Os** osmium 190.2	77 **Ir** iridium 192.2	78 **Pt** platinum 195.1	79 **Au** gold 197.0	80 **Hg** mercury 200.6	81 **Tl** thallium [204.3, 204.4]	82 **Pb** lead 207.2	83 **Bi** bismuth 209.0	84 **Po** polonium	85 **At** astatine	86 **Rn** radon
87 **Fr** francium	88 **Ra** radium	89-103 actinoids	104 **Rf** rutherfordium	105 **Db** dubnium	106 **Sg** seaborgium	107 **Bh** bohrium	108 **Hs** hassium	109 **Mt** meitnerium	110 **Ds** darmstadtium	111 **Rg** roentgenium	112 **Cn** copernicium	113 **Nh** nihonium	114 **Fl** flerovium	115 **Mc** moscovium	116 **Lv** livermorium	117 **Ts** tennessine	118 **Og** oganesson

Key: atomic number / **Symbol** / name / standard atomic weight

57 **La** lanthanum 138.9	58 **Ce** cerium 140.1	59 **Pr** praseodymium 140.9	60 **Nd** neodymium 144.2	61 **Pm** promethium	62 **Sm** samarium 150.4	63 **Eu** europium 152.0	64 **Gd** gadolinium 157.3	65 **Tb** terbium 158.9	66 **Dy** dysprosium 162.5	67 **Ho** holmium 164.9	68 **Er** erbium 167.3	69 **Tm** thulium 168.9	70 **Yb** ytterbium 173.0	71 **Lu** lutetium 175.0
89 **Ac** actinium	90 **Th** thorium 232.0	91 **Pa** protactinium 231.0	92 **U** uranium 238.0	93 **Np** neptunium	94 **Pu** plutonium	95 **Am** americium	96 **Cm** curium	97 **Bk** berkelium	98 **Cf** californium	99 **Es** einsteinium	100 **Fm** fermium	101 **Md** mendelevium	102 **No** nobelium	103 **Lr** lawrencium